Analysis and modeling of manufacturing systems

Analysis and Modeling of Manufacturing Systems

INTERNATIONAL SERIES IN OPERATIONS RESEARCH & MANAGEMENT SCIENCE

Frederick S. Hillier, Series Editor Stanford University

Liu, B. & Esogbue, A.O. / *DECISION CRITERIA AND OPTIMAL INVENTORY PROCESSES*
Gal, T., Stewart, T.J., Hanne, T. / *MULTICRITERIA DECISION MAKING: Advances in MCDM Models, Algorithms, Theory, and Applications*
Fox, B.L. / *STRATEGIES FOR QUASI-MONTE CARLO*
Hall, R.W. / *HANDBOOK OF TRANSPORTATION SCIENCE*
Grassman, W.K. / *COMPUTATIONAL PROBABILITY*
Pomerol, J-C. & Barba-Romero, S. / *MULTICRITERION DECISION IN MANAGEMENT*
Axsäter, S. / *INVENTORY CONTROL*
Wolkowicz, H., Saigal, R., & Vandenberghe, L. / *HANDBOOK OF SEMI-DEFINITE PROGRAMMING: Theory, Algorithms, and Applications*
Hobbs, B.F. & Meier, P. / *ENERGY DECISIONS AND THE ENVIRONMENT: A Guide to the Use of Multicriteria Methods*
Dar-El, E. / *HUMAN LEARNING: From Learning Curves to Learning Organizations*
Armstrong, J.S. / *PRINCIPLES OF FORECASTING: A Handbook for Researchers and Practitioners*
Balsamo, S., Personé, V., & Onvural, R./ *ANALYSIS OF QUEUEING NETWORKS WITH BLOCKING*
Bouyssou, D. et al. / *EVALUATION AND DECISION MODELS: A Critical Perspective*
Hanne, T. / *INTELLIGENT STRATEGIES FOR META MULTIPLE CRITERIA DECISION MAKING*
Saaty, T. & Vargas, L. / *MODELS, METHODS, CONCEPTS and APPLICATIONS OF THE ANALYTIC HIERARCHY PROCESS*
Chatterjee, K. & Samuelson, W. / *GAME THEORY AND BUSINESS APPLICATIONS*
Hobbs, B. et al. / *THE NEXT GENERATION OF ELECTRIC POWER UNIT COMMITMENT MODELS*
Vanderbei, R.J. / *LINEAR PROGRAMMING: Foundations and Extensions, 2nd Ed.*
Kimms, A. / *MATHEMATICAL PROGRAMMING AND FINANCIAL OBJECTIVES FOR SCHEDULING PROJECTS*
Baptiste, P., Le Pape, C. & Nuijten, W. / *CONSTRAINT-BASED SCHEDULING*
Feinberg, E. & Shwartz, A. / *HANDBOOK OF MARKOV DECISION PROCESSES: Methods and Applications*
Ramík, J. & Vlach, M. / *GENERALIZED CONCAVITY IN FUZZY OPTIMIZATION AND DECISION ANALYSIS*
Song, J. & Yao, D. / *SUPPLY CHAIN STRUCTURES: Coordination, Information and Optimization*
Kozan, E. & Ohuchi, A. / *OPERATIONS RESEARCH/ MANAGEMENT SCIENCE AT WORK*
Bouyssou et al. / *AIDING DECISIONS WITH MULTIPLE CRITERIA: Essays in Honor of Bernard Roy*
Cox, Louis Anthony, Jr. / *RISK ANALYSIS: Foundations, Models and Methods*
Dror, M., L'Ecuyer, P. & Szidarovszky, F. / *MODELING UNCERTAINTY: An Examination of Stochastic Theory, Methods, and Applications*
Dokuchaev, N. / *DYNAMIC PORTFOLIO STRATEGIES: Quantitative Methods and Empirical Rules for Incomplete Information*
Sarker, R., Mohammadian, M. & Yao, X. / *EVOLUTIONARY OPTIMIZATION*
Demeulemeester, R. & Herroelen, W. / *PROJECT SCHEDULING: A Research Handbook*
Gazis, D.C. / *TRAFFIC THEORY*
Zhu, J. / *QUANTITATIVE MODELS FOR PERFORMANCE EVALUATION AND BENCHMARKING*
Ehrgott, M. & Gandibleux, X. / *MULTIPLE CRITERIA OPTIMIZATION: State of the Art Annotated Bibliographical Surveys*
Bienstock, D. / *Potential Function Methods for Approx. Solving Linear Programming Problems*
Matsatsinis, N.F. & Siskos, Y. / *INTELLIGENT SUPPORT SYSTEMS FOR MARKETING DECISIONS*
Alpern, S. & Gal, S. / *THE THEORY OF SEARCH GAMES AND RENDEZVOUS*
Hall, R.W./*HANDBOOK OF TRANSPORTATION SCIENCE - 2nd Ed.*
Glover, F. & Kochenberger, G.A. / *HANDBOOK OF METAHEURISTICS*
Graves, S.B. & Ringuest, J.L. / *MODELS AND METHODS FOR PROJECT SELECTION: Concepts from Management Science, Finance and Information Technology*
Hassin, R. & Haviv, M./ *TO QUEUE OR NOT TO QUEUE: Equilibrium Behavior in Queueing Systems*

Analysis and Modeling of Manufacturing Systems

edited by

Stanley B. Gershwin
Massachusetts Institute of Technology
Cambridge, MA, U.S.A.

Yves Dallery
École Centrale Paris
Chatenay-Malabry, France

Chrissoleon T. Papadopoulos
University of the Aegean
Chios Island, Greece

J. MacGregor Smith
University of Massachusetts at Amherst
Amherst, MA, U.S.A.

KLUWER ACADEMIC PUBLISHERS
Boston / Dordrecht / London

Distributors for North, Central and South America:
Kluwer Academic Publishers
101 Philip Drive
Assinippi Park
Norwell, Massachusetts 02061 USA
Telephone (781) 871-6600
Fax (781) 681-9045
E-Mail < kluwer@wkap.com>

Distributors for all other countries:
Kluwer Academic Publishers Group
Post Office Box 322
3300 AH Dordrecht, THE NETHERLANDS
Telephone 31 78 6576 000
Fax 31 78 6576 254
E-Mail < services@wkap.nl>

Electronic Services < http://www.wkap.nl>

Library of Congress Cataloging-in-Publication Data

Title: ***Analysis and Modeling of Manufacturing Systems***
Edited by: Stanley B. Gershwin, Yves Dallery, Chrissoleon Papadopoulos, and J. MacGregor Smith
ISBN: 1-4020-7303-8

Printed on acid-free paper.

Printed in the United States of America.

Contents

List of Figures

List of Tables

Preface

This book is an outgrowth of the Second Aegean International Conference on Analysis and Modeling of Manufacturing Systems, which was held on May 16-20, 1999 at Tinos Island, Greece. Christos Papadopoulos has been the organizer of these conferences (the first in 1997 and the third in 2001), and has done a miraculous job of management and community-building. The papers roughly correspond to many of those presented at the conference, but there have been additions, subtractions, and each paper has had the improvements that the reviewing process brings. The conference participants and their families are illustrated on the next page.

We have decided to dedicate the book to Professor John A. Buzacott, who is retiring this year. John Buzacott has been a huge influence on the field, which is demonstrated by the fact that his name appears in nearly every reference list in this volume. His landmark 1967 *IJPR* paper was a career-clarifying inspiration to me and to many, many others.

We are grateful to Christos Papadopoulos and his staff for the productive and enjoyable time we had at the Aegean conference. We are also grateful to the sponsors.

We would like to express our sincere thanks to all the people of Tinos Island who generously provided us with both financial support and their hospitality. Special thanks are due to the members of the Administrative Committee of the Panhellenic Religious Foundation "Evangelistria of Tinos" and to the Municipalities of Tinos, Exomvourgon and Panormos of Tinos Island. Without their financial and other contributions there would have been no conference. We are grateful to the following organizations, who were the sponsors of the conference: INTRACOM, Unilever, Αθηναική Ζυθοποιία, ΔΕΛΤΑ Πρότυπος Βιομηχανία Γάλακτος, BSP, Tinos Beach Hotel, Johnson-Johnson, Metaxa, C^3T American Computers & Engineers Services, ΚΩΤΣΟΒΟΛΟΣ, ΕΛΑΪΣ, MAILIS, Perfetti Hellas.

We thank the many anonymous reviewers for their contributions of quality control and improvements to the papers compiled here. We are grateful to Karuna Mohindra and Christos Tsagaris for their assistance in the preparation of this volume. And we thank Fred Hillier for his role as Editor-in-Chief of

Kluwer's *International Series in Operations Research and Management Science*, and Gary Folven and Roberts Apse of Kluwer for their guidance in the nuts and bolts of this process, and mostly for their patience.

STANLEY B. GERSHWIN

Members of the 1999 Tinos Workshop, their families, students, and friends. *(Photo provided by Heinrich Kuhn)*

This book is dedicated to
John A. Buzacott for his
pioneering leadership of and
outstanding contributions to
the field of Manufacturing
Systems Engineering.

Chapter 1

CAPACITATED TWO-ECHELON INVENTORY MODELS FOR REPAIRABLE ITEM SYSTEMS

Zeynep Muge Avsar
Department of Industrial Engineering
Middle East Technical University
06531 Ankara Turkey

W. Henk Zijm
Faculty of Mathematical Sciences
University of Twente, P.O. Box 217, 7500 AE Enschede, The Netherlands

Abstract In this paper, we consider two-echelon maintenance systems with repair facilities both at a number of local service centers (called bases) and at a central location. Each repair facility may be considered to be a job shop and is modeled as a (limited capacity) open queuing network, while any transport from the central facility to the bases (and vice versa) is modeled as an ample server. At all bases as well as at the central repair facility, ready-for-use spare parts are kept in stock. Once an item in the field fails, it is returned to one of the bases and replaced by a ready-for-use item from the spare parts stock, if available. The returned failed item is either repaired at the base or shipped to and repaired at the central facility. In the case of local repair, the item is added to the local spare parts stock as a ready-for-use item after repair. If a repair at the central facility is needed, the base orders an item from the central spare parts stock to replenish its local stock, while the failed item is added to the central stock after repair. Orders are satisfied on a first-come-first-serve basis while any requirement that cannot be satisfied immediately either at the bases or at the central facility is backlogged.

We assume that failed items are returned to the bases according to a Poisson process, and that each repair shop (at the bases as well as at the central facility) can be modeled as a Jackson network. Under these conditions, we propose a special near-product-form solution that provides an excellent approximation for the steady-state distribution of a slightly aggregated system, that permits us to calculate all relevant performance measures (such as fill rates and stockout probabilities) at the bases as well as at the central facility, as a function of target inventory levels at each location. Errors of these performance measures are generally less than one percent, when compared with simulation results. Finally, we show how these approximations can be used to determine optimal inventory levels at both the central and local facilities.

Keywords: multi-echelon systems, repairable items, finite repair capacities, queuing-inventory models, base-stock policy

1. Introduction

Since the pathbreaking work of Sherbrooke [19], multi-echelon models for repairable items inventory control have received considerable attention. The current paper can be seen as an extension of Sherbrooke's METRIC (Multi Echelon Technique for Recoverable Item Control) models by integrating inventory systems with finite capacity repair centers (modeled as open Jackson networks) at both the local bases and a central depot. In this section, we first describe the general structure of two-echelon repairable item systems. Next, we briefly discuss some key references on multi-echelon repairable item systems, and indicate the main contributions of this paper.

Let us first outline the general structure of capacitated multi-echelon, multi-indenture repairable item models. Suppose items are placed in the field to operate, possibly as part of a large technical system. If an item fails, it is returned to a local, nearby repair facility (called a base) which at the same time ships a ready-for-use spare item, taken from a local spare parts stock, to the technical system in order to minimize the non-operating time of that system. The failed item is either repaired at the base or, in case the repair turns out to be hard and requires special engineering expertise, at a central repair shop. Note that there may also be cases where repair is possible only at the central location but not at the bases. In such cases, failed items are always returned to the central repair facility. In the case of a local repair, upon completion of the repair the item is added to the local stock. If, however, the failed item have to be repaired at the central facility, the base immediately orders a ready-for-use item from a central stock at this repair facility, to replenish the item shipped to the field installation. Once the repair at the central facility is completed, the revised item is added to the central stock as a ready-for-use item again. Hence, as long as no condemnation occurs (i.e., each item can be repaired either at one of the bases or at the central repair shop) the total number of items (operating in the field, being in repair or in transport between depot and bases, or stocked as ready-for-use items) is constant in principle. We assume that all requirements are fulfilled on a first-come-first-serve basis and that each demand that cannot be satisfied immediately either at the bases or at the central facility is backlogged.

Models similar to the one described above have initially been considered by Sherbrooke [19] and have become known as METRIC models. As in almost all (multi-echelon) inventory models, Sherbrooke focuses on the determination of optimal order-up-to levels at both the local and the central stocking centers, and ignores any limitation on the available repair capacities at any facility. On the other hand, he considers multiple items that may operate together in complex systems (such as aircrafts, ships or production facilities) and attempts to maximize the overall system availability under a given budget constraint. Although, as mentioned already, the total number of rotating items of each type is fixed, the number of items operating in the field is assumed to be sufficiently large to allow for the field demand (due to failures) to be approximated by a Poisson process.

During the last three decades several important improvements of METRIC have been proposed. Muckstadt [18] was the first to recognize the importance of the product structure with respect to recoverable item control. He extended the existing METRIC model, which may be characterized as a two-echelon, single-indenture model, to a two-echelon, two-indenture model, which is also referred to as MOD-METRIC. Another variant of METRIC is VARI-METRIC, a two-echelon, single-indenture model developed by Slay [22]. In the core part of the analysis of the initial METRIC model, it is assumed that, for each product, the number of items in repair follows a Poisson distribution (of which the variance equals the mean). In his VARI-METRIC method, Slay derives an approximate expression for the variances of the number of items in repair. Next, for each product, he fits a negative binomial distribution on the first two moments of these items in order to obtain a more accurate approximation. Graves [11] independently developed a slightly simpler approximation for the variance of the number of items in repair. Next, he also continues with fitting a negative binomial distribution on the first two moments. Sherbrooke [20] generalized the original VARI-METRIC method and developed a two-indenture, two-echelon version of VARI-METRIC. By simulation, it has been shown that the results produced by this method are fairly accurate. An overview of METRIC type models is given in Sherbrooke [21]. Extensions to more flexible models allowing for emergency repair or emergency supply (but still assuming no resource constraints) have been studied by Verrijdt [25].

Another important line of research was initiated by Gross [12]. The main difference between the VARI-METRIC model and the models of Gross and others (see, e.g., Gross et al. [13], [14] and Albright [1]) is in the constraints of the repair process. In VARI-METRIC, it is assumed that all repair leadtimes are independent variables, which corresponds to an infinite repair capacity. In the models by Gross and others, a limited repair capacity is assumed, however at the cost of other rather restrictive assumptions such as a dedicated repair capacity and fixed repair routings. In these models, the circulation of products through repair, distribution and use is usually modeled as a closed queuing network, and hence, in the spare parts literature, these models are also known as the "closed queuing network models for spare parts management". Approximations for general single stage queuing stations at both the central and local repair centers have been proposed by Diaz and Fu [8]. A previous work that is to be directly compared with our study is due to Lee and Zipkin [15], [16]. In [15], they approximated multi-echelon make to stock systems with a single exponential server at every echelon based on an independence assumption in the spirit of network decomposition methods of Whitt [26], [27], [28] employed previously for pure tandem queues. A later article of Lee and Zipkin [16] extended [15] for the cases of component commonality.

For this paper, another line of research is of interest. This line starts with the classical multi-echelon inventory systems under periodic review studied initially by Clark and Scarf [7] for serial systems, and later extended to inverse aborescent or distribution structures (one central depot and multiple local warehouses) by numerous authors, see Federgruen [10], Axsäter [2], Van Houtum et al. [24] and Diks et al. [9] for reviews of the literature. Again, almost all authors assume unlimited production capacities at any facility and model all supply lead times as being either fixed or an independent random variable. Only recently, models have been proposed to integrate serial base-stock systems with limited capacities at each facility where these facilities are modeled as either open or population constrained queuing network models, see, e.g., Buzacott et al. [5] and Buzacott and Shanthikumar [6]. Note, however, that in all these models items are not circulating but instead are procured from an external supplier, subsequently go through one production stage, next are stocked, subsequently go through a second downstream production stage, are stocked again, and finally leave the system to satisfy external demand. Hence, these models do not allow for distribution structures nor for the complex routings that occur in METRIC models (with either local or central repair).

In this paper, we attempt to integrate models of resource-constrained repair facilities and multi-echelon inventory models for repairable item systems. In order to keep description of the development of our approach simple and clear, we first consider a system with a single base having a local spare parts stock (but not a repair facility) and just one repair facility which is at the central location. Then, we proceed with the use of the same approximation idea for investigating the case of multiple local bases and a central repair facility, each modeled as a finite capacity open Jackson queuing network. Repair may take place either at a base or at the central facility. As in most papers, we assume that demand (due to failed items in the field) occurs at each base according to a Poisson process. The main contribution of this paper is described as follows: under a slight modification of the steady state equations of the total system, the resulting equations can be shown to have a product form solution. Based on the latter solution, several performance measures (such as fill rates, stock-out probabilities and expected shortages) can be calculated both at the bases and at the central facility, as a function of target inventory levels. Numerical experiments show that the calculated performance measures deviate in general less than one percent from those determined through simulation. This in turn allows the use of the modified models to determine optimal base stock inventory levels, e.g., in order to achieve target service levels, taking into account all repair constraints.

We wrap up this section with an outline of this paper. In the next section, we consider a very simple two-echelon system, consisting of one base without a repair facility and a single central repair node. This model mainly serves to explain the essential elements of the modification discussed above. We present the modified system with product form solution and numerical results to show the accuracy of the approximation. Next, we turn to more general repairable item network structures and more general repair facilities in Section 3, present the main results for the modified systems and again discuss numerical results. In Section 4, we show how to use the modified models for the purpose of optimizing the target inventory levels of ready-for-use items at both the central repair facility and the bases. In Section 5, we summarize our results and discuss a number of extensions that are currently being investigated.

2. Analysis of a simple two-echelon system with central repair

In this section, we first discuss a highly simplified repairable item system, to explain how a slight modification turns this system into a near-product-form network that can be completely analyzed. The results of this analysis are shown to serve as excellent approximations for key performance measures of the original system.

Consider the system displayed in Figure 1 with a single base and a central repair facility. Failed items that arrive at the base are always shipped to the central repair facility, hence no repair at the base occurs. Both the central repair facility and the base hold a number of ready-for-use items in stock. Each failed item generates a demand for a new item at the base, while at the same time the base orders an item from the stock at the central repair facility to update its local stock again. Each demand not immediately fulfilled at either the base or the depot is backordered. The central repair facility is modeled as an exponential server with repair rate μ_0 while transport of an item from the depot to the base stock is modeled as an ample server with exponential service rate μ_1. Transport of a failed item from the base to the depot is not modeled explicitly. In the next section, the single server repair center will be replaced by a product form network, including possible ample servers, hence then transport from base to depot is easily included in the network structure.

We assume that the system operates according to a base-stock policy. Let S_0 and S_1 be the specified target spare part inventory levels at the central repair facility and at the base, respectively. The stock location of the central repair facility is often referred to as the depot. There is one synchronization station associated with both the central depot and the base. Each request (which may be a backorder) is merged with ready-for-use item in stock if there is one available for each through these synchronization stations. The number of items to be repaired or being in repair at the central repair shop is denoted by the random variable N while the number of items in transport from the depot to the base is denoted by M. The number of ready-for-use items (i.e., items already repaired, now being stocked) at the depot (base) is denoted by $\bar{N}$ ($\bar{M}$). There is not a queue formed in front of the transportation node because it is an ample server. The number of items backordered in case there is no spare part in stock is denoted by K_0 for the depot and by K_1 for the base. Note that each failure of an item in the field results in a request both at the depot and at the base, due to the fact that any demand fulfillment at the base leads to a replenishment order from the base to the depot at the same time. We assume that items fail according to a Poisson process with rate λ where naturally $(\lambda/\mu_0) < 1$ in order for the system to be ergodic.

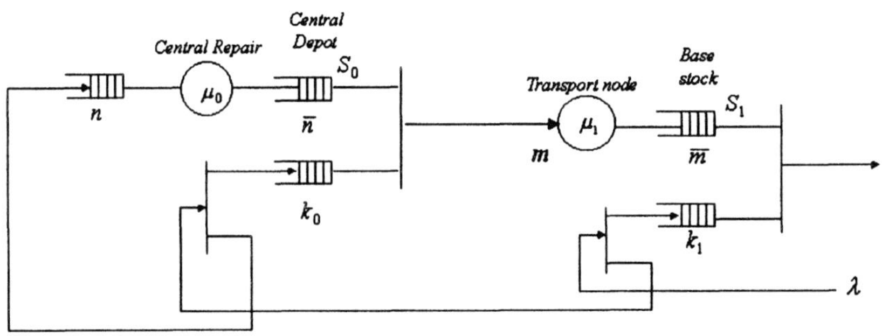

Figure 1. A single item, two-echelon model

For $N = n$, $\bar{N} = \bar{n}$, $M = m$, $\bar{M} = \bar{m}$, $K_0 = k_0$ and $K_1 = k_1$, the following equations hold as a result of the operating inventory control policies:

$$n + \bar{n} - k_0 = S_0,$$

$$m + k_0 + \bar{m} - k_1 = S_1.$$

Since any request is satisfied immediately if there are available spare parts at the depot (base), the request and the spare part are merged just upon the arrival of the request. Also, if requests are being backordered at the depot (at the base), then just after the repair completion of a part (arrival of a part shipped from the depot to the base) it is merged with the longest waiting request. So, the values of both $\bar{n}$ ($\bar{m}$) and k_0 (k_1) at the upper (lower) echelon can never be positive at the same time. More precisely,

$$If \;\; n \leq S_0, \;\; then \;\; \bar{n} = S_0 - n \;\; and \;\; k_0 = 0;$$

$$If \;\; n > S_0, \;\; then \;\; \bar{n} = 0 \;\; and \;\; k_0 = n - S_0;$$

$$If \;\; m + k_0 \leq S_1, \;\; then \;\; \bar{m} = S_1 - (m + k_0) \;\; and \;\; k_1 = 0;$$

$$If \;\; m + k_0 > S_1, \;\; then \;\; \bar{m} = 0 \;\; and \;\; k_1 = (m + k_0) - S_1.$$

From these relations, it follows immediately that n and m completely determine the state of the system, including the values of $\bar{n}$, k_0, $\bar{m}$ and k_1. Thus, this repairable item system can be modeled as a continuous time Markov chain with state description (n, m). The corresponding transition diagram is displayed in Figure 2. $P(N = n, M = m)$ is the steady-state probability of being in state (n, m). Note that for any $n > S_0$ the system behaves as an open tandem queuing system, since in this case a backlog occurs at the depot, causing each completed item at the central repair facility to be transferred immediately to the base. The more difficult part of the transition diagram (similar to that of a fork-join queue system since indeed here one demand generates both an additional repair request and the start of a transport activity) arises for $n < S_0$. Fortunately, that part of the state space can be very naturally aggregated since the states with $0 \leq n \leq S_0$ are precisely those states with no backlog at the depot, i.e., with $k_0 = 0$, while any $k_0 > 0$ corresponds to the set of states with $n = (S_0 + k_0)$.

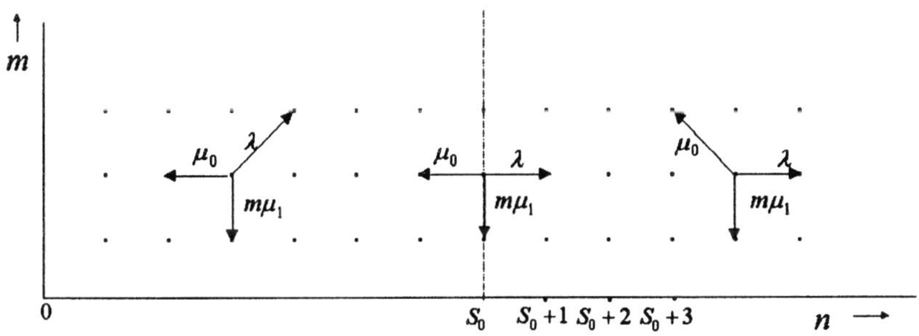

Figure 2. Transition diagram for state description (n, m)

Therefore, a natural aggregation is a description of the system through the states (k_0, m). Denote the steady-state probabilities of this new model by $\tilde{P}$. For simplification of the notation, P_{nm} and $\tilde{P}_{k_0m}$ will be used for $P(N = n, M = m)$ and $P(K_0 = k_0, M = m)$, respectively. Then, for any m,

$$\tilde{P}_{0m} = \sum_{n \leq S_0} P_{nm}, \tag{1}$$

$$\tilde{P}_{k_0m} = P_{S_0+k_0,m} \qquad for\ \ k_0 > 0. \tag{2}$$

Let $q(m)$ be the steady-state probability that an arriving request for an item at the depot has to wait, given that it finds no other waiting requests in front of it and $M = m$. Using the fact that Poisson arrivals see time averages, we have

$$q(m) = \frac{P_{S_0m}}{\sum_{n \leq S_0} P_{nm}} \qquad for\ \ every\ \ m,$$

while the transition diagram corresponding to the alternative state space description is displayed by Figure 3.

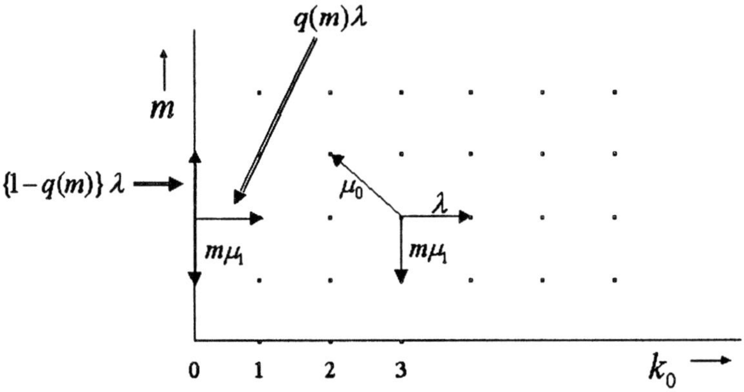

Figure 3. Transition diagram for state description (k_0, m)

The difficulty in the description of the aggregate model is of course in the dependence of $q(m)$ on m. Basically, the modification to be discussed below comes down to ignoring this dependence. Let $P_0(N = n)$ denote the marginal probability distributions of the states at the central repair facility. Let further $\rho_0 = (\lambda/\mu_0)$ and $\rho_1 = (\lambda/\mu_1)$. From Figure 1, it is immediately observed that $P_0(N = n) = (1 - \rho_0)\rho_0^n$.

The modification of the aggregate model with state description (k_0, m) proposed in this article is based on the ignorance of the dependence of $q(m)$ on m, meaning that in the balance equations of the states (k_0, m) the conditional probabilities $q(m)$ are all replaced by

$$q = \frac{P_0(N = S_0)}{P_0(N \leq S_0)} = \frac{(1-\rho_0)\rho_0^{S_0}}{(1-\rho_0^{S_0+1})}.$$

Note that q is the expected probability that a request arriving at the depot has to be backordered when this request does not see any backordered item at the depot at the time of its arrival, while not observing the value of m, i.e.,

$$q = \sum_{m=0}^{\infty} q(m)P(M = m|N \leq S_0).$$

Hence, q can be seen as a weighted average of the values $q(m)$. The following result is essential for the analysis in this paper. It follows immediately by substituting the given distribution in the balance equations of the aggregate model when $q(m)$ is replaced by q.

Theorem 1. *For the modified aggregate model, the steady-state distribution is given by*

$$\check{P}(K_0 = k_0, M = m) = \begin{cases} (1-\rho_0^{S_0+1})\,\frac{e^{-\rho_1}\rho_1^m}{m!} & \text{for } k_0 = 0, \\ (1-\rho_0)\rho_0^{S_0+k_0}\,\frac{e^{-\rho_1}\rho_1^m}{m!} & \text{for } k_0 > 0. \end{cases}$$

Note that $\left(\frac{e^{-\rho_1}\rho_1^m}{m!}\right)$ is the steady-state probability of having m jobs in an $M/M/\infty$ system with traffic intensity ρ_1. As mentioned earlier, for $k_0 > 0$ the system behaves as a tandem queuing system and hence the product form seems natural, given the one-to-one correspondence between k_0 and n in this case. The states with $k_0 = 0$ represent the aggregation and hence satisfy different transition rates. The solution given in the above theorem will be referred to as a near-product-form solution, for obvious reasons. The reader may note that for $S_0 = 0$ the modified model is equivalent to the original model, since in this case $q(m) = q = 1$ for all m.

Basically, the modification suggested above is similar to the approximations suggested by Buzacott et al. [5] and Buzacott and Shanthikumar [6]. However, by viewing the approximation in terms of the conditional probabilities $q(m)$ and q, and by explicitly considering the balance equations of the aggregate system with states (k_0, m), the near-product-form immediately follows, which in turn eliminates the need to consider the marginal distribution of the base explicitly. The above references decompose the system completely, and therefore require an explicit description of the arrival process of items at the second stage (the base) which is definitely *not* a Poisson process (see, e.g., the derivation in the first appendix of Buzacott et al. [5]). The derivation of the near-product-form solution eliminates the need for this decomposition.

At this point, we should note that the approximate near-product-form solution proposed in this article also results immediately from the stronger assumptions given in the following remark.

Remark: *Assume*

1) *the random variables N and M (in the original model shown by Figure 2) are independent,*

2) *the arrival process for the transport node is Poisson with rate λ.*

These assumptions imply that

$$P(N = n, M = m) = P_0(N = n)\left(\frac{e^{-\rho_1}\rho_1^m}{m!}\right)$$

for every (n, m) pair. Then, since the probability distribution of the partially aggregated model satisfies (1) and (2), the approximate probability distribution obtained for (k_0, m) is the same as the near-product-form solution given by theorem 1. □

Comparing the approach presented by this remark with the '$q(m)$ to q' argument, the key result of this article is observed as follows: the near-product-form solution is still the approximate distribution when the less restrictive assumption of constant q is considered instead of (1) and (2) in the remark. The following explanation is to clarify why constant q assumption isn't that strong: Independence of N and M (assumption (1) in the remark) leads to the independence of K_0 and M. However, the use of constant q means that when $K_0 = 0$, we impose

$$P(N = S_0, M = m) = P_0(N = S_0)P_1(M = m),$$

$$P(N \leq S_0, M = m) = P_0(N \leq S_0)P_1(M = m),$$

(defining P_1 as the marginal distribution of M), and nothing more, in particular constant q assumption does not imply the independence of N and M. Furthermore, with the approach in the remark the transport node is assumed as $M/M/\infty$ with arrival rate λ (assumption (2) in the remark) in addition to the independence of N and M. On the other hand, when '$q(m)$ to q' argument is used, it is just this argument leading to the near-product-form distribution with the approximate marginal steady-state probability of the random variable M turning out to be of $M/M/\infty$, i.e., $\left(\frac{e^{-\rho_1}\rho_1^m}{m!}\right)$.

As for the purpose of placing the contribution of this paper among others in the literature, we compare our approximation with the one due to Lee and Zipkin [15]. This is because once the ample node of the model analyzed in this section is replaced with a single exponential server, we come across with the two-echelon model studied by Lee and Zipkin [15] for which the arguments of our approach would still hold and give the approximate near-product-form probability distribution

$$\check{P}(K_0 = k_0, M = m) = \begin{cases} (1-\rho_0^{S_0+1})\,(1-\rho_1)\rho_1^m & for\ \ k_0 = 0, \\ (1-\rho_0)\rho_0^{S_0+k_0}\,(1-\rho_1)\rho_1^m & for\ \ k_0 > 0. \end{cases}$$

The point to which attention should be drawn is that assumption (2) in the remark is considered by Lee and Zipkin also in addition to an independence assumption for the sojourn times of a unit at different stages (echelons) they make for their approximation. Now, what is shown in lemma 1 below is that the probability distribution for $(K_0 + M)$, which is computed with the use of the near-product-form solution as

$$\begin{aligned} \check{P}(K_0 + M = z) &= (1-\rho_0^{S_0+1})\,(1-\rho_1)\rho_1^z \\ &\quad + I_{\{z>0\}} \sum_{m=0}^{z-1} (1-\rho_0)\rho_0^{S_0+(z-m)}(1-\rho_1)\rho_1^m \end{aligned}$$

for $z \geq 0$, is the same as the one proposed by Lee and Zipkin [15]. Lee and Zipkin's computations to approximate the probability distribution of $(K_0 + M)$ are based on a method due to Svoronos and Zipkin [23]. The approximate distribution they propose is

$$\bar{P}(K_0 + M = z) = \begin{cases} 1 - \pi A^0 \boldsymbol{e} & for\ \ z = 0, \\ \pi A^{z-1} \boldsymbol{\tau} & for\ \ z > 0, \end{cases}$$

where $\boldsymbol{e}$ is a vector with all of its entries being 1, $\pi = (\rho_0^{S_0+1}, (1-\rho_0^{S_0+1})\rho_1)$, $\boldsymbol{\tau} = ((1-\rho_0)(1-\rho_1), (1-\rho_1))^T$, $A = \begin{bmatrix} \rho_0 & (1-\rho_0)\rho_1 \\ 0 & \rho_1 \end{bmatrix}$. Lemma 1 results from the investigation of these formulas derived by Lee and Zipkin [15] as compared to the distribution of $(K_0 + M)$ approximated using our approach. The importance of the observation in lemma 1 is due to the fact that the performance measures are all expressed in terms of the probability distribution of $(K_0 + M)$. In order to see the validity of this observation numerically, we computed $E(K_0 + M)$ using near-product-form solution and observed that they are all the same as the ones listed in table I of [15].

Lemma 1.

a) For $z > 0$,

$$A^{z-1}\boldsymbol{\tau} = ((1-\rho_0)(1-\rho_1)\rho_0^{z-1}\sum_{m=0}^{z-1}\left(\frac{\rho_1}{\rho_0}\right)^m, (1-\rho_1)\rho_1^{z-1})^T.$$

b) $\bar{P}(K_0 + M = z) = \check{P}(K_0 + M = z)$ *for* $z \geq 0$.

Proof:

a) Proof is given by induction. Denote

$$((1-\rho_0)(1-\rho_1)\rho_0^{z-1}\sum_{m=0}^{z-1}\left(\frac{\rho_1}{\rho_0}\right)^m, (1-\rho_1)\rho_1^{z-1})^T$$

by $\boldsymbol{c}_z$ for $z > 0$. For $z = 1$, $A^0\boldsymbol{\tau} = I\boldsymbol{\tau} = \boldsymbol{\tau}$ is equal to $\boldsymbol{c}_1$. Assume $A^{z-1}\boldsymbol{\tau} = \boldsymbol{c}_z$. Then, it is observed that $A^z\boldsymbol{\tau} = A(A^{z-1}\boldsymbol{\tau}) = A\boldsymbol{c}_z$ is equal to $\boldsymbol{c}_{z+1}$.

b) For $z = 0$, $\bar{P}(K_0 + M = 0) = 1 - \pi A^0\boldsymbol{e} = 1 - \pi I\boldsymbol{e} = 1 - (\rho_0^{S_0+1} + (1 - \rho_0^{S_0+1})\rho_1)$, which is equal to $\check{P}(K_0 + M = 0)$. For $z > 0$,

$$\begin{aligned}\bar{P}(K_0 + M = z) &= \pi A^{z-1}\boldsymbol{\tau} = \pi\boldsymbol{c}_z \\ &= \rho_0^{S_0+1}(1-\rho_0)(1-\rho_1)\rho_0^{z-1}\left(\sum_{m=0}^{z-1}\left(\frac{\rho_1}{\rho_0}\right)^m\right) \\ &\quad +(1-\rho_0^{S_0+1})(1-\rho_1)\rho_1^z \\ &= \check{P}(K_0 + M = z).\end{aligned}$$

□

The advantage of our approach over the one of Lee and Zipkin[15] is that the $q(m)$-based near-product-form analysis can easily be extended to more complex systems (as shown in Section 3).

The rest of this section is devoted to the assessment of the proposed approximation. Performance measures typically considered for repairable item systems are the stockout probability, the fill rate and the expected shortage at the base. The fill rate is defined as the probability that any demand arriving at the base can be fulfilled immediately. To obtain these performance measures, the distribution of the random variable K_1 must be computed. This distribution satisfies

$$Pr(K_1 = 0) = \sum_{k_0+m\leq S_1} \tilde{P}_{k_0 m},$$

$$Pr(K_1 = k_1) = \sum_{k_0+m=S_1+k_1} \tilde{P}_{k_0 m} \qquad for \;\; k_1 > 0.$$

Then, in terms of this distribution the performance measures at the base are

Stockout Probability (SP): $Pr(K_1 > 0)$,

Fill Rate (FR): $Pr(K_0 + M < S_1) = Pr(K_1 = 0) - Pr(K_0 + M = S_1)$,

Expected Shortage (ES): $E(K_1)$.

We compute the measures regarding K_1 listed above by simulation and by using the analytical near product form solution of the modified system. The simulation results are based on an experimental setting with 15 replications and a simulation time of 10^6 for each replication. The confidence intervals are given for a confidence level of 0.95. Service rates μ_0 and μ_1 are 10. Tables 1, 2 in appendix are for the failure rate λ being 8, Table 3 is for $\lambda = 9.5$ and Table 4 is for $\lambda = 5$. Also, for the purpose of investigating performance of the proposed near-product-form solution when the ample transport node has a deterministic service time (as an extreme of a phase type transport time considering the decrease in the coefficient of variation), Table 5 is included in appendix for $\lambda = 9$, $\mu_0 = 10$ and the deterministic transport time being equal to 0.10. Note that the deterministic transportation time does not allow us to construct the model as in this section, but our aim in including Table 5 is to see how the near-product-form solution would work in such a case. The results in Tables 1, 2, 3, 4, 5 clearly show that the approximate solution performs excellently. A comparison of the results based on the solution of the modified model with the simulation results for different target stock levels shows errors less than 1 percent. Hence, it seems natural to investigate a similar modification for more complex repairable item systems, to see whether similar analytical results and equally accurate approximations can be derived. That will be the topic of the next section.

3. General capacitated two-echelon repairable item systems

In this section, we investigate complex but more realistic two-echelon repairable item systems, derive analytical results for a slightly modified aggregate system and study the performance of the solution of this modified system as an approximation for the exact performance measures. The basic characteristics of the systems studied in this section are listed below.

- *Repair units both at the depot and at each base:* Failed items arrive at each base according to a (possibly base-dependent) Poisson rate. The base is responsible for replacing the failed item with a ready-for-use spare part. A failed item is either repaired at a local base repair shop, after which it is stocked at the base as a ready-for-use item again, or sent to the central repair facility. In the latter case, at the same time a ready-for-use item is requested by the base from the depot to replenish its local base stock, while the item repaired at the central repair facility is placed in the depot upon completion.

- *Multiple bases:* The complete system consists of several bases and one central repair facility with a depot of ready-for-use spare parts. Requests from the bases for items from the depot stock are fulfilled on a first-come-first-serve basis.

- *Product form repair networks:* Each repair shop is modeled as an open Jackson network. Failed items that are returned from a base to the central repair facility follow a (probabilistic) routing in the central repair facility, with given process time distributions at each node being independent of the base that shipped them (see Section 5 for a relaxation of this assumption). However, requests from the different bases for new items to replenish their local stocks are distinguished in order to make sure that each shipped item arrives at the correct base. Transport times from the central depot to the bases may be base-dependent.

The most complicated maintenance system, for which the approximation approach proposed in this article is investigated, is obtained by combining all the extensions above. Such a system with two bases is displayed in Figure 4.

3.1 Modification and analysis of general repairable item systems

In this subsection, an extension of the analysis of modified aggregate systems is obtained and the existence of near-product form steady-state distribu-

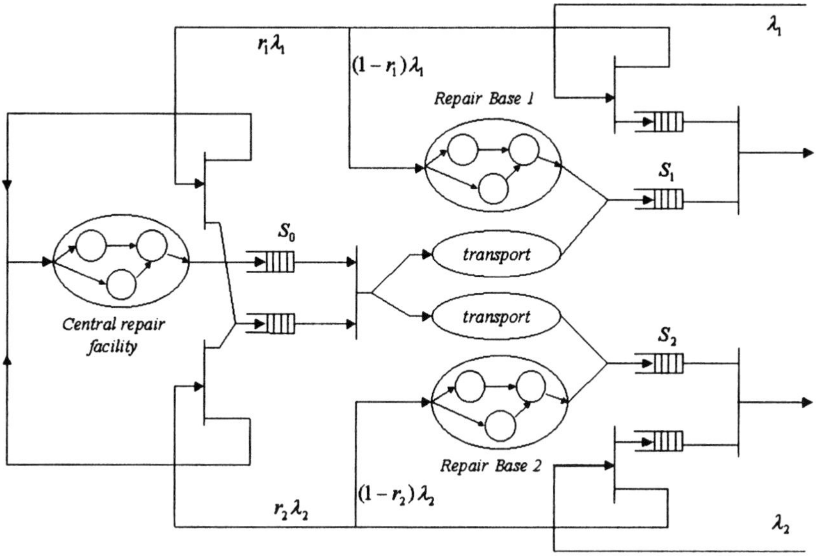

Figure 4. A two-echelon, single indenture model

tions is proven for the complex systems introduced above. First, we have to revise the notations introduced in the previous section. Let L be the number of bases and let M_l be the random variable describing the number of items in transfer from the depot to base l ($l = 1, 2, ..., L$). The target spare part inventory level is denoted by S_0 at the depot and by S_l at base l, the random variable denoting the number of requests not filled (backlogs) at that base is K_l while $\bar{M}_l$ is the random variable denoting the number of available spare parts that are ready-for-use, for $l = 1, 2, ..., L$. Suppose there are J nodes within the central repair network including an ample server node that represents transportation from the bases to the depot. The number of items being served or waiting to be served at node j of this network is denoted by random variable N_j for $j = 1, ..., J$. The states of the central repair network are represented by vectors $\boldsymbol{n} = (n_1, n_2, ..., n_J)$, where n_j is the number of items being in repair at node j ($j = 1, ..., J$) while $\bar{n}$ denotes the number of ready-for-use spare parts stocked at the depot. Let Z_l be the number of nodes at the repair network of base l, then similarly the states of the local repair network of base l are represented by vectors $\boldsymbol{h}_l = (h_{l1}, h_{l2}, ..., h_{lZ_l})$, for $l = 1, 2, ..., L$. Also, let $\boldsymbol{h} = (\boldsymbol{h}_1, \boldsymbol{h}_2, ..., \boldsymbol{h}_L)$. The vector $\boldsymbol{m} = (m_1, m_2, ..., m_L)$ describes the number of spare parts being transported from the depot to each base by base-specific ample servers, while $\bar{m}_l$ denotes the number of ready-for-use spare parts at base l ($l = 1, 2, ..., L$). Finally, let k_{0l} denote the number of requests from base l ($l = 1, 2, ..., L$), waiting to be fulfilled at the central depot, and let $\boldsymbol{k}_0 = (k_{01}, k_{02}, ..., k_{0L})$. For reasons to be explained below, we also define $k_0 = \sum_{l=1}^{L} k_{0l}$ as the total number of backordered requests at the central depot. $|\boldsymbol{n}|$ denotes the cardinality of $\boldsymbol{n}$, i.e., $\sum_{j=1}^{J} n_j$. Then, the state definition for the complete repairable item system is denoted by $(\boldsymbol{k}_0, \boldsymbol{n}, \boldsymbol{m}, \boldsymbol{h})$, while furthermore due to employment of the base-stock policies employed at the stock points

$$|\boldsymbol{n}| + \bar{n} - k_0 = S_0,$$

$$|\boldsymbol{h}_l| + m_l + k_{0l} + \bar{m}_l - k_l = S_l \qquad for \;\; l = 1, ..., L,$$

which imply the following relations:

$$If \;\; |\boldsymbol{n}| \leq S_0, \;\; then \;\; \bar{n} = S_0 - |\boldsymbol{n}| \;\; and \;\; k_0 = 0;$$

$$If \;\; |\boldsymbol{n}| > S_0, \;\; then \;\; \bar{n} = 0 \;\; and \;\; k_0 = |\boldsymbol{n}| - S_0;$$

$$For \;\; l = 1, 2, ..., L,$$

$$If \;\; |\boldsymbol{h}_l| + m_l + k_{0l} \leq S_l,$$
$$then \;\; \bar{m}_l = S_l - (|\boldsymbol{h}_l| + m_l + k_{0l}) \;\; and \;\; k_l = 0;$$

$$If \;\; |\boldsymbol{h}_l| + m_l + k_{0l} > S_l,$$
$$then \;\; \bar{m}_l = 0 \;\; and \;\; k_l = (|\boldsymbol{h}_l| + m_l + k_{0l}) - S_l.$$

The total failure rate is $\lambda = \sum_{l=1}^{L} \lambda_l$ with λ_l being the failure rate at base l, $l = 1, 2, ...L$. For each node j (z) of the central network (base l), the exponential service rate is denoted by μ_{0j} (μ_{lz}). Defining c_{0j} as the number of parallel servers at node j and $\mu_{0j}(n_j)$ being the service rate when there are n_j items at the node, we have

$$\mu_{0j}(n_j) = \begin{cases} n_j \mu_{0j} & if \;\; n_j < c_{0j}, \\ c_{0j} \mu_{0j} & if \;\; n_j \geq c_{0j}. \end{cases}$$

For each base l, c_{lz} and $\mu_{lz}(h_{lz})$ are defined similarly. Let r_l denote the probability that a failed item arriving at base l has to be sent to the central repair facility, hence $(1 - r_l)$ denotes the probability that the item can be repaired at the local base repair shop. Then, $\lambda_{CR} = \sum_{l=1}^{L} r_l \lambda_l$ denotes the total arrival rate of failed items (and hence also of replenishment requests from the bases) at the central repair shop while $(1 - r_l)\lambda_l$ is the arrival rate of items to be repaired at local repair shop l, for $l = 1, 2, ..., L$.

Let p_{ij} be the routing probability for an item being repaired at the central network to go from node i to node j. Denote the routing probability that an arriving item is sent to node j by p_{0j} and the routing probability of leaving the network after completion of the service at node j by p_{j0}. The total arrival rate $\hat{\lambda}_j$ into each node j of the central network is computed from

$$\hat{\lambda}_j = \left(\sum_{l=1}^{L} r_l \lambda_l \right) p_{0j} + \sum_{i=1}^{J} \hat{\lambda}_i \, p_{ij} \qquad for \;\; j = 1, ..., J. \tag{3}$$

Clearly the system is ergodic if $\hat{\lambda}_j < c_{0j}\mu_{0j}$ for $j = 1, ..., J$. The routing matrices and the effective arrival rates for the networks at the bases are defined similarly and similar ergodicity conditions hold.

Before continuing, we will slightly simplify the state space description by making an early aggregation step. Note that for a full system description the vector $\boldsymbol{k}_0 = (k_{01}, k_{02}, ..., k_{0L})$ is needed in order to determine the number of ready-for-use items and the backlogs at each base l. However, the system structure does not change at all if, as soon as a shipment from the central depot to a local base is initiated, we assume that with probability $r_l\lambda_l/\lambda_{CR}$ the item is shipped to base l. Note that physically all requests can be seen as standing in one queue in the order of arrival, since they are handled on a first-come-first serve basis. Hence, if we know the total number k_0 of backordered requests at the depot, the probability distribution of the values k_{0l} is easily determined by the binomial formula

$$P(K_{0l} = k_{0l}) = \binom{k_0}{k_{0l}} (r_l\lambda_l/\lambda_{CR})^{k_{0l}} (1 - (r_l\lambda_l/\lambda_{CR}))^{k_0 - k_{0l}},$$

while the joint probability distribution of the vectors $\boldsymbol{k}_0$ is given by the well-known multinomial distribution

$$P(K_{01} = k_{01}, ..., K_{0L} = k_{0L}) = \frac{k_0!}{\left(\prod_{l-1}^{L} k_{0l}!\right)} \left(\prod_{l-1}^{L} (r_l\lambda_l/\lambda_{CR})^{k_{0l}} \right).$$

Since k_0 is completely determined by the cardinality of the vector $\boldsymbol{n}$ (see the above state relations), it follows that we may restrict ourselves to the state description $(\boldsymbol{n}, \boldsymbol{m}, \boldsymbol{h})$ without loosing any generality, since the routing probabilities $(r_l\lambda_l/\lambda_{CR})$ can be used to derive the more detailed probability distributions on the initial state space described by $(\boldsymbol{k}_0, \boldsymbol{n}, \boldsymbol{m}, \boldsymbol{h})$, at least if some sort of product form solution arises again.

Now, we first concentrate on a special parameter setting for the above system in which $r_l = 1$ for all l, i.e., we ignore for the moment the local base repair shops and assume that all failed items have to be repaired at the central facility. Hence, for the moment we assume $\lambda_{CR} = \lambda$, i.e., we concentrate on the two-echelon system consisting of the central repair facility, its associated depot stock and possible requests from the bases, and the transport nodes to all bases. For this system, it can be shown that a modification similar to the one made in Section 2 leads to a set of balance equations with a near-product-form solution (see [3] for the details of this analysis). Next, the results can easily be extended to the system with possible repairs at the local base shops as well.

For the system with $r_l = 1$ for all l, the state description becomes $(\boldsymbol{n}, \boldsymbol{m})$. Analogous to the development in section 2, we define an aggregate model with states described by $(k_0, \boldsymbol{m})$ as follows:

$$
\begin{aligned}
&(0, \boldsymbol{m}) \quad for \ |\boldsymbol{n}| \leq S_0, \\
&(k_0, \boldsymbol{m}) \quad for \ |\boldsymbol{n}| = S_0 + k_0, \ \ k_0 > 0.
\end{aligned}
$$

The relation between the two models is investigated below. For any $\boldsymbol{m}$,

$$
\begin{aligned}
\tilde{P}_{0\boldsymbol{m}} &= \sum_{|\boldsymbol{n}| \leq S_0} P_{\boldsymbol{nm}}, \\
\tilde{P}_{k_0\boldsymbol{m}} &= \sum_{|\boldsymbol{n}| = S_0 + k_0} P_{\boldsymbol{nm}} \qquad for \ k_0 > 0.
\end{aligned}
$$

Note that now also for $k_0 > 0$ the state $(k_0, \boldsymbol{m})$ is an aggregate state. Using notations similar to those introduced in Section 2 for the steady-state probabilities, we define $q(\boldsymbol{m})$ by:

$$
q(\boldsymbol{m}) = \frac{\sum_{|\boldsymbol{n}| = S_0} P_{\boldsymbol{nm}}}{\sum_{|\boldsymbol{n}| \leq S_0} P_{\boldsymbol{nm}}} \qquad for \ every \ \boldsymbol{m}.
$$

Again, $q(\boldsymbol{m})$ denotes the probability that a request arriving at the central depot for a ready-for-use item finds no other requests waiting to be fulfilled while nevertheless the depot stock is depleted.

Now, as in the preceding section, the modification of the aggregate model with state description $(k_0, \boldsymbol{m})$ is based on the ignorance of the dependence of $q(\boldsymbol{m})$ on $\boldsymbol{m}$. Hence, (in the balance equations of the states $(k_0, \boldsymbol{m})$) the conditional probabilities $q(\boldsymbol{m})$ are all replaced by

$$
q = \frac{\sum_{|\boldsymbol{n}| = S_0} P_0(\boldsymbol{N} = \boldsymbol{n})}{\sum_{|\boldsymbol{n}| \leq S_0} P_0(\boldsymbol{N} = \boldsymbol{n})}.
$$

In [3], it is shown that for the special situation that $r_l = 1$ for all l this modified aggregate model has a near-product-form solution. However, it is easy to see that this result can be generalized for the situation of arbitrary $r_l, l = 1, 2, ..., L$. First of all, the result does not change at all if we replace λ_l by $r_l\lambda_l$ and λ by $\lambda_{CR} = \sum_{l=1}^{L} r_l\lambda_l$. Second, it is well-known that splitting a Poisson process into two streams with probabilities r_l and $(1 - r_l)$, respectively, results in two independent Poisson processes. Thirdly, merging L independent Poisson processes (as occurs at the central repair facility) again leads to a Poisson process. Finally, since each base repair shop is modeled as a Jackson network and hence has a product form steady-state solution on its own, and since the modified aggregate system (consisting of the central repair facility followed by the base transportation nodes) also has a product form solution (section 2), and all arrival processes are Poisson, the system as a whole (aggregated and modified in exactly the same way as before with all conditional probabilities $q(\boldsymbol{m})$ replaced by q) also has a product form solution (in which the initial product form solutions arise as marginal probabilities). This is summarized in the following theorem.

Theorem 2. *Consider the general capacitated repairable item system with one central repair facility and L local bases, as introduced in the beginning of this section. Suppose that each local repair shop and the central repair facility can be modeled as a Jackson network, while the transportation nodes are modeled as ample exponential servers. Then, the modified aggregate system that is obtained by considering only the backlogs at the central repair facility and replacing the conditional probabilities $q(\boldsymbol{m})$ by q for all $\boldsymbol{m}$, has a product form steady-state distribution. In particular, $P(K_0 = k_0, \boldsymbol{M} = \boldsymbol{m}, \boldsymbol{H} = \boldsymbol{h})$ is equal to*

$$\left(\sum_{|\boldsymbol{n}|\leq S_0} P_0(\boldsymbol{N} = \boldsymbol{n})\right)\left(\prod_{l=1}^{L} P_l(M_l = m_l)\right)\left(\prod_{l=1}^{L} P_l(\boldsymbol{H}_l = \boldsymbol{h}_l)\right)$$

and

$$\left(\sum_{|\boldsymbol{n}|=S_0+k_0} P_0(\boldsymbol{N} = \boldsymbol{n})\right)\left(\prod_{l=1}^{L} P_l(M_l = m_l)\right)\left(\prod_{l=1}^{L} P_l(\boldsymbol{H}_l = \boldsymbol{h}_l)\right)$$

for $k_0 = 0$ and $k_0 > 0$, respectively, where $P_0(\boldsymbol{N} = \boldsymbol{n})$ is the product form solution of the central repair shop with arrival rate $\lambda_{CR} = \sum_{l=1}^{L} r_l\lambda_l$, $P_l(M_l = m_l)$ denotes the steady-state probability of transportation node l with Poisson arrival rate $r_l\lambda_l$, and $P_l(\boldsymbol{H}_l = \boldsymbol{h}_l)$ is the product form solution of local repair shop l.

Again, we are to point out that this near-product-form solution is also obtained based on the assumption of Poisson arrival to each transport node l with rate $r_l\lambda_l$ and complete independence of N and M, as for the case of the simple model considered in Section 2 (recall the remark in Section 2). All the related discussions and comparisons for the '$q(m)$ to q' approach and the use of Poisson arrivals to transport nodes and complete independence could similarly be repeated here.

Since Theorem 2 holds, we can now exploit the multinomial expression to obtain the detailed probabilities $P(\boldsymbol{K}_0 = \boldsymbol{k}_0, \boldsymbol{M} = \boldsymbol{m}, \boldsymbol{H} = \boldsymbol{h})$. By using appropriate convolutions and the balance relations between the various state variables discussed in the beginning of this section, all relevant performance measures are now easily calculated. Note, however, that all these values are only exact for the modified aggregate system. In the next subsection, we show that they also provide excellent approximations for the performance measures (obtained through simulation) of the original system.

3.2 Numerical Experiments

In order to assess the performance of the approximation for a complicated maintenance system, numerical experiments are performed for a two-base system as displayed in Figure 4 where the central network consists of an ample transport node (to represent transportation from the bases to the depot) followed by a single repair node and there are ample nodes for transportation from the central depot to the bases and the repair units at the bases are single servers. Service rates for the nodes are all exponential. In Table 6 and 7 in appendix, simulation results and approximate values are given for the case $r_1 = r_2 = 0.5$ and the exponential service rates for local and central repairs and for the transportation nodes all being equal to 10. This system is referred as the symmetric system. Simulation runs are as designed for the simple maintenance system studied in Section 2. One other system for which the experiments are performed is of the same type with the repair rate at the bases being higher, 15, to make the system more realistic, $r_1 = 0.7$, $r_2 = 0.1$ and $\lambda_1 = 10$, $\lambda_2 = 8$, and other parameters being the same as before. Different r and λ values make this system asymmetric. The results are given by Table 8 in appendix.

All experiments carried out suggest that the approximation works very well for various sets of system parameters. Errors are generally less than one percent, with the largest error occurring in the determination of the expected backlogs at the bases.

4. Optimization

In the preceding section, we have observed that the performance measures as calculated for the modified model serve as an excellent approximation for those of the original model. This suggests that the modified model can be used to determine an optimal investment strategy, given that a target service level has to be attained, or an optimal allocation of stock between the depot and the local bases, given a certain budget restriction. As an example, we present in this section an approach to the first problem. Hence, we are interested in the minimum number of SKU's (stock keeping units or spare parts), and the optimal allocation of this number between depot and bases, such that a target fill rate is met. The realistic assumption here is that the cost per unit of stock to be allocated is the same for the depot and the bases.

Before answering this question we first have to define what is an overall fill rate. We have chosen this fill rate as a weighted average of the fill rates at the bases, i.e., we define the overall fill rate FR_b by

$$FR_b(\boldsymbol{S}) = \frac{\sum_{l=1}^{L} \lambda_l FR_l(\boldsymbol{S})}{\lambda},$$

where FR_l is the fill rate at base l, $l = 1, 2, ...L$, and $\boldsymbol{S} = (S_0, S_1, ..., S_L)$. Note that in this way FR_b is indeed the expected total proportion of requests that are fulfilled immediately. Next, we define a greedy, step-wise approach to find the minimum number of SKU's and the optimal allocation such that for some given $\alpha, 0 < \alpha < 1$, we have $FR_b(\boldsymbol{S}) \geq \alpha$.

The greedy approach works as follows: First, we determine the minimum order-up-to levels at the bases, $\widehat{S}_1, ..., \widehat{S}_L$, say, such that, at each base, the service level is at least α, while assuming that the central depot is never out of stock (equivalent to $S_0 = \infty$). It is clear that in an optimal allocation in which the stock position of the depot is taken into account, these values represent a lower bound for the base order-up-to levels. Next, we initialize a step-wise improvement procedure on $S_0 = 0$ and $S_l = \widehat{S}_l$, for $l = 1, 2, ..., L$. At each next step of the greedy approach, one more item is assigned to either of the bases or to the depot, based on the maximum improvement on the fill rate $FR_b(\boldsymbol{S})$. We stop as soon as a fill rate α is reached.

As an example, we have performed a number of numerical experiments for the same repairable item system as studied in Section 3.2, where we have chosen $\alpha = 0.95$. In appendix, the steps of the greedy approach are displayed in Tables 9 and 10 for the symmetric and asymmetric systems, respectively. At each step, one more item is assigned to either of the bases or to the depot based on the improvements implied at the current (S_0, S_1, S_2) by each assignment. For the symmetric system, the minimum resource to be utilized for FR_b to be at 0.95 is $S_0 + S_1 + S_2 = 6 + 7 + 6 = 19$. If a budget constraint allows for an investment of maximally 20 spare parts to be allocated to the depot and the bases, then the allocation obtained by the greedy approach is $S_0 = 6$, $S_1 = 7$, $S_2 = 7$. Note that, as observed in Tables 6 and 7, this allocation gives the best fill rate for the bases among all possible symmetric allocations, i.e., $S_1 = S_2$, of 20 spare part stocks.

Although a formal proof cannot be given here, extensive numerical experiments suggest that the fill rate defined above behaves as a multi-dimensional concave function as soon as $S_l \geq \widehat{S}_l$, for $l = 1, 2, ..., L$. Therefore, we believe that the resulting allocations are generally close to optimal. Apart from this, the optimization shows the power of the approximations based on the modified system analyzed in this paper. Without these approximations, extensive simulation studies would be needed to determine optimal allocations, which seem to be rather impractical for realistic systems.

5. Summary and possible extensions

In this paper, we have analyzed a fairly general two-echelon repairable item system with limited repair capacities at both a central repair facility and a number of local bases. Both the bases and the central facility are able to keep a number of ready-for-use items in stock. Items that have failed in the field are returned to the closest base and are replaced immediately by a ready-for-use item, if available. The returned item is either repaired at the base, in which case it is put in the local stock after repair, or sent to the central repair facility, in which case a ready-for-use item is immediately shipped to the base to update its stock level. Each repair shop (bases and central facility) can be modeled as an open Jackson queuing network. We have shown that a slight modification of the overall model reveals a product form solution which allows for a relatively easy calculation of several performance measures. Numerical experiments indicate that these functions excellently approximate the performance measures of the original system, given the order-up-to levels of the stocks at both the depot and the bases. Based upon this result, a greedy approach is defined to determine optimal order-up-to levels, in order to meet a given target fill rate.

The approach presented here can be extended easily for more complicated models. Note that the only approximations that are made in the paper concern the replacement of the conditional probabilities $q(m)$ by q for all m. We have presented the analysis under the assumption that all nodes (in either the base repair shop or the central facility, or the transportation nodes) show exponential behavior. In fact, all results remain valid as long as the repair networks are product form networks, i.e., each open BCMP network will do. In particular, this allows for general (non-exponential) transportation servers since in any BCMP network an ample server may have a general service time distribution without destroying the product form property.

One other extension to be mentioned here is for multi-echelon systems. Aggregate models of such systems would include conditional probabilities, namely q's, to be defined for each echelon as a function of the state of the downstream echelons. The approximation is then based upon the assumption of independence of the q values at each echelon from the states of the downstream echelons. The exact analysis of this modified model is similar to the one performed in this article for two-echelon systems.

If we skip the local repair shops and instead replace each transportation node between the central facility and a base by a product form network, then again the complete analysis can be repeated, with similar results. Again, by replacing $q(m)'s$ by q, the whole system turns out to be a product form network. The resulting model is the full equivalent of a two-echelon production-inventory system with limited production capacities at each site. Hence, such systems can be completely analyzed as long as the demand process is Poisson and no batching in production or shipping is allowed. Buzacott et al. [5] analyze serial systems (hence without the inverse aborescent or distribution structure). A further extension concerns the modeling of systems where in each stage only a limited number of products can be in production simultaneously, leading to so-called generalized Kanban systems (see, e.g., Buzacott [4], Buzacott and Shanthikumar [6] or Di Mascolo et al. [17]). However, as soon as the amount of work-in-process in a stage is restricted, the replacement of $q(m)'s$ by q no longer leads to product form solutions.

Clearly, although computational complexities increase, the extension of the models to multiple products presents no essential theoretical problems, as long as the repair shops can be modeled as product form networks. This allows in particular for base-dependent transportation times to the central repair facilities. Note that in the analysis in Section 3 this transportation time has been included in the central repair shop model, making these transportation times indistinguishable between bases. A more interesting application however concerns field operating systems that are built from several items where the complete system is down as soon as one item fails. The extension of the framework set up in this paper then leads to a complete generalization of METRIC models for capacitated systems.

In the same way, one may study multiple indenture levels, i.e., the case where each item is an assembly that may fail due to the failure of precisely one subassembly which may be replaced. This leads to questions on how many subassemblies and how many assemblies to store at either the central depot or each local base. In addition, one may study the impact of different repair policies. For instance, depending on the actual work load, one may decide to replace a complete assembly instead of only the component that caused the failure, to get the field system back to operation as soon as possible. Clearly, however, this leads to a shift of work because eventually the complete assembly has to be revised while in addition an assembly is obviously a more expensive SKU than a single component.

Finally, the impact of different levels of criticality of items in a system's operation may lead to different priority rules in the repair shops. This leads to the study of priority systems in a multi-echelon network. So far, no results are known to us on these systems.

In conclusion, we believe that the current analytical framework provides a powerful tool to assess the performance of fairly general capacitated multi-echelon repairable item systems, and subsequently to optimize these systems, while various extensions seem to be possible. However, the analysis of the models will be limited due to computational complexities and hence further numerical approximations may be needed (e.g., two moment approximations instead of a full characterization of probability distributions for stockouts, number of items in stock, etc.). These will be the topic of future research.

References

[1] S.C. Albright. An Approximation to the Stationary Distribution of a Multi-Echelon Repairable Item Inventory System with Finite Sources and Repair Channels. *Naval Research Logistics*, 36:179-195, 1989.

[2] S. Axsater. *Continuous Review Policies for Multi-Level Inventory Systems with Stochastic Demand*, volume 4. S. Graves, A. Rinnooy Kan and P. Zipkin: Logistics of Production and Inventory, Handbook of Operations Research, Elsevier-North-Holland, Amsterdam, The Netherlands, 1993.

[3] Z.M. Avşar and W.H.M. Zijm. *Resource-Constrained Two-Echelon Inventory Models for Repairable Item Systems.* Technical Report, Memorandum No.1521, University of Twente, Enschede, 2000.

[4] J.A. Buzacott. Generalized Kanban/MRP Systems. Technical Report, Department of Management Science, University of Waterloo, 1989.

[5] J.A. Buzacott, S.M. Price and J.G. Shantikumar. *Service Level in Multi-stage MRP and Base Stock Controlled Production Systems.* G. Fandel, Th. Gulledge and A. Jones (eds.): New Directions for Operations Research in Manufacturing, Springer, Berlin, 1992.

[6] J.A. Buzacott and J.G. Shantikumar. *Stochastic Models of Manufacturing Systems.* Prentice-Hall, EnglewoodCliffs, NJ, 1993.

[7] A.J. Clark and H. Scarf. Optimal Policies for a Multi-Echelon Inventory Problem. *Management Science*, 6:475-490, 1960.

[8] A. Diaz and M.C. Fu. Models for Multi-Echelon Repairable Item Inventory Systems with Limited Repair Capacity. *European Journal of Operational Research*, 97:480-492, 1997.

[9] E.B. Diks, A.G. de Kok and A.G. Lagodimos. Multi-Echelon Systems: A Service Measure Perspective. *European Journal of Operational Research*, 95:241-263, 1996.

[10] A. Federgruen. *Centralized Planning Models for Multi-Echelon Inventory Systems under Uncertainty*, volume 4. S. Graves, A. Rinnooy Kan and P. Zipkin: Logistics of Production and Inventory, Handbook of Operations Research, Elsevier-North-Holland, Amsterdam, The Netherlands, 1993.

[11] S.C. Graves. A Multi-Echelon Inventory Model for a Repairable Item with One-for-One Replenishment. *Management Science*, 31:1247-1256, 1985.

[12] D. Gross. On the Ample Service Assumption of Palm's Theorem in Inventory Modeling. *Management Science*, 28:1065-1079, 1982.

[13] D. Gross, L.C. Kioussin and D.R. Miller. A Network Decomposition Approach for Approximate Steady State Behavior of Markovian Multi-Echelon Repairable Item Inventory Systems. *Management Science*, 33:1453-1468, 1987.

[14] D. Gross, D.R. Miller and R.M. Soland. A Closed Queuing Network Model for Multi-Echelon Repairable Item Provisioning. *IIE Transactions*, 15:344-352, 1983.

[15] Y.J. Lee and P. Zipkin. Tandem Queues with Planned Inventories. *Operations Research*, 40:936-947, 1992.

[16] Y.J. Lee and P. Zipkin. Processing Networks with Inventories: Sequential Refinement Systems. Operations Research, 43:1025-1036, 1995.

[17] M. Di Mascolo, Y. Frein and Y. Dallery. An Analytical Method for Performance Evaluation of Kanban-Contralled Production Systems. *Operations Research*, 44 (1):50-64, 1996.

[18] J. Muckstadt. A Model for Multi-Item, Multi-Echelon, Multi-Indenture Inventory Systems. *Management Science*, 20:472-481, 1973.

[19] C.C. Sherbrooke. METRIC: A Multi-Echelon Technique for Recoverable Item Control. *Operations Research*, 16:122-141, 1968.

[20] C.C. Sherbrooke. VARI-METRIC: Improved Approximations for Multi-Indenture, Multi-Echelon Availability Models. *Operations Research*, 34:311-319, 1986.

[21] C.C. Sherbrooke. *Optimal Inventory Modeling of Systems: Multi-Echelon Techniques*. Wiley, New York, 1992.

[22] F.M. Slay. VARI-METRIC: An Approach to Modeling Multi-echelon Resupply When the Demand Process is Poisson with a Gamma Prior. Report AF301-3, Logistics Management Institute, Washington D.C., 1984.

[23] A. Svoronos and P. Zipkin. Evaluation of One-for-One Replenishment Policies for Multiechelon Inventory Systems. *Management Science*, 37:68-83, 1991.

[24] G.J. van Houtum, K. Inderfurth and W.H.M. Zijm. Materials Coordination in stochastic Multi-Echelon Systems. *European Journal of Operational Research*, 95:1-23, 1996.

[25] J.H.C.M. Verrijdt. *Design and Control of Service Part Distribution Systems*. PhD thesis, Eindhoven University of Technology, Faculty of Technology Management, Eindhoven, 1997.

[26] W. Whitt. Performance of the Queuing Network Analyzer. *Bell Syst. Tech. J.*, 62:2817-2843, 1983.

[27] W. Whitt. The Queuing Network Analyzer. *Bell Syst. Tech. J.*, 62:2779-2815, 1983.

[28] W. Whitt. Approximations for Departure Processes and Queues in Series. *Naval Research Logistics Quarterly*, 31:499-524, 1984.

Appendix

Table 1. Performance measures for single-base case, $\rho_0 = \rho_1 = 8/10$

$S_0 - S_1$		SP	FR	ES
	sim	0.4999	0.3774	2.50545
0-3		(.4997,.5002)	(.3773,.3775)	(2.4998,2.5092)
	app	0.4998	0.3774	2.5010
	sim	0.4003	0.5000	2.0046
0-4		(.4000,.4007)	(.4998,.5003)	(2.0000,2.0091)
	app	0.4002	0.5001	2.0011
	sim	0.4036	0.4900	2.0092
1-3		(.4033,.4039)	(.4898,.4902)	(2.0046,2.0137)
	app	0.40171	0.4925	2.0029
	sim	0.3203	0.5997	1.6042
0-5		(.3199,.3207)	(.5994,.6000)	(1.6000,1.6085)
	app	0.3202	0.5998	1.6009
	sim	0.3214	0.5965	1.6056
1-4		(.3210,.3217)	(.5962,.5967)	(1.6013,1.6100)
	app	0.32042	0.59829	1.60119
	sim	0.3256	0.5816	1.6114
2-3		(.3252,.3259)	(.5813,.5818)	(1.6073,1.6155)
	app	0.3232	0.5845	1.6045
	sim	0.2563	0.6797	1.2839
0-6		(.2560,.2567)	(.6793,.6800)	(1.2799,1.2879)
	app	0.2562	0.6798	1.2807
	sim	0.2566	0.6786	1.2842
1-5		(.2563,.2570)	(.6783,.6790)	(1.2802,1.2883)
	app	0.2562	0.6796	1.2808
	sim	0.2578	0.6744	1.2858
2-4		(.2575,.2582)	(.6741,.6748)	(1.2820,1.2897)
	app	0.2566	0.6768	1.2813
	sim	0.2626	0.6556	1.2921
3-3		(.2622,.2629)	(.6553,.6559)	(1.2882,1.2961)
	app	0.26036	0.65811	1.28571
	sim	0.2051	0.7437	1.0275
0-7		(.2048,.2055)	(.7433,.7440)	(1.0238,1.0313)
	app	0.2049	0.7439	1.0246

Table 2. Performance measures for single-base case, $\rho_0 = \rho_1 = 8/10$ (cont'd)

$S_0 - S_1$		SP	FR	ES
	sim	0.2052	0.7434	1.0276
1-6		(.2049,.2056)	(.7430,.7437)	(1.0238,1.0314)
	app	0.2049	0.7438	1.0246
	sim	0.2055	0.7422	1.0280
2-5		(.2052,.2059)	(.7418,.7425)	(1.0244,1.0316)
	app	0.2050	0.7434	1.0247
	sim	0.2067	0.7374	1.0296
3-4		(.2064,.2070)	(.7371,.7378)	(1.0259,1.0333)
	app	0.2056	0.7396	1.0254
	sim	0.2120	0.7149	1.0367
4-3		(.2117,.2124)	(.7146,0.7152)	(1.0331,1.0404)
	app	0.2101	0.7170	1.0307
	sim	0.9101	0	4.8036
0-0		(.9101,.9102)		(4.7988,4.8084)
	app	0.9101	0	4.8000
	sim	0.0114	0.9858	0.0572
10-10		(.0113,.0115)	(.9857,.9859)	(0.0557,0.0586)
	app	0.0113	0.9859	0.0563
	sim	0.1055	0.8670	0.5275
5-5		(.1052,.1058)	(.8667,.8673)	(0.5245,0.5305)
	app	0.1050	0.8679	0.5247
	sim	0.1449	0.7125	0.5769
8-2		(.1446,.1452)	(.7122,.7126)	(0.5740,0.5798)
	app	0.1439	0.7123	0.5724
	sim	0.1051	0.8687	0.5269
2-8		(.1048,.1054)	(.8683,.8690)	(0.5240,0.5296)
	app	0.1049	0.8689	0.5246

Table 3. Performance measures for single-base case, $\rho_0 = \rho_1 = 9.5/10$

$S_0 - S_1$		SP	FR	ES
0-4	sim	0.8135	0.1439	16.2786
		(.8132,.8138)	(.1437,.1441)	(16.1737,16.3836)
	app	0.8134	0.1439	16.2618
0-9	sim	0.6298	0.3372	12.5975
		(.6290,.6305)	(.3365,.3378)	(12.4939,12.7012)
	app	0.6294	0.3374	12.5816
2-7	sim	0.6298	0.3371	12.5968
		(.6291,.6306)	(.3364,.3378)	(12.4929,12.7008)
	app	0.6294	0.3374	12.5816
4-5	sim	0.6302	0.3360	12.5977
		(.6295,.6309)	(.3353,.3366)	(12.4928,12.7027)
	app	0.6294	0.3369	12.5816
5-4	sim	0.6310	0.3327	12.5984
		(.6303,.6318)	(.3320,.3334)	(12.4937,12.7030)
	app	0.6301	0.3339	12.5824
0-20	sim	0.3587	0.6224	7.1624
		(.3572,.3602)	(.6210,.6239)	(7.0680,7.2568)
	app	0.3580	0.6231	7.1538
5-15	sim	0.3587	0.6224	7.1613
		(.3572,.3602)	(.6210,.6239)	(7.0664,7.2563)
	app	0.3580	0.6231	7.1538
10-10	sim	0.3587	0.6224	7.1625
		(.3572,.3602)	(.6210,.6238)	(7.0681,7.2570)
	app	0.3580	0.6231	7.1538
16-4	sim	0.3607	0.6130	7.1644
		(.3592,.3621)	(.6116,.6144)	(7.0691,7.2596)
	app	0.3596	0.6142	7.1557
20-30	sim	0.0765	0.9194	1.5336
		(.0750,.0780)	(.9180,.9210)	(1.4870,1.5802)
	app	0.0768	0.9191	1.5312

Table 4. Performance measures for single-base case, $\rho_0 = \rho_1 = 5/10$

$S_0 - S_1$		SP	FR	ES
0-2	sim	0.2039	0.6065	0.4098
		(.2038,.2041)	(.6064,.6066)	(.4092,.4104)
	app	0.2039	0.6065	0.4098
2-4	sim	0.0134	0.9718	0.0264
		(.0133,.0134)	(.9717,.9719)	(.0261,0.0266)
	app	0.0130	0.9730	0.0259
2-3	sim	0.0282	0.9356	0.0545
		(.0281,.0283)	(.9355,.9358)	(.0542,.0549)
	app	0.0270	0.9382	0.0529
3-3	sim	0.0151	0.9605	0.0284
		(.0150,0.0152)	(.9604,.9606)	(.0282,.0287)
	app	0.0144	0.9619	0.0274
1-4	sim	0.0262	0.9464	0.0521
		(.0261,.0263)	(.9463,.9465)	(0.0518,.0524)
	app	0.0258	0.9477	0.0516
4-1	sim	0.1095	0.5892	0.1517
		(.1094,.1095)	(.5891,.5892)	(.1514,0.1520)
	app	0.1092	0.5876	0.1517
6-2	sim	0.0176	0.9050	0.0228
		(.0175,.0176)	(.9049,.9051)	(.0227,0.0229)
	app	0.0174	0.9051	0.0225
6-1	sim	0.0950	0.6022	0.1179
		(.0950,.0951)	(.6021,.6023)	(.1177,.1180)
	app	0.0949	0.6018	0.1174
15-1	sim	0.0899	0.6061	0.1062
		(.0891,.0906)	(.6052,.6070)	(.1053,.1071)
	app	0.0902	0.6065	0.1066
5-5	sim	0.0008	0.9982	0.0016
		(.0008,.0009)	(.9981,.9982)	(.0016,.0017)
	app	0.0008	0.9982	0.0016

Table 5. Performance measures for single-base case, $\lambda = 9$, $\mu_0 = 10$ and constant transport time 0.10

$S_0 - S_1$		SP	FR	ES
0-10	sim	0.3472	0.6144	3.4727
		(.3466,.3477)	(.6139,.6149)	(3.4571,3.4884)
	app	0.3468	0.6147	3.4681
2-10	sim	0.2813	0.6875	2.8131
		(.2807,.2819)	(.6869,.6881)	(2.7981,2.8280)
	app	0.2809	0.6879	2.8092
2-14	sim	0.1848	0.7948	1.8453
		(.1842,.1854)	(.7942,.7954)	(1.8320,1.8586)
	app	0.1843	0.7952	1.8431
4-8	sim	0.2813	0.6875	2.8131
		(.2807,.2819)	(.6869,.6881)	(2.7981,2.8280)
	app	0.2809	0.6879	2.8092
6-6	sim	0.2814	0.6873	2.8131
		(.2808,.2819)	(.6868,.6879)	(2.7982,2.8280)
	app	0.2809	0.6877	2.8092
8-4	sim	0.2826	0.6800	2.8146
		(.2821,.2832)	(.6794,.6805)	(2.7997,2.8295)
	app	0.2822	0.6803	2.8107
10-2	sim	0.3216	0.5425	2.8640
		(.3211,.3222)	(.5421,.5430)	(2.8488,2.8791)
	app	0.3212	0.5428	2.8599
12-0	sim	0.6969	0	3.4455
		(.6967,.6972)		(3.4308,3.4603)
	app	0.6968	0	3.4418
0-5	sim	0.5874	0.3474	5.8787
		(.5871,.5878)	(.3471,.3477)	(5.8618,5.8956)
	app	0.5873	0.3475	5.8733
1-4	sim	0.5876	0.3463	5.8789
		(.5873,.5880)	(.3460,.3466)	(5.8620,5.8958)
	app	0.5875	0.3463	5.8735
2-3	sim	0.5897	0.3369	5.8815
		(.5894,.5901)	(.3366,.3372)	(5.8646,5.8984)
	app	0.5896	0.3370	5.8761
3-2	sim	0.6031	0.2923	5.8985
		(.6028,.6034)	(.2921,.2925)	(5.8816,5.9154)
	app	0.6030	0.2923	5.8931
4-1	sim	0.6597	0.1665	5.9767
		(.6595,.6600)	(.1664,.1666)	(5.9599,5.9936)
	app	0.6597	0.1665	5.9714

Table 6. Performance measures for two-base case, the symmetric system

$S_0 - S_1 - S_2$		SP	FR	ES
	sim	base 1 0.0258	0.9614	0.0771
		(.0256,.0260)	(.9611,.9616)	(.0763,.0780)
0-10-10		base 2 0.0257	0.9615	0.0771
		(.0255,.0260)	(.9612,.9618)	(.0762,.0780)
	app	base 0.0257	0.9614	0.0763
	sim	base 1 0.0248	0.9628	0.0741
		(.0246,.0250)	(.9626,.9631)	(.0733,.0750)
2-9-9		base 2 0.0247	0.9629	0.0741
		(.0245,.0250)	(.9626,.9632)	(.0731,.0750)
	app	base 0.0247	0.9628	0.0732
	sim	base 1 0.0240	0.9639	0.0714
		(.0238,.0242)	(.9637,.9642)	(.0706,.0722)
4-8-8		base 2 0.0239	0.9640	0.0714
		(.0237,.0241)	(.9637,.9643)	(.0705,.0723)
	app	base 0.0239	0.9639	0.0705
	sim	base 1 0.0235	0.9641	0.0694
		(.0233,.0237)	(.9638,.9643)	(.0686,.0702)
6-7-7		base 2 0.0235	0.9642	0.0694
		(.0233,.0237)	(.9639,.9644)	(.0685,.0703)
	app	base 0.0234	0.9641	0.0685
	sim	base 1 0.0240	0.9618	0.0691
		(.0238,.0242)	(.9616,.9621)	(.0682,.0699)
8-6-6		base 2 0.0240	0.9619	0.0691
		(.0238,.0242)	(.9616,.9622)	(.0682,.0700)
	app	base 0.0240	0.9619	0.0681
	sim	base 1 0.0271	0.9533	0.0730
		(.0269,.0273)	(.9531,.9536)	(.0722,.0738)
10-5-5		base 2 0.0271	0.9534	0.0730
		(.0268,.0273)	(.9531,.9537)	(.0721,.0739)
	app	base 0.027	0.9534	0.0720
	sim	base 1 0.0366	0.9289	0.0877
		(.0364,.0368)	(.9286,.9292)	(.0869,.0885)
12-4-4		base 2 0.0366	0.9289	0.0877
		(.0363,.0368)	(.9287,.9292)	(.0868,.0886)
	app	base 0.0365	0.9290	0.0867
	sim	base 1 0.0623	0.8650	0.1296
		(.0621,.0625)	(.8647,.8652)	(.1287,.1304)
14-3-3		base 2 0.0622	0.8650	0.1295
		(.0620,.0624)	(.8647,.8652)	(.1286,.1304)
	app	base 0.0621	0.8650	0.1285

Table 7. Performance measures for two-base case, the symmetric system (cont'd)

$S_0 - S_1 - S_2$		SP	FR	ES
	sim	base 1 0.1279	0.7104	0.2389
		(.1277,.1281)	(.7102,.7106)	(.2380,.2397)
16-2-2		base 2 0.1279	0.7105	0.2387
		(.1277,.1281)	(.7103,.7107)	(.2378,.2396)
	app	base 0.1278	0.7104	0.2376
	sim	base 1 0.2847	0.3964	0.5070
		(.2845,.2849)	(.3963-.3965)	(.5062-.5078)
18-1-1		base 2 0.2847	0.3964	0.5068
		(.2845,.2849)	(.3962,.3966)	(.5059,.5076)
	app	base 0.2846	0.3963	0.5057
	sim	base 1 0.6015	0	1.0948
		(.6014,.6016)		(1.0941,1.0955)
20-0-0		base 2 0.6015	0	1.0947
		(.6014,.6017)		(1.0939,1.0955)
	app	base 0.6015	0	1.0936
	sim	base 1 0.0084	0.9873	0.0253
		(.0083,.0086)	(.9871,.9874)	(.0248,.0258)
5-10-5		base 2 0.0673	0.8959	0.1976
		(.0670,.0677)	(.8955,.8963)	(.1960,.1992)
	app	base 1 0.0084	0.9872	0.0245
		base 2 0.0673	0.8960	0.1967

Table 8. Performance measures for two-base case, the asymmetric system

$S_0 - S_1 - S_2$		SP	FR	ES
0-10-10	sim	base 1 0.0833	0.8905	0.3478
		(.0829,.0837)	(.8901,.8910)	(.3444,.3511)
		base 2 0.0006	0.9987	0.0012
		(.0006,.0006)	(.9987,.9988)	(.0011,.0012)
	app	base 1 0.0834	0.8904	0.3477
		base 2 0.0006	0.9987	0.0012
0-13-5	sim	base 1 0.0367	0.9518	0.1529
		(.0364,.0370)	(.9515,.9522)	(.1505,.1553)
		base 2 0.0236	0.9514	0.0456
		(.0235,.0237)	(.9513,.9516)	(.0454,.0459)
	app	base 1 0.0367	0.9517	0.1527
		base 2 0.0236	0.9514	0.0456
2-11-5	sim	base 1 0.0385	0.9493	0.1607
		(.0382,.0389)	(.9490,.9497)	(.1583,.1631)
		base 2 0.0197	0.9594	0.0380
		(.0196,.0198)	(.9593,.9595)	(.0377,.0382)
	app	base 1 0.0386	0.9493	0.1605
		base 2 0.0197	0.9594	0.0379
4-9-5	sim	base 1 0.0405	0.9467	0.1689
		(.0402,.0408)	(.9464,.9471)	(.1664,.1714)
		base 2 0.0172	0.9644	0.0331
		(.0171,.0173)	(.9643,.9646)	(.0329,.0334)
	app	base 1 0.0405	0.9467	0.1687
		base 2 0.0172	0.9644	0.0331
6-7-5	sim	base 1 0.0426	0.9438	0.1777
		(.0423,.0430)	(.9434,.9442)	(.1751,.1803)
		base 2 0.0157	0.9675	0.0302
		(.0156,.0158)	(.9674,.9679)	(.0300,.0305)
	app	base 1 0.0426	0.9438	0.1774
		base 2 0.0159	0.9674	0.0302
8-5-5	sim	base 1 0.0457	0.9371	0.1879
		(.0454,.0460)	(.9367,.9374)	(.1853,.1905)
		base 2 0.0148	0.9693	0.0284
		(.0147,.0149)	(.9692,.9694)	(.0282,.0287)
	app	base 1 0.0456	0.9374	0.1875
		base 2 0.0148	0.9693	0.0284
10-3-5	sim	base 1 0.0650	0.8703	0.2198
		(.0647,.0654)	(.8700,.8707)	(.2171,.2224)
		base 2 0.0142	0.9705	0.0273
		(.0141,.0143)	(.9704,.9706)	(.0271,.0276)
	app	base 1 0.0647	0.8706	0.2188
		base 2 0.0142	0.9704	0.0273

Table 9. Results of the greedy optimization procedure for the symmetric system

$S_0 - S_1 - S_2$	FR_b	(S_0+1, S_1, S_2)	(S_0, S_1+1, S_2)	(S_0, S_1, S_2+1)
0-4-4	0.6003	FR_1=.6688 FR_2=.6688 FR_b=.6688	FR_1=.7223 FR_2=.6003 FR_b=.6613	FR_1=.6003 FR_2=.7223 FR_b=.6613
1-4-4	0.6688	FR_1=.7253 FR_2=.7253 FR_b=.7253	FR_1=.7733 FR_2=.6688 FR_b=.7211	FR_1=.6688 FR_2=.7733 FR_b=.7211
2-4-4	0.7253	FR_1=.7709 FR_2=.7709 FR_b=.7709	FR_1=.8147 FR_2=.7253 FR_b=.7700	FR_1=.7253 FR_2=.8147 FR_b=.7700
3-4-4	0.7709	FR_1=.8074 FR_2=.8074 FR_b=.8074	FR_1=.8480 FR_2=.7709 FR_b=.8094	FR_1=.7709 FR_2=.8480 FR_b=.8094
3-5-4	0.8094	FR_1=.8747 FR_2=.8074 FR_b=.8410	FR_1=.8990 FR_2=.7709 FR_b=.8349	FR_1=.8480 FR_2=.8480 FR_b=.8480
3-5-5	0.8480	FR_1=.8747 FR_2=.8747 FR_b=.8750	FR_1=.8990 FR_2=.8480 FR_b=.8735	FR_1=.8480 FR_2=.8990 FR_b=.8735
4-5-5	0.8747	FR_1=.8960 FR_2=.8960 FR_b=.8960	FR_1=.9177 FR_2=.8747 FR_b=.8962	FR_1=.8747 FR_2=.9177 FR_b=.8962
4-6-5	0.8962	FR_1=.9326 FR_2=.8960 FR_b=.9143	FR_1=0.9456 FR_2=.8747 FR_b=.9101	FR_1=.9177 FR_2=.9177 FR_b=.9177
4-6-6	0.9177	FR_1=.9326 FR_2=.9326 FR_b=.9326	FR_1=.9456 FR_2=.9177 FR_b=.9316	FR_1=.9177 FR_2=.9456 FR_b=.9316
5-6-6	0.9326	FR_1=.9446 FR_2=.9446 FR_b=.9446	FR_1=.9559 FR_2=.9326 FR_b=.9443	FR_1=.9326 FR_2=.9559 FR_b=.9443
6-6-6	0.9446	FR_1=.9542 FR_2=.9542 FR_b=.9542	FR_1=.9641 FR_2=.9446 FR_b=.9544	FR_1=.9446 FR_2=.9641 FR_b=.9544
6-7-6	0.9544	FR_1=.9707 FR_2=.9542 FR_b=.9625	FR_1=.9765 FR_2=.9446 FR_b=.9606	FR_1=.9641 FR_2=.9641 FR_b=.9641
6-7-7	0.9641	FR_1=.9707 FR_2=.9765 FR_b=.9765	FR_1=.9707 FR_2=.9641 FR_b=.9641	FR_1=.9707 FR_2=.9703 FR_b=.9703

Table 10. Results of the greedy optimization procedure for the asymmetric system

$S_0 - S_1 - S_2$	FR_b	(S_0+1, S_1, S_2)	(S_0, S_1+1, S_2)	(S_0, S_1, S_2+1)
0-3-5	0.5920	FR_1=.4310 FR_2=.9558 FR_b=.6643	FR_1=.4496 FR_2=.9514 FR_b=.6726	FR_1=.3046 FR_2=.9764 FR_b=.6032
0-4-5	0.6726	FR_1=.5623 FR_2=.9558 FR_b=.7372	FR_1=.5742 FR_2=.9514 FR_b=0.7418	FR_1=.4496 FR_2=.9764 FR_b=0.6837
0-5-5	0.7418	FR_1=.6655 FR_2=.9558 FR_b=.7945	FR_1=.6740 FR_2=.9514 FR_b=.7973	FR_1=.5742 FR_2=.9764 FR_b=.7530
0-6-5	0.7973	FR_1=.7452 FR_2=.9558 FR_b=.8388	FR_1=.7515 FR_2=.9514 FR_b=.8403	FR_1=.6740 FR_2=.9764 FR_b=.8084
0-7-5	0.8403	FR_1=.8060 FR_2=.9558 FR_b=.8726	FR_1=.8108 FR_2=.9514 FR_b=.8733	FR_1=.7515 FR_2=.9764 FR_b=.8514
0-8-5	0.8733	FR_1=.8524 FR_2=.9558 FR_b=.8983	FR_1=.8560 FR_2=.9514 FR_b=.8984	FR_1=.8108 FR_2=.9764 FR_b=.8844
0-9-5	0.8984	FR_1=.8877 FR_2=.9558 FR_b=.9179	FR_1=.8904 FR_2=.9514 FR_b=.9175	FR_1=.8560 FR_2=.9764 FR_b=.9095
1-9-5	0.9179	FR_1=.9124 FR_2=.9594 FR_b=.9333	FR_1=.9145 FR_2=.9558 FR_b=.9329	FR_1=.8877 FR_2=.9785 FR_b=.9281
2-9-5	0.9333	FR_1=.9317 FR_2=.9622 FR_b=.9452	FR_1=.9333 FR_2=.9594 FR_b=.9449	FR_1=.9124 FR_2=.9803 FR_b=.9426
3-9-5	0.9452	FR_1=.9467 FR_2=.9644 FR_b=.9546	FR_1=.9480 FR_2=.9622 FR_b=.9543	FR_1=.9317 FR_2=.9817 FR_b=.9539

Chapter 2

DISTRIBUTION RESOURCE PLANNING SYSTEMS: A CRITIQUE AND ENHANCEMENT

Gerald E. Feigin *
SmartOps
One North Shore Center, Suite 400
12 Federal Street, Pittsburgh, PA 15212
GFeigin@smartops.com

Kaan Katircioglu
IBM Research Division
T.J. Watson Research Center
Yorktown Heights, NY 10598
kaan@us.ibm.com

David D. Yao †
IEOR Department
Columbia University
New York, NY 10027
yao@ieor.columbia.edu

*Research undertaken while affiliated with IBM Research Division, T.J. Watson Research Center.
†Research undertaken while an academic visitor at IBM Research Division, T.J. Watson Research Center.

Abstract

Distribution Resource Planning (DRP) is a general framework for planning and managing inventory in distribution networks. The DRP framework can be applied to complex distribution networks with thousands of unique stock-keeping units and hundreds of stocking locations. It allows for non-stationary (e.g. seasonal) demand patterns and a wide variety of user-specified inventory control rules including all standard inventory policies such as (S, s) and fixed order quantity rules. A number of software implementations of DRP are commercially available and are widely used in industry. In this paper, we describe the logic underlying DRP and point out some of its limitations. The inner workings of DRP are not always familiar to the research/academic community. On the other hand, practitioners may be unaware of some of the shortfalls and limitations of the system. Our objective here is to bridge this gap. In particular, we show how the performance evaluation capability of DRP can be substantially enhanced by some simple analytical formulas, derived as approximations from base-stock and (S, s) control schemes.

1. Introduction

In the retail and distribution sectors of many industries, inventory management activities are made difficult by the sheer enormity of the number of distinct SKUs (stock-keeping units) that are sold. For example, it is not unusual for a large department store to stock more than 100,000 items. In such an environment, inventory management can be reasonably carried out only with the aid of decision support software which automates many of the planning activities, and enables decision makers to view at a glance what their inventory requirements are over time at different levels of product aggregation and at different points in the network. In addition, the software gives estimates of future on-hand inventory, backorders, and service levels, and provides the decision maker with the ability to evaluate quickly how changes in planning parameters such as safety stock levels and order quantities will affect inventory performance, measured primarily by inventory turnover and order fill rate. Because this software can typically be linked directly with order entry and purchasing systems, the entire replenishment decision process, if desired, can be automated. The role of the inventory analyst then becomes one of tuning parameters governing the replenishment decisions to achieve desired business performance, as well as responding to exceptions that the software brings to the attention of the analyst.

A class of inventory decision support software widely used in industry is based on Distribution Resource Planning (DRP), which refers to a general framework for planning and managing inventory in distribution networks. In particular, it allows for the setting of inventory control parameters and the calculation of time-phased inventory requirements, based on user-specified goals, such as service level targets or total inventory investment. It can take into consideration a host of different constraints, including supply capacity constraints, minimum and maximum order quantity constraints, etc. Since the early 1980's, DRP systems have been implemented in a variety of commercial software packages, offering a wide range of decision support capabilities to inventory managers and procurement personnel, and have been widely used in industry. For descriptions of DRP as well as examples of how it is used, refer to Martin [2], Mlot et al. [3], Silver and Peterson [6], Stenger [7], and Smith [8]. Software packages that implement DRP have in recent years become the main products of many business software companies.

One of the advertised strengths of DRP is its ability to project future on-hand inventory and future replenishment requirements at all echelons of a distribution network. The projection of future on-hand inventory and replenishment requirements is accomplished using standard DRP logic, described below. However, as we shall explain, standard DRP logic provides these estimates without taking into consideration the uncertainty of future demands. Specifically, DRP logic, in its calculation of projected on-hand inventory and replenishment requirements, assumes that the demand that will be realized in the future is exactly equal to the forecast and this can lead to misleading performance projections.

Our objectives in this paper are as follows. First, we present a detailed analysis on the inner workings of DRP so as to pinpoint its shortfalls and limitations, in particular, in estimating future performance measures as mentioned above. Second, we want to illustrate how the performance evaluation capability of DRP can be substantially enhanced by incorporating into its logic some simple analytical formulas. We shall derive these formulas through analyzing certain familiar inventory systems, such as the base-stock control and the (S, s) control schemes. Third, in the context of a multi-echelon distribution network, we want to illustrate how safety stock levels can be set at upstream locations so as to meet a target service level at the retail level using an echelon inventory mechanism and how this mechanism can be implemented within the DRP framework. Throughout the paper, wherever exact analysis becomes intractable or requires heavy computation, we will not hesitate to resort to approximation, seeking to derive closed-form formulas that are easy to evaluate.

The use of simple approximate formulas are justified by the enormous volume of SKUs in any inventory system for which DRP-based software is meant to generate fast decision alternatives. An alternative would be to use discrete-event simulation, but experiments indicate that the required computation time for simulation, even on a powerful computer, makes analysis of large numbers of SKUs infeasible. Approximate formulas are also justified simply because in many industrial applications, demand streams fluctuate significantly over time, and exact analytical performance estimates in the case of non-stationary demands are mostly intractable and unavailable in the inventory literature. For notational simplicity, however, we shall not distinguish approximate formulas by the $\approx$ sign. What specific approximation is involved in a formula is often clear from the context. (For instance, in §4.2, $\mathsf{E}(X)$, the average inventory position, in (16) is based on X following a uniform distribution, which follows from the linear asymptote of the renewal function; the average backorder and inventory levels in (17) and (19) are based on approximating demand by a normal distribution.)

The remainder of the paper is organized as follows. In §2 we review DRP logic and point out its weaknesses. In §3, we derive analytical approximations that enhance DRP's projections on inventory performance, under the lot-for-lot ordering rule. In §4, we review the stationary (S, s) model, focusing on bringing out approximate formulas that can be adapted to the non-stationary setting and are easy to evaluate. In §5, we show that in general, DRP ordering rules can be unified as a dynamic (S, s) rule — with time-dependent S and s values. We then develop improved estimates on inventory performance based on the approximations in §4. In §6 we present some numerical results comparing our analytical formulas with standard DRP logic and with simulation. Finally, extensions to multi-echelon networks are briefly discussed in the concluding §7.

2. How Does DRP Work?

2.1 The DRP Logic

By DRP logic, we mean the mathematical algorithm used in standard DRP procedures to calculate future on-hand inventory and future replenishment requirements. While descriptions of standard DRP procedures are widely available in professional references such as Stenger [7], as well as in manuals of commercial software, we present a formal approach below. We explain the logic in terms of a single product at a single location; later, we provide an example illustrating how the logic carries over to distribution networks with multiple locations and multiple products.

DRP logic is based on a set of recursive equations that characterize the inventory dynamics over time. To be more precise, the principal logic of DRP involves the recursive calculation of the replenishment requirements, the on-hand inventory, and the backordered demand in the future for a specific product at a specific location. The first step in describing the recursion is to divide the future time horizon into a set of discrete periods of equal length, indexed by $t = 1, 2, \ldots, n$, where n is the time horizon of interest (As a practical matter, DRP logic is typically implemented using weekly or monthly periods with a time horizon of 6 months to one year.) So, assuming that we are currently at the beginning of period 1, the objective of DRP logic is to project, for the product and location of interest, the following quantities for all future periods, t, $t = 1, 2, \ldots, n$:

I_t := the on-hand inventory at the *end* of time period t.

B_t := the backordered demand at the *end* of time period t.

A_t := the required quantity of product needed at the beginning of period t.

Q_t := the constrained (or feasible) quantity of product needed at the beginning of period t.

R_t := the recommended order quantity at the beginning of period t.

The distinction between the requirements, A_t, and the constrained order quantity, Q_t, is important. Whereas A_t represents the quantity that is needed at the beginning of period t, Q_t reflects what is feasible, taking into account lead time constraints and order quantity restrictions. For example, if $A_t = 40$ and the maximum order quantity is 30, then Q_t would be set to 30.

To carry out the recursive calculation of these quantities, DRP assumes that the following information is known (again, for the specific product and location) at the start of time period 1:

D_t := the demand in future time period t, $t = 1, 2, \ldots, n$. Although the actual future demand is unknown, DRP assumes that the future demand in period t is known through its mean, $\mathsf{E}[D_t]$.

T_t := the scheduled receipts at the *beginning* of time period t. This is the quantity of the product that is currently (i.e., at the beginning of period 1) scheduled to arrive at the location at the start of period t.

K_t := the safety stock requirement for period t. This is the quantity of the product that should be kept at this location in this period to protect against uncertainty in demand. (Most DRP software implementations provide a variety of methods to automatically calculate safety stock requirements for each period. These methods vary in degree of sophistication but, just as with forecast generation, the particular approach used is not relevant to DRP logic.)

I_0 := the on-hand inventory at the *beginning* of time period 1. This is the *current* on-hand inventory.

B_0 := the backordered demand at the *beginning* of time period 1.

The scheduled receipts, T_t, should not be confused with the constrained order quantity, Q_t. The former is an input to DRP and represents what has been already ordered or in transit, while the latter is an output of DRP and reflects an assessment of what is additionally needed at the start of period t subject to order constraints. The scheduled receipts are sometimes referred to as "in-transit" to suggest that the goods are physically en route to the location in question.

At the beginning of each period t, the following sequence of events takes place. First, replenishments (if any) arrive, including both Q_t – the constrained order quantity, which relates to A_t and will be specified below – and T_t. These are used to first satisfy the backorders, if any, left from the previous period. Next, demand of the period, D_t, is realized and filled, which brings us to the *end* of the period, when I_t (on-hand inventory) and B_t (backorders) are updated.

To project future requirements, A_t, $t = 1, \ldots, n$, DRP makes use of the following recursive relation:

$$A_t = [B_{t-1} - I_{t-1} - T_t + D_t + K_t]^+, \tag{1}$$

where $[x]^+$ denotes $\max\{x, 0\}$. The above equation says that the replenishment requirement in each period along with any on-hand inventory from the last period and scheduled receipts due at the beginning of this period, should be able to supply the demand of the current period and any backlog from the last period, and still result in a surplus that is equal to the required safety stock for this period.

The next step in DRP logic is to modify A_t to reflect what can feasibly be obtained by the start of period t. To this end, the constrained order quantities, Q_t, are derived from A_t, by applying a set of prespecified order-size restrictions, which we refer to as *order policies* or *order rules*. Several commonly used rules are listed below. Note that all of the rules apply if and only if $A_t > 0$; otherwise, we set $Q_t = A_t = 0$.

(i) *Lot-for-Lot*: $Q_t = A_t$, i.e., there is no restriction on the order quantity.

(ii) *Min-Max*: with $Q_{\min}$ and $Q_{\max}$ being the (given) lower and upper limit on order quantities,

$$Q_t = \min\{\max(Q_{\min}, A_t), Q_{\max}\},$$

becomes the order quantity planned for period t.

(iii) *DOS* ("days of supply"): Q_t is equal to the (projected) demand over a given number of periods.

(iv) *EOQ* (economic order quantity): $Q_t = \sqrt{2\mu C/h}$, the classical EOQ formula (e.g., [6]), where μ is the average demand per period, C is a fixed cost for placing a replenishment order, and h is the inventory holding cost per period.

Note that all the rules above with the exception of (i) impose some restrictions on the order quantity. In general, we can assume there is some prespecified Q_t, which is a function of A_t (and other parameters).

The recursive equations governing the on-hand inventory, I_t, and the backorders, B_t, are simple to describe as well. First, denote the *net* inventory level at the beginning of period t as

$$Y_t = I_{t-1} - B_{t-1} + T_t, \quad t = 1, \ldots, n. \tag{2}$$

That is, Y_t accounts for the (net) balance left over from the previous period and the scheduled receipts for the current period. This amount, if positive, can be used to supply demand in the current period, along with the constrained order quantity, Q_t, which is derived from the requirement A_t following a prespecified order policy. If Y_t is negative, i.e., $|Y_t|$ is the number of backordered units, then $|Y_t|$ must be first deducted from Q_t, before any remaining units can be used to supply demand. Hence, $Q_t + Y_t$ is the net inventory at the beginning of period t, *after* the replenishment; and combining this quantity with the demand in period t yields the on-hand inventory and backorder levels:

$$I_t = [Q_t + Y_t - D_t]^+, \quad \text{and} \quad B_t = [D_t - Q_t - Y_t]^+. \tag{3}$$

The final calculation performed by DRP is the recommended orders. Suppose that the order lead time for a specific product at a specific location is L periods. In DRP, the recommended order quantity in period t corresponds to the calculated order quantity $L + t$ periods later:

$$R_t = Q_{t+L}, \quad t = 1, \ldots, n - L. \tag{4}$$

So, the recommended order quantity at the start of period t is whatever the constrained order quantity is for period $t + L$.

Since the future demand D_t is known only through its forccasted mean and standard deviation, to implement the recursions defined in (1, 2, 3, 4), DRP logic uses the forecasted mean demand, $\mathsf{E}[D_t]$, in place of D_t. Therefore, these quantities should be viewed as DRP's estimates of unknown future quantities. For example, R_t is DRP's estimate of the quantity that will be ordered at the start of period t standing at the beginning of period 1. In particular, R_1 is the current recommended order quantity; whereas R_5, for instance, is DRP's best guess at the quantity that is most likely to be ordered at the start of period 5.

2.2 An Example

To help illustrate DRP logic, we consider a simple distribution network, illustrated schematically in Figure 1. The network consists of three retailers supplied by one warehouse location. The retailers obtain replenishments from the warehouse, which in turn obtains replenishments from a supplier. For the purpose of this example, we focus on one product that is sold by all three retailers. We assume that the replenishment lead time for this product at each of the retailers is 5 weeks and that the replenishment lead time at the warehouse is 15 weeks. This means, for example, that if a retailer places a replenishment order at the start of week 1, then it will arrive at the start of week 6. Suppose the weekly demand at each retailer for this product follows a discrete uniform distribution on $[0, 20]$ – that is, the weekly demand equals n with probability $p = 1/21$, $n = 0, 1, \ldots 20$. Thus, the weekly demand at each retail location has a mean of 10 unites. Also suppose that at the start of week 1, there are 60 units of on-hand inventory available at each retailer and 100 units available at the warehouse.

Next, suppose that there are currently no scheduled receipts; and that the retailers wish to hold 10 units of safety stock and the warehouse wishes to hold 15 units of safety stock. Also, assume that the retailers place orders in quantities of 30 (i.e., three weeks of average demand) and that the warehouse places orders in quantities of 200. (Note that in most industrial applications, order quantities are often constrained by concerns that are independent of the demand process; and this is reflected in the difference between A_t and Q_t in the DRP logic highlighted above.)

Let us suppose that we are currently at the beginning of week 1 (i.e. $t = 1$) and we are interested in using DRP to project forward 16 weeks into the future. DRP logic assumes that the following information is known at this time:

- The mean and standard deviation of each retailer's future weekly demand over the horizon in question.
- The current on-hand inventory at each location, namely at all three retailers and at the warehouse.
- The scheduled receipts – i.e. a list of all shipments scheduled to arrive at each location by week over the horizon in question.
- The safety stock to be held at each location.

To project on-hand inventory and replenishment requirements for the next 16 weeks at each location in this network, DRP logic first starts with the retail locations and constructs a table for each retailer that reflects the weekly operation of the retailer over the next 16 weeks. The table consists of 16 columns (one for each week) and the following rows:

1 Expected (i.e. mean) demand ($\mathsf{E}[D_t]$)

2 Scheduled receipts (T_t)

3 Safety stock requirements (K_t)

4 Replenishment requirements (A_t)

5 Constrained order quantity (Q_t)

6 On-hand inventory (I_t)

7 Backordered demand (B_t)

8 Recommended orders (R_t)

Table 1 shows a completed DRP table for one retailer in our example. (Since all the retailers are identical in this example, the tables generated for the retailers are identical). As the reader can verify, all of the projected quantities, A_t, Q_t, I_t, B_t and R_t, have been calculated using the DRP recursions specified above while respecting the order quantity restriction.

To continue with our example, let us assume that the DRP tables have been constructed for all three retailers following the DRP logic described above. The next step is to construct an analogous DRP table for the warehouse. To do this, DRP sums up the recommended order quantities by week from all three retailers and this becomes the weekly demand forecast used to drive the DRP logic at the warehouse level. That is, row 1 of the DRP table at the warehouse corresponding to the expected weekly demand at the warehouse is constructed by summing the corresponding recommended order quantities from the three retail DRP tables. Once the weekly demand is specified, the DRP recursive calculations are carried out exactly as for the retailers, resulting in a completed DRP table for the warehouse. Table 2 shows a completed DRP table for the warehouse in our example.

To summarize, the end result of the DRP calculations is a set of tables, one for each product and location in the distribution network. The key quantities that these tables contain are *projections* (equivalently, *estimates*) about the future on-hand inventory and replenishment requirements. While we have illustrated the construction of these tables using a simple example, the logic we have described carries over in a straightforward manner to much more complex networks.

2.3 Critique of DRP Logic

The primary criticism of DRP logic is that it ignores the randomness of future demand in its calculations by substituting the mean forecasted demand in the recursive equations. In our example, the mean demand at the retail level is 10 units per week and this is the number that is used in carrying out the DRP calculations. However, the standard deviation of demand is 6.06 and nowhere in the DRP logic is this fact used. (The calculation of safety stock may make use of the standard deviation of demand, but this calculation falls outside the scope of DRP logic. As stated earlier, the safety stock is assumed to be given as an input to the DRP calculations. See the Appendix for a description of how safety stock levels are calculated to achieve a target fill rate within DRP.)

In fact, the implicit assumption in DRP logic is that the standard deviation of demand is 0.

Another way to view the situation is that the formal recursive equations governing DRP logic as defined in (1), (2), (3), and (4), are statements about *random* quantities. Viewed in this light, it is reasonable to ask if the DRP projections specified in the recursive equations correspond to the correct *expected values* of these random quantities. In fact, since $f(x) = (x)^+$ is a convex function of x, it is easy to see (following Jensen's inequality, or the so-called "variability ordering" — [4]) that in replacing D_t by its mean, the DRP logic will *underestimate* the expected values of A_t, I_t and B_t.

Furthermore, assuming a lot-for-lot order policy, i.e., $Q_t = A_t$, we can rewrite (3) as follows:

$$I_t = \max\{A_t + Y_t - D_t, 0\} = \max\{K_t, Y_t - D_t, 0\},$$

and

$$B_t = -\min\{A_t + Y_t - D_t, 0\}, = -\min\{\max\{K_t, Y_t - D_t\}, 0\}.$$

Note that the safety stock, K_t, need *not* always be positive. (A negative safety stock level could happen when either the required service level is low or the inventory cost is low, or both.)

Therefore, if the safety stock is positive, then DRP always estimates positive on-hand inventory (i.e., with zero backlog), which is at or above the level of safety stock, since $I_t \geq K_t$ from the above. On the other hand, when the safety stock is negative, DRP prescribes a backlog strategy of not keeping any on-inventory, but instead keeping a level of backlog that is at least as much as the absolute value of the (negative) safety stock. That is,

$$I_t = K_t \qquad \text{or} \qquad B_t = -K_t$$

according to $K_t > 0$ or $K_t < 0$. These predictions, of course, can be quite remote from the reality of any typical inventory system.

3. Improved Projections: Lot-for-Lot Rule

In this section, we discuss how to improve the DRP performance estimates under the lot-for-lot ordering rule. Assume that the demand in period t can be expressed as

$$D_t = \mu_t + \sigma_t Z_t$$

where μ_t is the mean demand in period t, σ_t is the standard deviation of demand in period t and $Z_t, t = 1, \ldots, n$ are i.i.d. random variables following any distribution with zero mean and unit variance. For ease of discussion, we focus on the case in which Z_t follows a standard normal distribution. The results extend readily to more general demand distributions.

To start with, assume zero lead time: $L = 0$. Suppose we are at the beginning of period t, with Y_t available as a realized (deterministic) value. We want to know what should be the requirement A_t, which once determined and ordered will arrive immediately (due to zero lead time). We want to make this decision so that B_t, the backorder at the end of the period, will satisfy the desired service level. In addition, we also want to have a good estimate on I_t, the on-hand inventory at the end of the period.

What we do *not* have access to, in determining A_t, is D_t, the demand to be realized over the period: it's actual value will only become available by the end of the period. At our decision epoch (the beginning of the period), we only know the distribution of D_t, which, as specified above, follows a normal distribution with mean μ_t and standard deviation σ_t.

Substituting (2) into (1), we can rewrite the latter as follows:

$$A_t = [D_t + K_t - Y_t]^+. \tag{5}$$

Since D_t is treated as deterministic, and replaced by μ_t in the DRP logic, we have $A_t = [\mu_t + K_t - Y_t]^+$. Therefore, with a lot-for-lot ordering policy, DRP logic basically follows a base-stock policy, with $\mu_t + K_t$ as the base-stock level: if the net inventory at the beginning of t, Y_t, falls below the base-stock level, $\mu_t + K_t$, then $A_t > 0$, and an order quantity of $Q_t = A_t$ will bring the inventory position back to the base-stock level. On the other hand, since Y_t can never exceeds the base-stock level (this can be shown inductively), $A_t = 0$ implies that Y_t is equal to the base-stock level (which could happen when there is no demand in period t 1). In either case, after the review (at the beginning of period t), since $L = 0$, with the instantaneous arrival of A_t (possibly zero), the on-hand inventory is brought back to the base-stock level. The inventory will then be depleted to supply the demand D_t. Write

$$K_t = k_t\sigma_t,$$

where k_t is the so-called *safety factor* (whereas K_t is the safety stock). At the end of the period, we have

$$\mathsf{E}[B_t] = \mathsf{E}[D_t - \mu_t - k_t\sigma_t]^+ = \sigma_t\mathsf{E}[Z_t - k_t]^+ = \sigma_t G(k_t);$$

and

$$\mathsf{E}[I_t] = \mathsf{E}[\mu_t + k_t\sigma_t - D_t]^+ = \sigma_t\mathsf{E}[k_t - Z_t]^+ = \sigma_t H(k_t).$$

where

$$G(k) := \mathsf{E}[Z - k]^+ = \int_k^\infty (x - k)d\Phi(x),$$

with $\Phi(x)$ denoting the distribution function of Z, and $H(k) := k + G(k)$.

Observe that $H(x)$ is an increasing function, $G(x)$ is decreasing, and both are convex. When Z is the standard normal variate, these two functions are widely used in inventory theory; refer to e.g., [10]. When x is reasonably large such that $\Phi(x) \approx 1$ (say, $x \geq 3$), we have, approximately, $H(x) \approx x$, $H(-x) \approx 0$; and $G(x) \approx 0$, $G(-x) \approx x$. These properties will become useful in the analysis below.

Next, consider the more general case of $L \geq 0$. Suppose again, we want to determine the requirement A_t for period t. We need to make this decision at the beginning of period $t - L$ so as to place the order in time for it to arrive in period t.

Position ourselves at the beginning of period $t - L$. What we know are the following: (a) the net inventory left from the previous period, Y_{t-L}; and (b) the scheduled arrivals, $A_{t-L}, A_{t-L+1}, ..., A_{t-1}$, which had already been decided and ordered. What we do *not* know is the sequence of demands over the lead time, $D_{t-L}, D_{t-L+1}, ..., D_t$.

Denote:

$$D(s,t) := D_s + \cdots + D_t, \qquad A(s,t) := A_s + \cdots + A_t;$$

$$\mu(s,t) := \mathsf{E}[D(s,t)], \qquad \sigma(s,t) := \mathsf{sd}[D(s,t)].$$

We claim that the right A_t value should be the following:

$$A_t = [\mu(t-L,t) + k_t\sigma(t-L,t) - Y_{t-L} - A(t-L,t-1)]^+. \tag{6}$$

First, suppose the quantity inside $[\cdot]^+$ on the right hand side above is positive, and hence so is A_t. Then, at the beginning of period $t - L$, the net inventory is Y_{t-L}, which, along with the scheduled arrivals over the lead time — a total of $A(t - L, t)$, is going to supply all the demands over the lead time — a total of $D(t - L, t)$. Hence, at the end of period t, the expected backorder is:

$$\begin{aligned} & \mathsf{E}[B_t | A_t > 0] \\ = \; & \mathsf{E}[D(t-L,t) - Y_{t-L} - A(t-L,t)]^+ \\ = \; & \mathsf{E}[\mu(t-L,t) + Z_t\sigma(t-L,t) - \mu(t-L,t) - k_t\sigma(t-L,t)]^+ \\ = \; & \sigma(t-L,t)\, \mathsf{E}[Z_t - k_t]^+ \\ = \; & \sigma(t-L,t)\, G(k_t), \end{aligned} \tag{7}$$

where the second equality follows from (6), assuming $A_t > 0$:

$$Y_{t-L} + A(t-L,t) = \mu(t-L,t) + k_t\sigma(t-L,t).$$

Similarly,

$$\begin{aligned} & \mathsf{E}[I_t | A_t > 0] \\ = \; & \mathsf{E}[Y_{t-L} + A(t-L,t) - D(t-L,t)]^+ \\ = \; & \sigma(t-L,t)\, \mathsf{E}[k_t - Z_t]^+ \\ = \; & \sigma(t-L,t)\, H(k_t). \end{aligned} \tag{8}$$

Next, suppose $A_t = 0$ in (6). This implies

$$Y_{t-L} + A(t-L, t-1) \geq \mu(t-L,t) + k_t\sigma(t-L,t).$$

Hence,

$$\begin{aligned} & \mathsf{E}[B_t | A_t = 0] \\ = \; & \mathsf{E}[\mu(t-L,t) + Z \cdot \sigma(t-L,t) - Y_{t-L} - A(t-L,t-1)]^+ \\ = \; & \sigma(t-L,t)\, G(k_t'), \end{aligned} \tag{9}$$

with

$$k_t' := \frac{Y_{t-L} + A(t-L,t-1) - \mu(t-L,t)}{\sigma(t-L,t)} \geq k_t. \tag{10}$$

Hence, $\mathsf{E}[B_t]$ in (8) is no greater than $\mathsf{E}[B_t]$ in (6), since $G(\cdot)$ is a decreasing function. (Intuitively, in this case, since the available (net) inventory Y_{t-L} is higher, the effective safety factor k_t' is also higher; hence, the (projected) backorder is lower.)

Similarly, when $A_t = 0$, we have

$$\mathsf{E}[I_t | A_t = 0] = \sigma(t-L,t)\, H(k_t'), \tag{11}$$

with k_t' following (10); and $\mathsf{E}[I_t]$ above is larger than $\mathsf{E}[I_t]$ of (7).

To summarize, we have

Proposition 1 When the order rule is lot-for-lot, i.e., $Q_t = A_t$ for all t, the required quantity needed at the beginning of period t (A_t), and hence the recommended order quantity at the beginning of period $t-L$ (R_{t-L}), should be set as follows:

$$R_{t-L} = Q_t = A_t = [s_{t-L} - Y_{t-L} - A(t-L,t-1)]^+,$$

where

$$s_{t-L} = \mu(t-L,t) + k_t\sigma(t-L,t).$$

In this case, DRP follows a base-stock control mechanism, with s_{t-L} being the base-stock level (or, reorder point) for period $t-L$. The estimates for expected on-hand inventory and backorders (for period t) follow (7,11) and (6,8), respectively.

It is important to note that the reorder point, s_{t-L}, in the above proposition has a time index $(t-L)$ that is consistent with the time when the order is placed. On the other hand, we choose to index the safety factor k_t by t, since it relates more closely to the on-hand inventory and backorders in period t.

We should point out that the expression in (6) that specifies the requirement A_t can also be derived directly from the DRP recursion. From (2,3), with $Q_t = A_t$, and making use of the identity:

$$(a-b)^+ - (b-a)^+ = a - b,$$

we have

$$\begin{aligned} Y_t &= I_{t-1} - B_{t-1} + T_t \\ &= Y_{t-1} + A_{t-1} - D_{t-1} \\ &= Y_{t-2} + A_{t-2} - D_{t-2} + A_{t-1} - D_{t-1} \\ &= \cdots \\ &= Y_{t-L} + A(t-L, t-1) - D(t-L, t-1). \end{aligned} \tag{12}$$

Substituting the above into (5), we recover (6), noticing that since DRP treats demand as deterministic, $D_t + D(t-L, t-1) = \mu(t-L, t)$. In other words, if DRP uses (6) directly to generate the requirements A_t, instead of going through the recursions that involves I_t and B_t, the estimates on these requirements will be protected from the inaccuracies involved in estimating I_t and B_t. The expression in (6), however, is more explicit and flexible. For instance, when there are constraints on the requirements A_t, they can be readily included: any terms in the summation $A(t-L, t-1)$ in (6) can be modified to account for the order quantity constraints. In other words, these terms should be viewed as the actual orders that have been placed.

Following the expressions derived above for A_t, B_t and I_t, the DRP recursion can be carried over to future periods through $Y_{t+1} = I_t - B_t$. Specifically, starting from period 1, we first derive A_{L+1}, followed by B_{L+1} and I_{L+1}, and then Y_{L+1}. Continue this for $t = L+2, \ldots, 2L$, assuming that $Y_1, Y_2, \ldots, Y_L$ have all been prespecified, i.e., given at beginning of period 1. (Note that nothing in periods 1 through L can be affected by any replenishment decision, at the beginning of period 1, due to the lead time.) In return, we will derive $Y_{L+2}, \ldots, Y_{2L}$. Next, we will derive A_{2L+1}, at the beginning of period $L+1$. At that point, we will need the value of Y_{L+1}, which has already been derived; and in return, we will derive Y_{2L+1}. The recursion then continues.

4. The (S, s) Inventory System

Having analyzed the lot-for-lot rule, we want to study next other rules, which impose restrictions on the order quantities. It will become evident below (§5) that all these rules can be unified into a dynamic (S, s) control scheme — dynamic in the sense that the control parameters, S and s, in general change over time. In order to analyze the dynamic (S, s) control rule, here we first review the standard (S, s) inventory model under stationary demand, and develop some simple approximations for its performance.

The (S, s) policy is known to be optimal in a quite general setting, in terms of minimizing the total costs of placing orders, keeping inventory, and paying penalty of backorders. Refer to Scarf [5]. Our concern here is quite different, however; our focus is on performance evaluation, rather than cost minimization. Our objective is to develop certain simple approximate formulas that are not only easy to evaluate but also useful in suggesting approximations in the non-stationary (i.e., dynamic) setting. More refined approximations for the stationary (S, s) systems are available in the literature, e.g., Tijms [9], although it is not clear how they can be adapted to the non-stationary setting.

4.1 Stationary Analysis

Consider an (S, s) inventory model with periodic review. A review takes place at the beginning of each period t, at which time the inventory position is updated and a replenishment decision is made following the (S, s) rule — order up to S if and only if the inventory position (on-hand plus on-order minus back-order) falls below s.

Observe that, defined as above, the inventory position, X_t, always takes values between s and S. In particular, if an arriving (batch) order brings the on-hand inventory level to s or below, the inventory position is instantaneously brought up to S.

Suppose, without loss of generality, that we start with $X_0 = S$. Then, we have a regenerative structure, with the cycle being the time between two consecutive replenishment orders. Given x ($x = s, s + 1, ..., S$), let

$$T_x = \min\{t : D_1 + \cdots + D_t > S - x\}$$

be the time, in a cycle, for the inventory position to drop from S to below x. In particular, T_s is the cycle length.

Since demands are i.i.d., we can view $T_x - 1$ as the number of renewals associated with a renewal process with "interarrival time" D_t. (Refer to, e.g., Ross [4], §3.4, in particular pp. 69-70.) Hence,

$$\mathsf{E}[T_x] = M(S - x) + 1, \qquad \mathsf{E}[T_s] = M(S - s) + 1,$$

where

$$M(x) = \sum_{n=1}^{\infty} F_{*n}(x)$$

is the renewal function, with $F_{*n}(\cdot)$ denoting the n-fold convolution of the (per-period) demand distribution, $F(\cdot)$. (Recall, the renewal function is the mean number of renewals up to a time point; refer to [4].) Furthermore, from standard renewal theory ([4], Chapter 3), we know

$$\mathsf{P}[X \geq x] = \lim_{t\to\infty} \mathsf{P}[X_t \geq x] = \frac{\mathsf{E}[T_x]}{\mathsf{E}[T_s]} = \frac{1 + M(S - x)}{1 + M(S - s)},$$

for $x \in [s, S]$. Hence,

$$\begin{aligned} \mathsf{P}[X = x] &= \frac{M(S - x) - M(S - x - 1)}{1 + M(S - s)}, \qquad x = s, ..., S - 1; \\ \mathsf{P}[X = S] &= \frac{1}{1 + M(S - s)}. \end{aligned}$$

The mean then follows:

$$\mathsf{E}[X] = \frac{S + sM(S - s) + \sum_{x=1}^{S-s-1} M(x)}{1 + M(S - s)}.$$

To characterize the on-hand inventory and the backorders, the key is to observe that by time t all orders that are placed before or at $t - L$ will have arrived. In other words, the on-order quantities included in the inventory position at $t-L$ will all have arrived by t. As before, let $D(t-L, t)$ denote the total demand over the $L + 1$ periods: $t - L, t - L + 1, ..., t$. We have

$$I_t = [X_{t-L} - D(t - L, t)]^+, \tag{13}$$

where $[x]^+ = \max\{x, 0\}$. Note that in (13), $D(t - L, t)$ is independent of X_{t-L}. Similarly, we have

$$B_t = [D(t - L, t) - X_{t-L}]^+. \tag{14}$$

Letting $t \to \infty$, and denoting the limits by omitting the time index, we have

$$I = [X - D(L + 1)]^+, \quad \text{and} \quad B = [D(L + 1) - X]^+, \tag{15}$$

where $D(L + 1)$ denotes a random variable that is equal in distribution to $D(t - L, t)$ and is independent of X.

4.2 Approximations

The renewal function is known to have a linear asymptote ([4]):

$$M(x) \sim \frac{x}{\mu} + \frac{m_2}{2\mu^2} - 1, \tag{16}$$

for large x, where $\mu = \mathsf{E}(D)$ and $m_2 = \mathsf{E}(D^2)$ are the first two moments of the (per period) demand distribution. (More precisely, the difference between $M(x)$ and the linear function on the right hand side above goes to zero when $x \to \infty$.)

Since in most applications the (S, s) values are large or moderately large, we can use the linear asymptote in (16) as an approximation for the renewal function. This way, the probability distribution of the inventory position X in §4 becomes:

$$\begin{aligned} \mathsf{P}[X = x] &= \frac{1}{\frac{m_2}{2\mu} + S - s}, \qquad x = s, ..., S - 1; \\ \mathsf{P}[X = S] &= \frac{\frac{m_2}{2\mu}}{\frac{m_2}{2\mu} + S - s}; \end{aligned}$$

where note that $\frac{m_2}{2\mu}$ is the average "undershoot" — the gap between the inventory position when an order is placed (which is necessarily smaller than s) and the reorder point s. Hence, from the above distribution of X, we can derive its mean:

$$\begin{aligned} \mathsf{E}[X] &= \frac{\frac{1}{2}(S + s - 1)(S - s) + \frac{m_2}{2\mu}S}{\frac{m_2}{2\mu} + S - s} \\ &= \frac{\mu(S - s)(S + s - 1) + m_2 S}{m_2 + 2\mu(S - s)}. \end{aligned} \tag{17}$$

In some special cases, the above reduces to a uniform distribution over the integers $\{s, ..., S\}$, and $\mathsf{E}[X] = (S + s)/2$. For instance, this is the case when the (per period) demand follows a truncated normal distribution with unit mean and unit coefficient of variation (hence, $\mu = 1$ and $m_2 = 2$). Note that for truncated normal distributions, just like normal distributions, assuming a unit coefficient of variation loses no generality; hence, unit mean is the only significant assumption here.

Next, making use of (15), along with the normal approximation of $D(L+1)$, we get the following approximation:

$$\mathsf{E}[B] = \sum_{x=s}^{S} \sigma\sqrt{L+1}G(\frac{x-(L+1)\mu}{\sigma\sqrt{L+1}})\mathsf{P}[X=x].$$

Substituting the uniform distribution of X into the above, we have

$$\begin{aligned} \mathsf{E}[B] &= \frac{1}{1+S-s}\sum_{x=s}^{S}\sigma\sqrt{L+1}G(\frac{x-(L+1)\mu}{\sigma\sqrt{L+1}}) \\ &\leq \sigma\sqrt{L+1}G(\frac{s-(L+1)\mu}{\sigma\sqrt{L+1}}), \end{aligned} \tag{18}$$

where the inequality follows from the fact that $G(x)$ is decreasing in x.

Once $\mathsf{E}[B]$ is derived, $\mathsf{E}[I]$ follows from (15), we have

$$\mathsf{E}[I] = \mathsf{E}[B] + \mathsf{E}[X] - \mu(L+1).$$

For instance, from (17), taking into account $G(k) = H(k) - k$, and approximating X with a uniform distribution, we have

$$\mathsf{E}[I] = \sigma\sqrt{L+1}H(\frac{s-(L+1)\mu}{\sigma\sqrt{L+1}}) + \frac{S-s}{2}. \tag{19}$$

Alternatively, $\mathsf{E}[X]$ can follow the approximation in (16).

5. Improved Projections: Dynamic (*S*, *s*) Rule

Observe that when following any policy other than lot-for-lot, the equation in (6) governing the requirements should be modified as follows:

$$A_t = [\mu(t-L,t) + k_t\sigma(t-L,t) - Y_{t-L} - Q(t-L,t-1)]^+ \tag{20}$$

where

$$Q(t-L,t-1) = Q_{t-L} + \cdots + Q_{t-1}$$

replaces $A(t-L,t-1)$. Hence, from $A_t > 0$, we have

$$\begin{aligned} A_t &= \mu(t-L,t) + k_t\sigma(t-L,t) - Y_{t-L} - Q(t-L,t-1) \\ &= s_{t-L} - Y_{t-L} - Q(t-L,t-1). \end{aligned}$$

That is, s_{t-L} is the reorder point, since following the DRP logic, an order (R_{t-L}) is placed (at $t-L$) if and only if $A_t > 0$. From the above expression, A_t is the required quantity to bring the inventory position (again, at $t-L$) back to s_{t-L}.

(This is consistent with the base-stock mechanism when there is no order-size restriction.)

Now, suppose the ordering policy is a dynamic (S, s) rule. Specifically, when $A_t > 0$, we want to bring the inventory position to S_{t-L} $(> s_{t-L})$, and an additional amount, $S_{t-L} - s_{t-L}$, is needed. Hence, the order quantity is:

$$Q_t = A_t + S_{t-L} - s_{t-L} = S_{t-L} - Y_{t-L} - Q(t-L, t-1), \tag{21}$$

when $A_t > 0$. ($Q_t = 0$ when $A_t = 0$).

On the other hand, given any order rule with a prespecified Q_t (as a function of $A_t > 0$, as discussed in §2.1), we can implement this rule as a dynamic (S, s) rule by letting [cf. (21)]

$$S_{t-L} = Q_t + s_{t-L} - A_t = Q_t + Q(t-L, t-1) + Y_{t-L},$$

when $A_t > 0$. When $A_t = 0$, we have $Q_t = 0$, and from the first equation above, we have $S_{t-L} = s_{t-L}$.

Implementation of a dynamic (S, s) policy within DRP is straightforward: once the parameters S_t and s_t for all t are specified, following (21) requires knowledge only of the quantities A_t, which are directly available within DRP. Therefore, here we focus on the dynamic (S, s) rule as a unified ordering rule for DRP when there are order-quantity restrictions.

We now turn to performance projections for DRP under the dynamic (S, s) rule. Note that the inventory position at $t - L$ is:

$$X_{t-L} = Y_{t-L} + Q(t-L, t), \tag{22}$$

i.e., the net inventory plus the on-order quantities (orders that have been placed but have yet to arrive). In particular, when $Q_t > 0$, from (21) and (22), we have,

$$S_{t-L} = Y_{t-L} + Q(t-L, t) = X_{t-L},$$

i.e., after the order is placed, the inventory position is brought up to S_{t-L}.

Just as in (11), we can iterate on (2,3) to obtain

$$Y_t = Y_{t-L} + Q(t-L, t-1) - D(t-L, t-1),$$

which, upon substitution back into (3), yields:

$$I_t = [Y_{t-L} + Q(t-L, t) - D(t-L, t)]^+ = [X_{t-L} - D(t-L, t)]^+,$$

and

$$B_t = [D(t-L, t) - Y_{t-L} - Q(t-L, t)]^+ = [D(t-L, t) - X_{t-L}]^+.$$

These are exactly the same formulas as in (13,14).

Therefore, we can adapt the approximations in the stationary (S, s) model. For instance, based on the backorder approximation in (17), and taking into account the formulas in (6,8) for the lot-for-lot case, we develop the following approximation:

$$\mathsf{E}[B_t|A_t > 0] = \sigma(t-L,t)\cdot G(k_t), \qquad \mathsf{E}[B_t|A_t = 0] = \sigma(t-L,t)\cdot G(k_t'), \tag{23}$$

where

$$k_t' := \frac{Y_{t-L} + Q(t-L,t-1) - \mu(t-L,t)}{\sigma(t-L,t)},$$

which, again, dominates k_t.

To approximate $\mathsf{E}[I_t]$, just as in the stationary case, we make use of the identity [from (13]

$$I_t - B_t = X_{t-L} - D(t-L,t).$$

Hence,

$$\mathsf{E}[I_t] = \mathsf{E}[B_t] + X_{t-L} - \mu(t-L,t),$$

where $\mathsf{E}[B_t]$ follows the approximations in (23), and X_{t-L} follows the expression in (22). According to the two cases in (23), and making use of the expression in (22) and the relation: $H(k) = k + G(k)$, we have:

$$\mathsf{E}[I_t|A_t > 0] = \sigma(t-L,t)\cdot H(k_t) + S_{t-L} - s_{t-L}, \tag{24}$$

(note in this case $X_{t-L} = S_{t-L}$); and

$$\begin{aligned} & \mathsf{E}[I_t|A_t = 0] \\ = \; & \sigma(t-L,t)\cdot H(k_t') - k_t'\sigma(t-L,t) + X_{t-L} - \mu(t-L,t) \\ = \; & \sigma(t-L,t)\cdot H(k_t'). \end{aligned} \tag{25}$$

Approximating $H(k_t')$ by k_t' in the above equation, we have

$$\mathsf{E}[I_t|A_t = 0] = k_t'\sigma(t-L,t) = X_{t-L} - \mu(t-L,t).$$

The above has the intuitive interpretation that if no order is placed at $t-L$ (i.e., $R_{t_L} = Q_t = A_t = 0$), then the expected on-hand inventory at t is simply the inventory position — of which all the on-order quantities will have arrived by t — minus the demand over the lead time, period $t-L$ through period t.

To implement the above approximations requires the derivation of Y_t, which is involved in both X_t and A_t. Recursively, Y_t can be approximated by its mean:

$$\mathsf{E}[Y_t] = \mathsf{E}[I_{t-1}] - \mathsf{E}[B_{t-1}].$$

Alternatively, a cruder approximation is to forego the distinction between the two cases $A_t > 0$ and $A_t = 0$. (Observe that this distinction is not present in the stationary case: it is averaged out in each regenerative cycle.) Specifically, ignore the k'_t case in (23), and approximate X_{t-L} by $(S_{t-L} + s_{t-L})/2$. This way, we have

$$\mathsf{E}[B_t] = \sigma(t-L,t) \cdot G(k_t), \tag{26}$$

and

$$\begin{aligned} & \mathsf{E}[I_t] \\ = \; & \sigma(t-L,t) \cdot H(k_t) - k_t\sigma(t-L,t) + (S_{t-L} + s_{t-L})/2 - \mu(t-L,t) \\ = \; & \sigma(t-L,t) \cdot H(k_t) + (S_{t-L} - s_{t-L})/2. \end{aligned} \tag{27}$$

Both are consistent with the stationary approximations in (17) and (19).

Proposition 2 With order-quantity restrictions, the implementation of DRP can be unified into a dynamic (S, s) control rule. Under this rule, the calculated quantity of product needed at the beginning of period t (Q_t), and hence the recommended order quantity at the beginning of period $t - L$ (R_{t-L}), should be set as follows:

$$R_{t-L} = Q_t = [A_t + S_{t-L} - s_{t-L}] \cdot 1[A_t > 0],$$

where $1[\cdot]$ denotes the indicator function, and

$$A_t = [s_{t-L} - Y_{t-L} - Q(t-L, t-1)]^+,$$

follows (20), with

$$s_{t-L} = \mu(t-L,t) + k_t\sigma(t-L,t).$$

The relationship between Q_t and S_{t-L} follows (21); in particular, when $A_t > 0$, $Q_t = S_{t-L} - Y_{t-L} - Q(t-L, t-1)$; and when $A_t = 0$, $Q_t = 0$ and $S_{t-L} = s_{t-L}$. The estimates on the expected on-hand inventory and backorders (for period t) follow (24,24) and (23), respectively; or alternatively, follow (26) and (26), respectively.

6. Numerical Examples

We present results from some numerical experiments designed to compare standard DRP estimates to those obtained using the approximations developed in §3 and §5. In all of the experiments, we use discrete event simulation as a benchmark for our comparison. The purpose of the experiments is to compare the accuracy of the DRP estimates as well as the computation time required to obtain the estimates. We conducted two sets of experiments. The first set is based on the example described in §2.2 in which demand is stationary. The second set explores the accuracy of the approximations in the case of non-stationary demand and is based on actual demand history for four different SKUs obtained from a large retailer.

For the first set of experiments, we modify the example of §2.2 so that the order quantity rule is lot-for-lot. We then compare estimates of on-hand inventory at the start of week 10 at one of the retailers obtained using three approaches: simulation, standard DRP, and improved DRP estimates based on approximations developed in §3 and §5. We carry out the comparison at several different safety stock levels. Figure 2 plots the estimates of on-hand inventory for all three methods. Note that the simulation and approximation values are essentially indistinguishable. For high safety stock values, the standard DRP estimates are close to those obtained from simulation. However, at lower values, the estimates deteriorate significantly. The reason for this behavior can be understood by referring back to the recursive equations governing DRP logic. When the safety stock levels are high, stockouts occur rarely and thus little distortion results when interchanging the expectation operation with the "max" operation in (5). At lower safety stock levels, this interchange leads to greater distortion, and, thus, poorer estimates of on-hand inventory. The situation for backorders, although not displayed here, is exactly analogous.

In the second set of experiments, we are largely interested in seeing how well our approximations for the dynamic (S, s) order policy work in the case of non-stationary demands. To this end, we compare estimates of on-hand inventory and backorders by week for four different SKUs at one retail location, each with a distinct non-stationary demand pattern. For all of these experiments, the coefficient of variation of demand is set to be 0.3 and demand per week is assumed to follow a truncated normal distribution. (We varied the coefficient of variation and found that the quality of the estimates is not significantly affected.) We use the method outlined in the Appendix to set s_t in order to meet a target fill rate of 95% each week. S_t is set to be s_t plus one week of demand. The review period is one week. Product 1 has a lead time of 5 weeks, Product 2 a leadtime of 1 week, Product 3 a lead time of 4 weeks, and Product 4 a lead time of 4 weeks. We apply the formula (26) in §5 to obtain estimates of on-hand inventory at the end of each week for 47 weeks and compare these to estimates obtained via simulation. Figures 3–6 plot week-ending on-hand inventory estimates for each week obtained via analytical approximation and simulation.

In general, the approximation appears "smoother" than the simulation results: When sharp shifts in demand occur from one week to the next the results from simulation indicate a more dramatic shift in on-hand inventory than the approximation estimates. One explanation for this phenomenon is that the approximation formula in (26) is based on the approximation in the stationary case, which contribute to the smoothing effect by averaging out the two cases $A_t > 0$ and $A_t = 0$. As for computational effort, the time required to run 30 replications for all four SKUs is approximately 0.5 seconds on an IBM RISC-System6000 Model 43P-140. To apply the approximation formulas for these SKUs required 0.009 seconds, approximately 55 times faster than simulation. Hence, for instance, in evaluating the performance of a system of 100,000 SKUs, the simulation would take about 3.5 hours, while the approximation needs less than 4 minutes.

7. Concluding Remarks

Our approach is readily extended to multi-echelon networks (e.g., Federgruen [1]). For instance, consider an assembly system. Suppose c components are stored at c different locations, indexed by $i = 1, ..., c$; and one unit from each is assembled into an end product at another location 0, where the end products supply a demand stream, $\{D_t, t = 1, 2, ...\}$. Let L_0, L_1, ..., L_c be the lead times at the stages. We can analyze the assembly stage 0 using the single-stage model. For each component stage i, we can combine it with stage 0 to form an echelon system, with a single demand stream $\{D_t\}$, and with $L_i + L_0$ being the lead time. This way, each component stage i follows an echelon control scheme; in particular, the reorder point is determined by the echelon inventory position that includes all the on-hand inventory in stage 0 and stage i, plus the on-order position of stage i, and minus the backorders waiting at stage 0.

Similarly, our approach also extends to a distribution network (essentially, a *disassembly* system) with a central stocking facility (e.g., a warehouse, or depot) supplying a set of local stocking facilities (e.g., retailers, or outlets), each of which, in turn, supplies its own customer demand. Our proposed approximation formulas can be applied to calculate safety stock levels at the different echelons so as to achieve a target fill rate for customer service.

8. Appendix: Setting Safety Stock Levels

In the preceding sections, we have assumed that the safety stock levels, K_t, are prespecified – that is, they are given as input to the DRP logic. Most DRP systems allow a number of methods for automatically setting the safety stock levels. One such method, which is of particular importance because of its wide use, sets safety stock levels based on achieving a target fill rate. For example, the DRP system may set safety stock levels based on a user-input target fill rate of 95% at the retail level. Here, for completeness, we present the formulas used to set this type of safety stock.

For ease of discussion, here we assume stationary demand, and omit the time index t wherever possible. Suppose the service requirement is that the fraction of demand backordered should be limited to $1-\beta$, where β is the required fill rate.

To characterize the fraction of backordered demand, we need to pick a "typical" time frame. In the standard inventory literature, this is taken to be the time between two consecutive orders — known as a regenerative cycle (when demands are independent and identically distributed). The required fraction is then the ratio of the average number of backorders to the average number of demand units, both over the regenerative cycle.

In the stationary (S, s) model, the average number of backorders, following (17), is approximated by

$$\sigma\sqrt{L+1}G\left(\frac{s-(L+1)\mu}{\sigma\sqrt{L+1}}\right);$$

whereas the average demand per cycle is

$$\left(\frac{S-s}{\mu}+\frac{m_2}{2\mu^2}\right)\mathsf{E}(D)=S-s+\frac{m_2}{2\mu},$$

where the first factor on the left hand side is the expected cycle length $\mathsf{E}[T_s] = 1+M(S-s)$ (refer to §4.1) with $M(S-s)$ approximated by the linear asymptote in (16). Note that the right hand side above is nothing but the expected order quantity; in particular, $\frac{m_2}{2\mu}$ is the expected undershoot.

Therefore, based on the above, under the dynamic (S, s) rule, k_t should be the solution to:

$$\sigma(t-L,t)G(k_t)=(1-\beta)\left[\Delta_{t-L}+\frac{\mathsf{E}(D_{t-L}^2)}{2\mathsf{E}(D_{t-L})}\right], \tag{28}$$

where $\Delta_{t-L} := S_{t-L} - s_{t-L}$ is assumed given.

The lot-for-lot rule is equivalent to $S_{t-L} = s_{t-L}$, or $\Delta_{t-L} = 0$. Also, since in every period an order is placed, the order quantity is simply equal to the demand. Hence, the equation in (28) is reduced to:

$$\sigma(t-L,t)G(k_t)=(1-\beta)\mathsf{E}(D_{t-L}). \tag{29}$$

Note that the equations in (28) and (29) are easily solved through the Newton's method. Write the equations in the form of $G(k) = c$. The Newton's iteration, indexed by the superscript (n), is as follows:

$$k^{(n+1)} = k^{(n)} - \frac{G(k^{(n)}) - c}{G'(k^{(n)})}.$$

When demand follows a normal distribution, $G'(k) = -1 + \Phi(k)$, where Φ denotes the distribution function of the standard normal variate.

Acknowledgments

Part of David Yao's research was supported by NSF grants ECS-9705392 and DMI-0085124, and by HK/RGC grant CUHK4376/99E while he was on leave at the Chinese University of Hong Kong, Dept. of Systems Engineering and Engineering Management.

References

[1] FEDERGRUEN, A., Centralized Planning Models for Multi-Echelon Inventory Systems under Uncertainty, in: S.C. Graves, A.H.G. Rinnooy Kan and P.H. Zipkin (eds.) *Logistics of Production and Inventory,* North-Holland, 133-173, 1993.

[2] MARTIN, A.J., *DRP Distribution Resource Planning: Distribution Management's Most Powerful Tool*, Prentice-Hall, Englewood Cliffs, NJ, and Oliver Wight Ltd Publications, INC., Essex Junction, VT, 2nd ed., 1990.

[3] MLOT, B., DIFRANCESCO, L., PERRY, D., LANDVATER, D. AND MARTIN, A., Distribution Resource Planning: The Critical Link from Final Point of Manufacture to Final Point of Sale, *The 39th APICS International Conference Proceedings,* 1996, pp. 294-297.

[4] ROSS, S.M. *Stochastic Processes,* Wiley, New York, 2nd edition, 1996.

[5] SCARF, H., The Optimality of (s, S) Policies in the Dynamic Inventory Problem, in: K. Arrow, S. Karlin and P. Suppes (eds.), *Mathematical Methods in the Social Sciences,* Stanford University Press, Stanford, CA, 1960.

[6] SILVER, E.A. AND PETERSON, R., *Decision Systems for Inventory Management and Production Planning,* Wiley, New York, 2nd ed., 1985.

[7] STENGER, A.J., Distribution Resource Planning, in: *The Logistics Handbook,* J.F. Robeson and W.C. Copacino (eds.), The Free Press, New York, 1994.

[8] SMITH, B.T., Using Distribution Resource Planning to Manage Inventories in Multiple Locations, *The 39th APICS International Conference Proceedings,* 1996, pp. 27-29.

[9] TIJMS, H., *Stochastic Modeling and Analysis: A Computational Approach*, Wiley, New York, 1994.

[10] ZIPKIN, P., *Foundations of Inventory Management,* Irwin/McGraw-Hill, New York, 2000.

Planning parameters

Order Quantity	30
Lead Time	5
Safety Stock	10

Retailer #1

Period	**0**	**1**	**2**	**3**	**4**	**5**	**6**	**7**	**8**	**9**	**10**	**11**	**12**	**13**	**14**	**15**	**16**
Demand		10	10	10	10	10	10	10	10	10	10	10	10	10	10	10	10
Scheduled Receipts		0	0	0	0	0	0	0	0	0	0	0	0	0	0	0	0
Safety Stock		10	10	10	10	10	10	10	10	10	10	10	10	10	10	10	10
On-Hand (pd end)	60	50	40	30	20	10	30	20	10	30	20	10	30	20	10	30	20
Requirements		0	0	0	0	0	10	0	0	10	0	0	10	0	0	10	0
Planned Receipts		0	0	0	0	0	30	0	0	30	0	0	30	0	0	30	0
Backorders		0	0	0	0	0	0	0	0	0	0	0	0	0	0	0	0
Rec. Orders		30	0	0	30	0	0	30	0	0	30	0	0	30	0	0	30

Table 1. Sample DRP Table for one retail location

Planning parameters

Order Quantity	200
Lead Time	15
Safety Stock	15

Warehouse

Period	0	1	2	3	4	5	6	7	8	9	10	11	12	13	14	15	16
Demand		90	0	0	90	0	0	90	0	0	90	0	0	90	0	0	90
Scheduled Receipts		0	0	0	0	0	0	0	0	0	0	0	0	0	0	0	0
Safety Stock		15	15	15	15	15	15	15	15	15	15	15	15	15	15	15	15
On-Hand (pd end)	200	110	110	110	20	20	20	0	0	0	0	0	0	0	0	0	0
Requirements		0	0	0	0	0	0	85	85	85	175	175	175	265	265	265	355
Planned Receipts		0	0	0	0	0	0	0	0	0	0	0	0	0	0	0	200
Backorders		0	0	0	0	0	0	70	70	70	160	160	160	250	250	250	140
Rec. Orders		200	200	0	200	0	0	0	0	0	200	0	0	0	0	0	200

Table 2. Sample DRP Table for warehouse

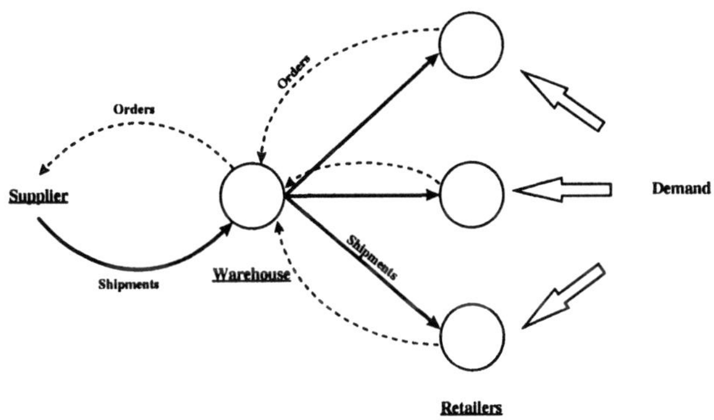

Figure 1. Layout of distribution network example

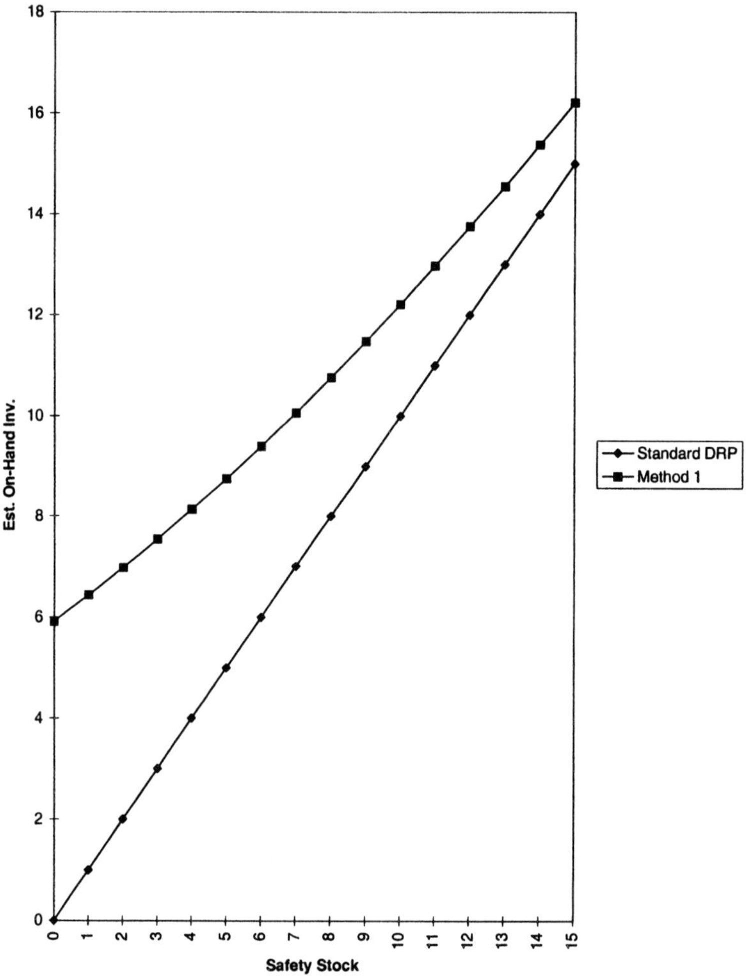

Figure 2. Estimates of Week 10 On-Hand Inventory as Function of Safety Stock for Stationary Demand Example using Lot-For-Lot Order Policy

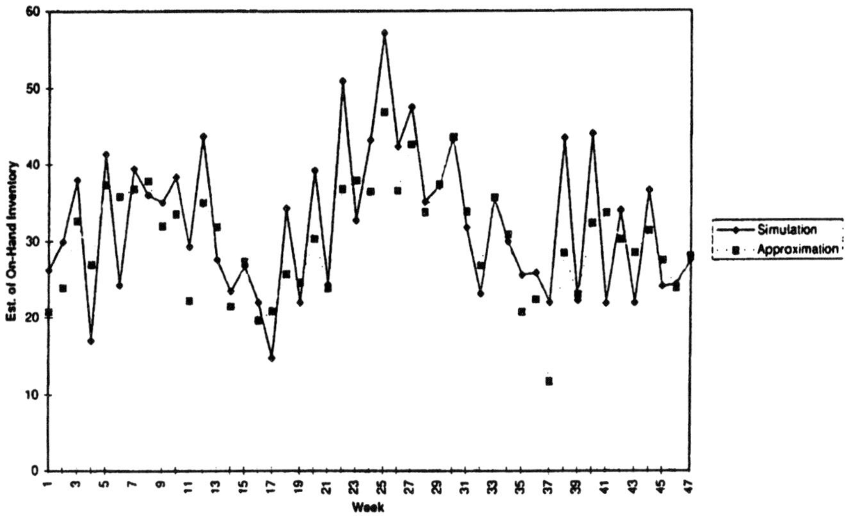

Figure 3. Estimates of Week-Ending On-Hand Inventory Levels for Product 1 with Non-Stationary Demand and Dynamic (S, s) Order Policy

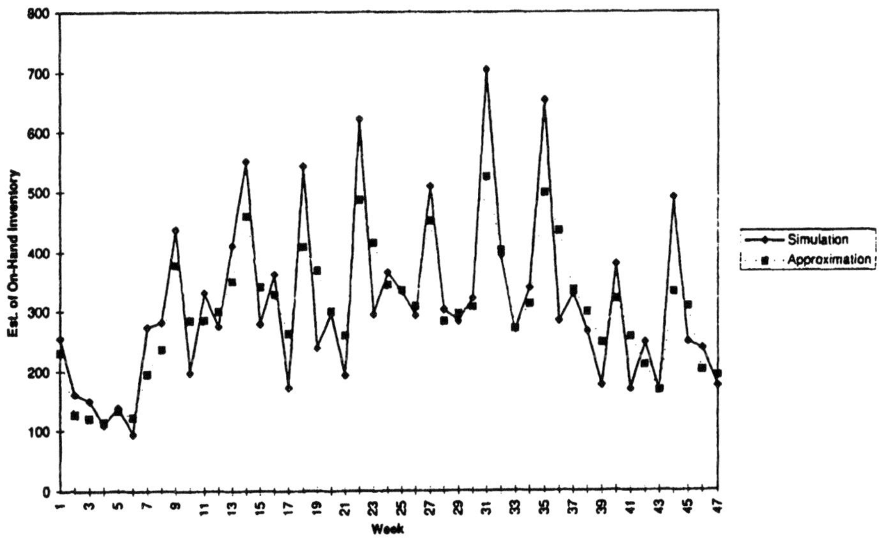

Figure 4. Estimates of Week-Ending On-Hand Inventory Levels for Product 2 with Non-Stationary Demand and Dynamic (S, s) Order Policy

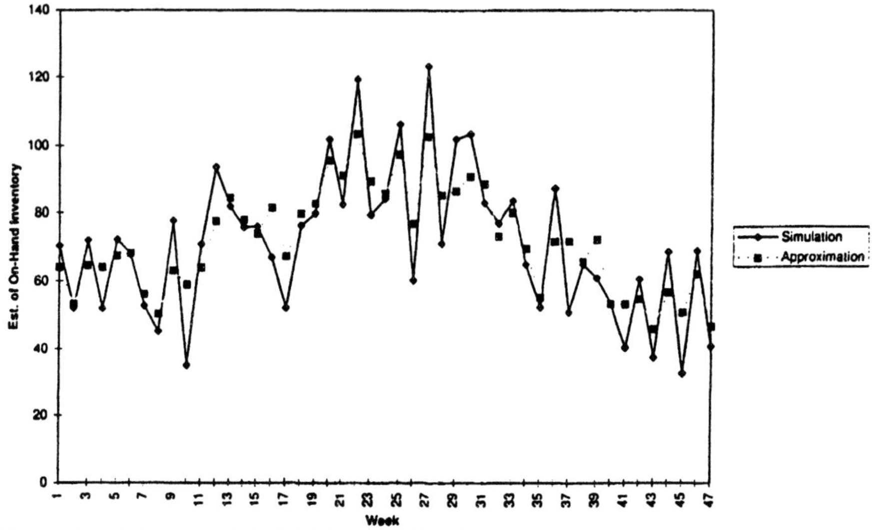

Figure 5. Estimates of Week-Ending On-Hand Inventory Levels for Product 3 with Non-Stationary Demand and Dynamic (S, s) Order Policy

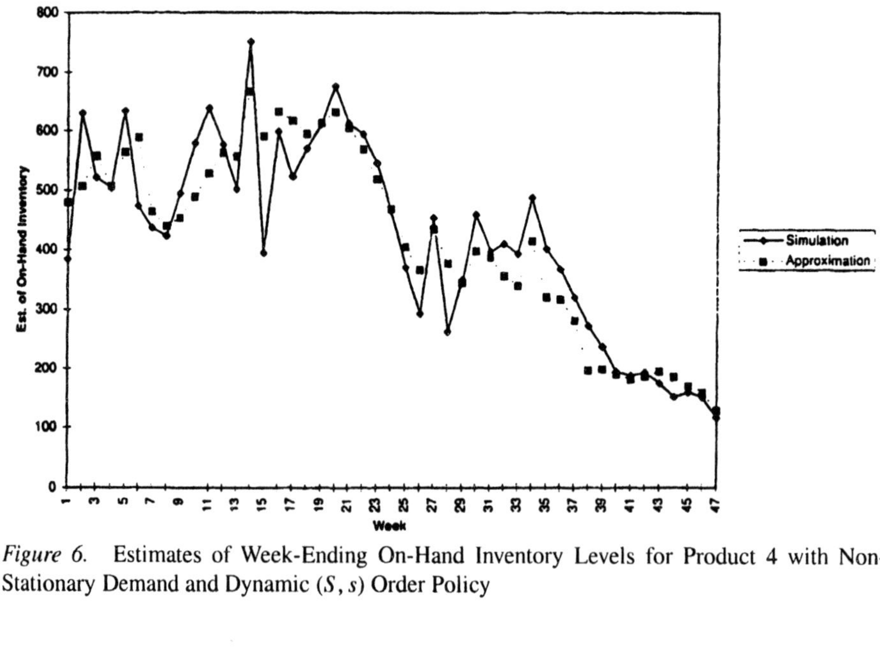

Figure 6. Estimates of Week-Ending On-Hand Inventory Levels for Product 4 with Non-Stationary Demand and Dynamic (S, s) Order Policy

Chapter 3

PROCESS ADJUSTMENT FOR ASSEMBLIES

Should Components, or Assemblies, Be Made to Specifications?

David W. Glenn *
University of British Columbia
Vancouver, BC, Canada V6T 1Z2
david.glenn@commerce.ubc.ca

Stephen M. Pollock
University of Michigan
Ann Arbor, MI 48109
Steve.Pollock@umich.edu

Abstract After the design and construction of manufacturing tools, adjustment is often necessary to reduce mean deviation from target of product dimensions. We formulate a stochastic dynamic program to find the optimal adjustment policy when the adjustment outcome is uncertain. The optimal policy minimizes the sum of adjustment cost during pre-production and quality cost during production. The formulation is developed for a single process and for two processes producing components for assembly. In the single process case, sufficient conditions are given for the optimal policy to have an upper and lower control limit form. In the two process case, we compare separate and combined policies and give an example where appreciable savings can be realized from a combined policy.

Keywords: Process develoment, die development, quality control, Markov decision process

*The authors wish to acknowledge the National Science Foundation (grant No. SBR-9712997) and the Tauber Manufacturing Institute of the University of Michigan for their support of their research.

Introduction

Tooling designed and constructed for manufacture of a new product often requires adjustment before use in commercial production. Product dimensions may vary excessively within a run, indicating the need for variance reduction; or dimensions may deviate systematically from target values, indicating the need for a mean correction. If these dimensional problems exist, the manufacturer must decide whether to spend time and resources on tool adjustment. Tool adjustment may range from slight tool modification to major reconstruction, and may even include changes in tool design. Furthermore, the outcome of an adjustment may be uncertain, requiring many cycles of testing and adjustment. Since in some situations tooling adjustment is extensive (and expensive) enough to impact overall process development cost and lead time, adjustment policies warrant careful consideration.

We present a model for adjustment of a group of tools that produce components for assembly. The simple system we consider consists of tools at two levels: component fabrication and product assembly. A group of *component tools* (tools used to produce components) operate in parallel. Each component tool is constructed separately and operated independently (perhaps at different geographic locations) to produce distinct components. The components are then transported and joined by an *assembly tool* to make the product. Because of error in tool design or construction, tools at both levels in the system may require adjustment to produce a dimensionally correct product. Adjustment uncertainty may cause the need for a sequence of adjustment attempts. An adjustment *policy* determines adjustment actions over time as a function of measured component and product quality.

One common approach in practice is to treat each component tool separately until the produced component dimensions conform to design intent. Assembly tools are then adjusted until assembled products conform to design. Policies that adjust component tools only as a function of component dimensions will be called *separate* policies.

Another approach is to treat the group of component tools jointly. Before making component tool adjustment decisions, assembled product quality is assessed. Then both component and/or assembly tools are adjusted, as necessary, to improve product quality. Such a policy may be worthwhile, for example, if assembly tools can be used to compensate for dimensional problems caused by component tools. Policies that adjust component and assembly tools as a function of product and component quality will be called *combined* policies because they use dimensional information from all adjoining components.

Consider an example from automotive sheet metal stamping and assembly. Large dies (component tools) are designed and constructed to stamp panels which are in turn assembled to produce a subassembly (e.g., door, fender assembly, etc.). Because sheet metal behavior in the stamping operation is highly complex and difficult to model, at the die design stage, dimensional quality of stamped panels is somewhat unpredictable. In particular, the mean dimensions of panels stamped from newly constructed dies typically deviate significantly from design target. Before panels can be used in production, dies often require costly adjustment, called *die rework*. The outcome of die rework, reflected in the panel dimensions, is also uncertain. Clark and Fujimoto (1991, pp. 182–203) show that die development accounts for a substantial fraction of total cost and lead time in the development of a new car program.

However, non-rigidity of sheet metal panels allows some latitude in assembly. Assembly tools and fixtures can often be adapted to pull mating surfaces together so as to produce dimensionally correct products from out-of-spec panels (see Takezawa (1980) for empirical evidence of the ability to alleviate stamping variation in assembly tooling). Using assembly fixtures to correct sheet metal problems is often much less expensive than die rework. A combined policy for die rework would ideally identify and distinguish those dimensional problems which require die (component tool) rework from those which can be mitigated in assembly. Hammett (1995) describes the advantages of assessing assembled product quality before making die rework decisions during auto-body process development.

Trietsch (1998) presents a method for improving mean deviation from target when the mean is unknown and estimated at each adjustment. This method is appropriate when adjustment cost is negligeable and the outcome is deterministic. Fine and Porteous (1989) model process improvement over an infinite horizon with uncertain outcome, where improvement is measured as a function of the reduction in the fraction of defective items, and a stochastic dynamic program is used to model uncertain improvement from adjustment. However, neither consider the two-stage system (component production followed by assembly) that we analyze here.

In the analysis that follows we make use of some simplifying assumptions in order to expose the underlying differences between separate and combined policies. These are discussed in detail in the Conclusion section.

1. Single Tool Adjustment with Uncertainty

We first consider tool adjustment with uncertainty in the context of manufacturing a single product (component) with a single critical dimension. Construction and adjustments of the tool take place at period k, where period $k = 0$ is immediately after die construction, and $k = N$ is the start of production. The critical dimension varies about the design target by the *deviation* X, a random variable with (known) probability distribution function $F_X(x|\theta_k)$. The mean of X is θ_k at the start of period k, $k = 0, 1, \ldots, N-1$. The variance of X is assumed to be a known constant for all k. At each period, the manufacturer may adjust the tool in an attempt to move the mean closer to zero.

Due to uncertainty during tool design and construction, the mean Θ_0 (after construction, before any adjustment) is also a random variable. At period k the manufacturer observes θ_k (to simplify the analysis we assume θ_k is measured perfectly at each period) and chooses an adjustment $d_k = \delta_k(\theta_0, \ldots, \theta_k)$, $d_k \in D_k$, where D_k is the set of feasible adjustments at period k. $\delta_k(\cdot)$ is called the *adjustment policy* at period k, and $\delta = (\delta_0, \delta_1, \ldots, \delta_{N-1})$ is the manufacturer's policy for all N adjustment periods.

The effect of choosing decision d_k is to change the distribution of Θ_{k+1}, the mean deviation at the next period. We assume that $\tilde{p}(\theta_{k+1}|\theta_k, d_k)$, the resulting distribution of Θ_{k+1}, has a Markovian property in that it depends only upon the deviation at period k and the decision at period k.

At each period the adjustment cost is a known function $R_k(\theta_k, d_k) \geq 0$. When the tool is finally put into production, after perhaps many adjustments, a terminal cost function $t_N(\theta_N) \geq 0$ represents the expected cost of operating in production with $\Theta_N = \theta_N$.

The total expected cost when using an arbitrary policy δ is then

$$J(\delta) = E\left\{\sum_{k=0}^{N-1} R_k(\Theta_k, \delta_k(\Theta_0, \ldots, \Theta_k)) + t_N(\Theta_N)\right\}, \tag{1}$$

where the expectation in Equation (1) is taken over all possible realizations of the sequence $\{\Theta_k\}_{k=0}^{N}$. The objective is to find an adjustment policy that minimizes J.

We use a standard dynamic programming approach by defining the function $V_k(\theta_k)$ to be the minimum expected cost at period k given the current deviation θ_k. The resulting recursive equations are

$$V_k(\theta_k) = \min_{d \in D_k}\left[R_k(\theta_k, d) + \int V_{k+1}(\theta_{k+1})\tilde{p}(\theta_{k+1}|\theta_k, d)d\theta_{k+1}\right] \tag{2}$$

for $k = 0, 1, \ldots, N-1$, with boundary condition

$$V_N(\theta_N) = t_N(\theta_N) \tag{3}$$

The solution $V_0(\theta)$ is the minimum expected cost of the adjustment process given an initial mean value θ. It is well known that $J(\delta) \geq E_\theta[V_0(\theta)]$ for arbitrary δ, and that the optimal policy at period k, δ_k^*, depends only on the current state θ_k, $k = 0, \ldots, N-1$ (see Kumar & Varaiya (1986)). Thus $J^* = J(\delta^*)$, the total expected cost using policy δ^*, is the minimum total expected cost; i.e., $J^* = E_\theta[V_0(\theta)]$.

1.1 Optimal policy for a restricted binary action space

Consider the restricted action space $D_k = \{0, -\theta_k\}$: either accept the process with no adjustment, or attempt to bring the mean to zero by and adjusting by an amount equal to the observed deviation. It seems reasonable that a good adjustment policy should be "accept" only when θ_k is in an interval that includes zero. That is, the adjustment policy should have the form

$$\delta_k(\theta_k) = \begin{cases} 0 & l_k \le \theta_k \le u_k \\ -\theta_k & \text{otherwise} \end{cases} \tag{4}$$

where $l_k \le 0 \le u_k$. Glenn and Pollock (1998) give a set of mild conditions on the cost functions and adjustment distributions under which the optimal policy has such a form. With this restricted action space, it becomes convenient to use the notation

$$p(\theta_{k+1}|\theta_k) \equiv \tilde{p}(\theta_{k+1}|\theta_k, -\theta_k) \tag{5}$$

1.2 Example 1: Die rework for a single dimension

1.2.1 Assumptions. A set of dies has been designed and constructed for fabrication of a specific sheet metal panel. The panel has a single critical dimension, X, measured in millimeters of deviation from target. The variable X is Gaussian with mean θ and fixed variance σ^2.

Prior to die construction, Θ_0 is a Gaussian random variable with $E[\Theta_0] = 0$ and *initial precision* $h_0 \equiv 1/Var[\Theta_0]$. After die construction a value θ_0 is observed.

At each period $k = 0, 1, \ldots, N-1$, the set of possible actions is $D_k = \{0, -\theta_k\}$: either make no adjustment, or attempt to remove the mean deviation from target.

The outcome Θ_{k+1}, given θ_k and action d_k, is

$$\Theta_{k+1} = \theta_k + W_k(1+b)d_k \tag{6}$$

where $0 \le b \le 1$, and the random variables $W_k \in [0, 1]$, $k = 0, 1, \ldots, N-1$, are independent and identically distributed with known cumulative distribution function $F_W(w)$ and precision $\rho = 1/Var(W)$. Equation (6) reflects the assumption that with no adjustment ($d_k = 0$) the expected value of the process mean for the next period is the current value (θ_k); with adjustment ($d_k = -\theta_k$) it is the current value multiplied by the random factor $[1 - W(1-b)]$. If $\theta_k > 0$, then $\Theta_{k+1} \in [-b\theta_k, \theta_k]$. Thus, the parameter b controls the degree to which the process is subject to possible over-correction.

The associated *rework precision* h_k can be readily shown to be:

$$h_k \equiv \frac{1}{Var[\Theta_{k+1}|\theta_k]} = \rho \frac{1}{(1+b)^2\theta_k^2}.$$

If $\theta_k < 0$, then $\Theta_{k+1} \in [\theta_k, -b\theta_k]$.

The die rework cost is constant for all periods; so that $R_k(\theta, -\theta) = r$ for all k.

The manufacturer's quality loss for production is quadratic in the deviation X, so that the terminal cost is

$$\begin{aligned} t_N(\theta) &= c\int x^2 dF_X(x|\theta) \\ &= c(\sigma^2 + \theta^2). \end{aligned}$$

where c is a constant reflecting the known quantity of parts to be produced.

Substituting the cost functions and the transition probability function into Equations (2) and (3) gives

$$V_k(\theta) = \min \begin{cases} r + \int V_{k+1}(\theta_{k+1})p(\theta_{k+1}|\theta, -\theta)d\theta_{k+1} & d_k = -\theta \\ V_{k+1}(\theta) & d_k = 0 \end{cases}$$

with boundary condition

$$V_N(\theta) = c(\sigma^2 + \theta^2) \tag{7}$$

Because rework cost is constant, it can be readily shown (see Glenn and Pollock (1998)) that the adjustment limit u_k of Equation (4) is a constant u for all k and can be obtained from the solution to

$$u = \left(\frac{r/c}{1-(1+b)^2 Var[W] - (1-(1+b)E[W])^2}\right)^{1/2}. \tag{8}$$

Note that the over-correction parameter b is explicitly included in the caclulation of u. Because the terminal cost function is symmetric in θ, it can also shown that $l = -u$.

1.2.2 Numerical Results. In the example that follows, W is a beta random variable with precision $\rho = 35$ and zero mode. Figure 1 shows the resulting $p(\theta_{k+1}|\theta_k)$ when $\theta_k = 4.0$ mm, for different values of b and ρ.

When $c = \$10^6$ (which is consistent with an average loss per part of \$40 when $x = \pm 2$ mm, and a total production volume of 100,000 parts) and $r = \$20,000$ (die rework cost for labor, equipment, and material). Equation (8) yields the optimal adjustment limit $u = .47$ mm.

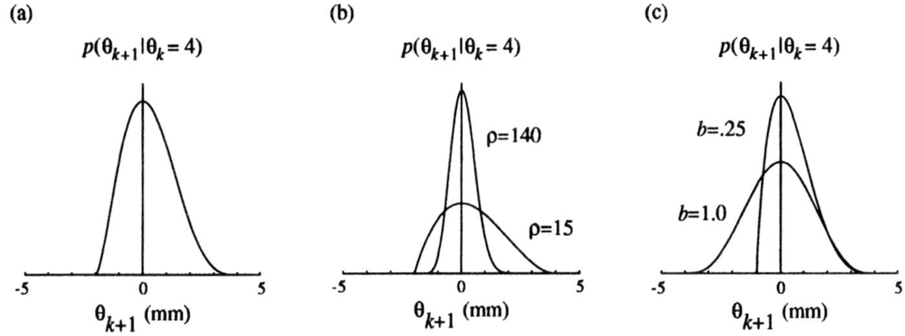

Figure 1. Die rework outcome p.d.f. for Θ_{k+1} given $\theta_k = 4$; (a) $b = 0.5$ and $\rho = 35$;(b) $b = .5$, $\rho = 25$ and $\rho = 140$; (c) $b = 0.25$, $b = 1.0$, and $\rho = 35$.

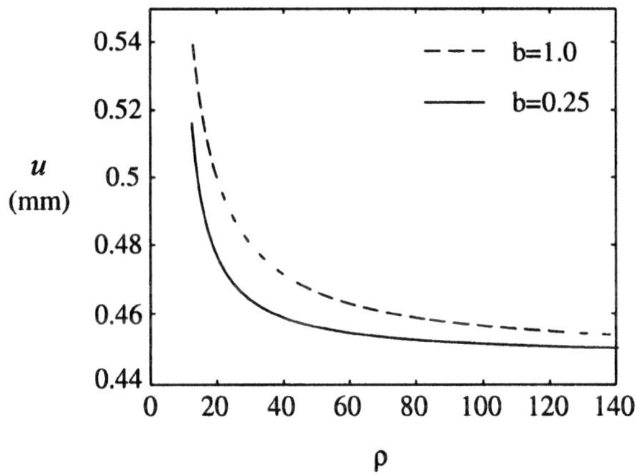

Figure 2. Optimal upper acceptance limit u as a function of ρ for two values of the over-correction parameter b.

Figure 2 shows u, as a function of ρ, for $b = 0.25$ and $b = 1.0$. As rework precision increases, due to either a decrease in b or an increase in ρ, the acceptance region narrows.

Figure 3 shows the minimum total expected cost J^* as a function of initial precision h_0 for $N = 10$, $b = .75$, and $\sigma = 0.1$. The two curves correspond to $\rho = 15$ and $\rho = 70$, which represent relatively low and high rework precisions, respectively.

In general, the expected cost decreases as either initial precision or rework precision increases. The cost reduction attained by improving rework precision diminishes as initial precision increases because Θ_0 is more likely to be close to

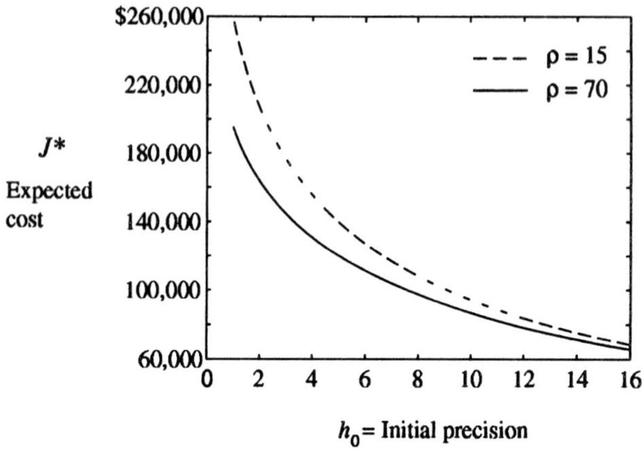

Figure 3. Minimum expected cost as a function of initial precision for two values of rework precision.

target. Figure 3 shows that the cost reduction from increasing initial precision is much greater per unit than the value of increasing rework precision.

2. Combined Policy for Adjustment of Two Tools

The previous analysis considered a single component with a single dimension requiring process adjustment. We now consider a system comprised of two component tools and an assembly process (see Figure 4). Each component has a single critical dimension. Deviations from design target of these dimensions are the independent random variables $X \sim F_X(x|\theta)$ and $Y \sim F_Y(y|\phi)$ with means θ and ϕ, respectively, and fixed variances. The mean deviations at $k = 0$ are independent random variables.

The assembly process joins the two components into a product having a single critical dimension. The resulting (random) deviation from target is $Z = g(X, Y, V)$, a known function of component dimensions X and Y and a random error variable V. As in the previous section attention will be restricted to only adjustment of the mean, $\eta \equiv E[Z]$.

At each period $k = 0, 1, \ldots, N-1$, a *combined* policy is defined to be one that specifies the amount to move *both* θ_k and ϕ_k, by adjustment of component tools, as a function of η. For each value of θ_k and ϕ_k, η is by definition

$$\eta(\theta_k, \phi_k) \equiv \int\int\int g(x, y, v) dF_X(x|\theta_k) dF_Y(y|\phi_k) dF_V(v)$$

Let $d_k = (d_k^1, d_k^2) \in D_k$ be the amount the manufacturer chooses to adjust θ_k and ϕ_k respectively, where D_k is the set of feasible adjustments in period k. Extending the dynamic program formulation in Equations (2) and (3), the

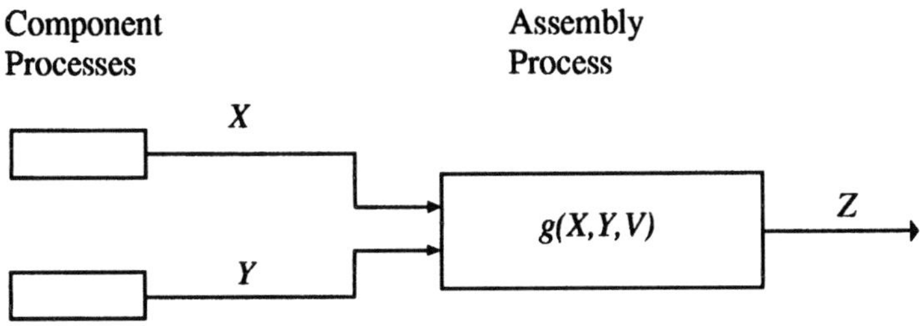

Figure 4. Manufacturing system consisting of two processes producing components for assembly.

state at each period k is (θ_k, ϕ_k). The joint p.d.f. $p(\theta_{k+1}|\theta_k, d_k)p(\phi_{k+1}|\phi_k, d_k)$ describes the random adjustment outcome under the assumption that the conditional random variables $(\theta_{k+1}|d_k)$ and $(\phi_{k+1}|d_k)$ are independent and have the Markov property. It can be shown that we need only consider the set of policies having the form $d_k = \delta_k(\theta_k, \phi_k)$.

If the cost of using decision d with values θ_k and ϕ_k is $R_k(\theta_k, \phi_k, d)$, the dynamic program becomes

$$V_k(\theta_k, \phi_k) = \min_{d \in D_k} \{R_k(\theta_k, \phi_k, d) + \int\int V_{k+1}(\theta_{k+1}, \phi_{k+1})p(\theta_{k+1}|\theta_k, d_k)p(\phi_{k+1}|\phi_k, d_k)d\theta_{k+1}d\phi_{k+1}\Big\}$$

for $k = 0, 1, \ldots, N-1$, with boundary condition

$$V_N(\theta_N, \phi_N) = t_N(\eta(\theta_N, \phi_N))$$

2.1 Example 2: Lap Joint Assembly

Lap joints are a common method of joining sheet metal components. Figure 5 (from Hu and Liu (1998)) shows the welding process for a lap joint between two components with one spot weld. Any deviation from nominal of the two components, labeled x and y (Figure 5(a)), is pressed out by the weld gun (Figure 5(b)). The result is an assembly deviation z due to spring back after the weld gun releases (Figure 5(c)).

The outcome z resulting from x, y, and weld gun displacement v is given by the mechanistic linear model

$$z = g(x, y, v) = s_x x + s_y y + s_v v \tag{9}$$

derived in Hu and Liu (1994). The sensitivity coefficients s_x, s_y, and s_v for the respective input variables are a function of metal thickness h_x and h_y, metal

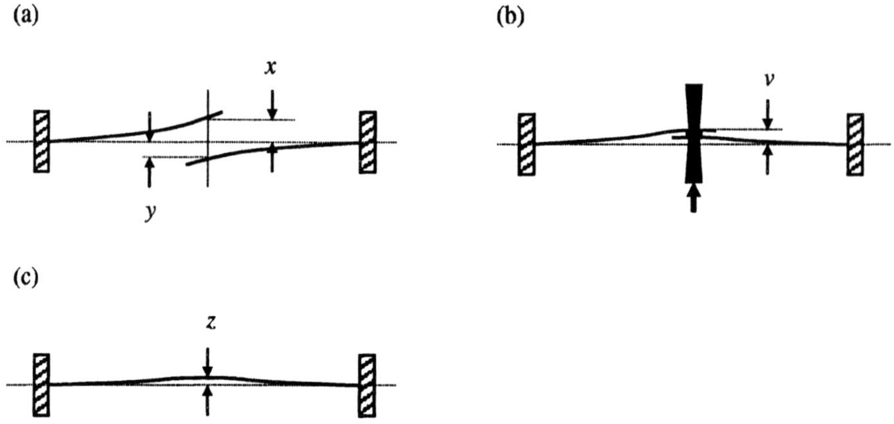

Figure 5. Lap joint. (a) Adjoining dimensions have displacement x and y; (b) the weld gun pulls metal together and has displacement v; (c) metal springs back after weld gun release and assembled dimension displacement is z. (Adapted from Hu and Liu (1998))

width, weld nugget properties, and distance of weld nugget from the clamps. Values for s_x, s_y, and s_v for joining two 10×100mm parts (as a function of the metal thickness ratio) are shown in Table 1.

Table 1. Sensitivity coefficients for joining two 10×100mm parts as a function of the metal thickness ratio. (Adapted from Hu and Liu (1994))

h_y/h_x	s_x	s_y	s_v
1.00	.1250	.1250	.7500
1.25	.0887	.1732	.7381
1.50	.0660	.2228	.7112
1.75	.0507	.2718	.6775
2.00	.0398	.3186	.6416

We now assume that:

- $X \sim N(\theta, \sigma_x^2)$, $Y \sim N(\phi, \sigma_y^2)$, and $V \sim N(0, \sigma_v^2)$; so that using Equation (9), $Z \sim F_Z(z|\theta, \phi) = N(\mu, \sigma_z^2)$ where $\mu = s_x\theta + s_y\phi$ and $\sigma_z^2 = s_x^2\sigma_x^2 + s_y^2\sigma_y^2 + s_v^2$
- After (θ_k, ϕ_k) is observed the options are to either rework one die, rework both dies, or accept both dies. The rework action is thus restricted to the set
  $D_k = \{(0,0), (0, -\phi_k), (-\theta_k, 0), (-\theta_k, -\phi_k)\}$. (To simplify presentation these regions are labelled (0,0), (0,1), (1,0), and (1,1) respectively.)
- $\Theta_0 \sim N(0, 1/h_0)$ and $\Phi_0 \sim N(0, 1/h_0)$ where h_0 is called the *initial precision.*

- When rework is performed on either die, the outcome has a transition probability of the same form used in example 1 (shown in Figure 1).

- The rework cost is a constant r, the same for either die.

- The loss function for random deviation is quadratic in the product dimension Z.

Under these assumptions the terminal cost function is

$$\begin{aligned} t_N(\theta,\phi) &= c\int Z^2 dF_Z(z|\theta,\phi) \\ &= c[\sigma_z^2+\mu^2], \end{aligned}$$

and the dynamic program becomes: For $k = 0,\ldots,N-1$,

$$V_k(\theta,\phi) = \min\begin{cases} r + E_{\theta_{k+1}}[V_{k+1}(\theta_{k+1},\phi_k)|(\theta,\phi),d] \\ r + E_{\phi_{k+1}}[V_{k+1}(\theta_k,\phi_{k+1})|(\theta,\phi),d] \\ 2r + E_{\theta_{k+1},\phi_{k+1}}[V_{k+1}(\theta_{k+1},\phi_{k+1})|(\theta,\phi),d] \\ V_{k+1}(\theta,\phi) \end{cases}$$

where the actions are $(1,0),(0,1),(1,1)$ and $(0,0)$ respectively. The boundary condition is

$$V_N(\theta,\phi) = t_N(\eta(\theta,\phi))$$

The minimum total expected cost is $J^* \equiv E_{\theta,\phi}[V_0(\theta,\phi)]$.

Figures 6 - 10 show numerically computed results for $r = 1$, $c = 50$, $s_x = s_y = 0.125$, and equal metal thicknesses (unless otherwise specified). Figure 6(a) shows the optimal policy for $k = N-1$ (only one rework possible before production). Note that whenever $\theta \approx -\phi$ the optimal policy is to accept both processes. For comparison, Figure 6(b) shows a (non-optimal) *separate* policy where each die is reworked if its component mean deviation is greater than 1.0 mm in absolute value (a typical process validation criterion for die rework for sheet metal components).

Figure 7 shows the difference between J^* and $J(\delta_S)$, obtained by using a (non-optimal) separate policy δ_s when $N = 10$ and $h_0 = 0.25$, as a function of l where

$$\delta_S(\theta) = \begin{cases} 0 & \text{if } |\theta| < u \\ -\theta & \text{if } |\theta| > u \end{cases}.$$

The best value for u in the separate policy is approximately 1.1 mm, yielding a cost about 17% more than the optimal policy.

Figure 8 shows $J(\delta_S)$ and J^* as functions of rework cost r, with $h_0 = 0.25$ and $N = 5$. The expected cost of the separate policy increases linearly with r, since it is non-optimal and thus does not reflect rework costs. However,

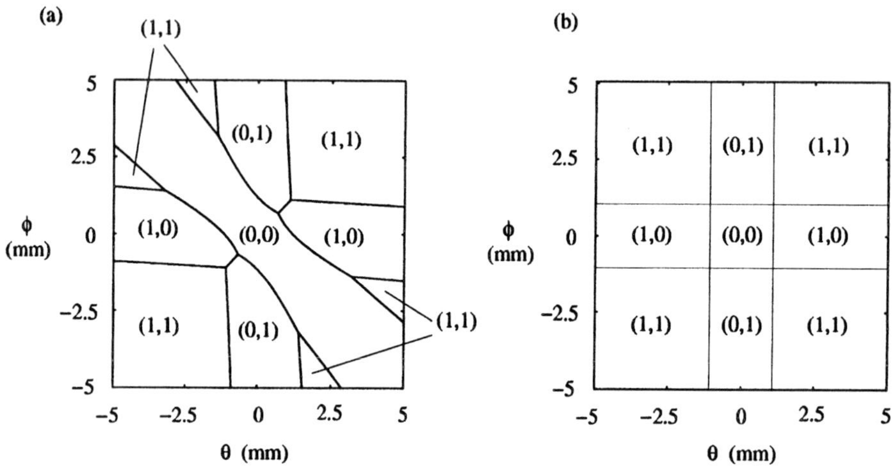

Figure 6. (a) The optimal combined policy for die rework when $N = 1$, metal thicknesses are equal, $r = 1$, and $c = 50$; (b) a (non-optimal) separate policy for reworking two sets of dies.

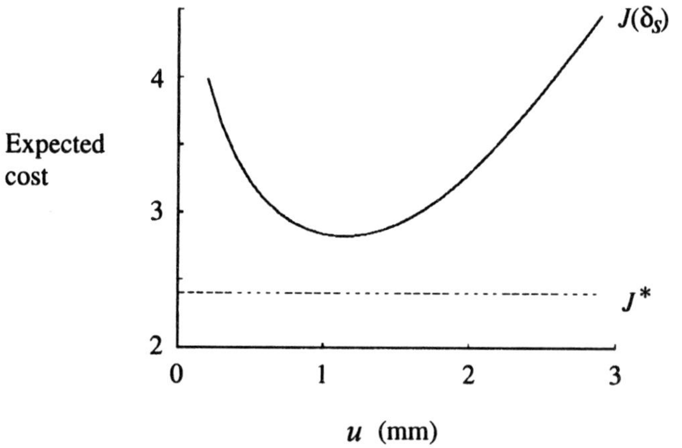

Figure 7. Minimum expected cost J^* using the optimal combined policy, with a horizon of $N = 10$, compared to the expected cost using a separate policy defined by the parameter u.

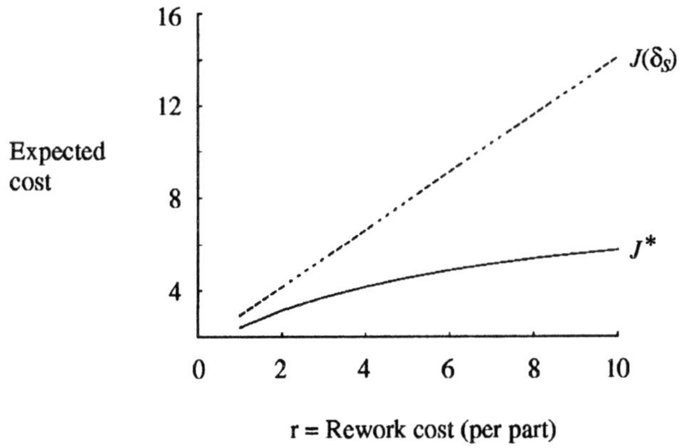

Figure 8. Expected cost of the separate policy and the optimal combined policy, with horizon $N = 5$, as a function of rework cost r.

J^* is asymptotic to an upper bound: the expected cost of accepting both processes without any rework. However, as $r \to 0$ the relative advantage of using a combined policy decreases. Since it is more costly to implement a combined policy, the analysis leading to Figure 8 can be used to find a value r' such that a combined policy is worthwhile when $r > r'$.

Figure 9 illustrates the value of a combined policy, with horizon $N = 5$, as a function of $h_0^{-1/2}$, the initial standard deviation, relative to the separate policy with $u = 1.0$ mm. As $h_0^{-1/2}$ decreases (i.e., as construction precision increases) the relative advantage of using a combined policy decreases. As $h_0^{-1/2} \to 0$ both $J(\delta_S)$ and J^* converge to $V_0(0, 0)$, the terminal cost when both process means are on target. This analysis can be used to find a value h_0' such that a combined policy is worthwhile when $h_0 < h_0'$.

In contrast to the policy illustrated in Figure 6(a), assembling components with unequal metal thicknesses will have non-symmetric die rework policies because the thicker component will dominate the assembly process output. Figures 10(a)-(c) show the general shapes of the optimal combined policy on the $\theta - \phi$ plane, with one adjustment period remaining for three ratios of metal thickness: $h_y = h_x$, $h_y = 1.25h_x$, and $h_y = 1.75h_x$. As h_y increases relative to h_x the acceptance region for ϕ narrows while that for θ expands. Furthermore, as h_y/h_x increases the expected cost of using the combined policy becomes less sensitive to θ, as shown in Figures 10(d)-(f).

3. Conclusion

The formulation and the examples explored in this paper provide useful insights into the problem of optimizing costly tool adjustment to reduce mean

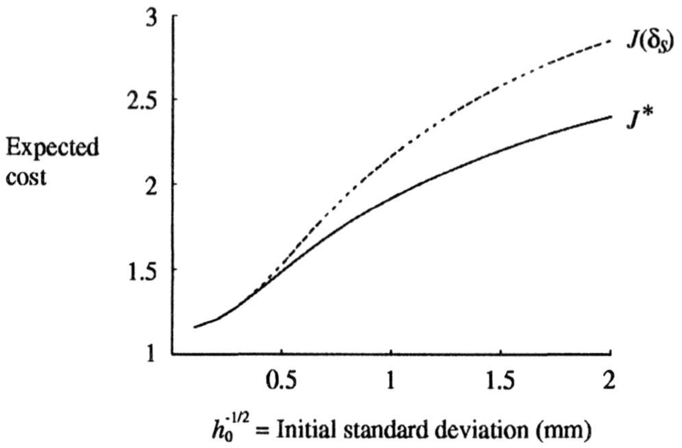

Figure 9. Expected cost of the separate policy and the optimal combined policy, with horizon $N = 5$, as a function of $h_0^{-1/2}$.

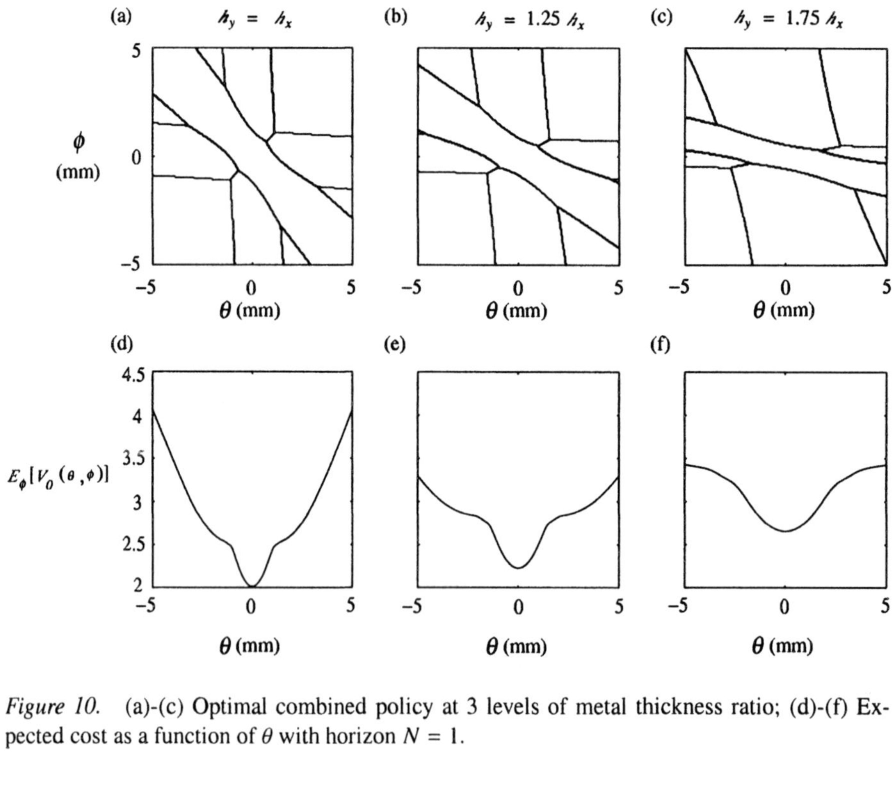

Figure 10. (a)-(c) Optimal combined policy at 3 levels of metal thickness ratio; (d)-(f) Expected cost as a function of θ with horizon $N = 1$.

deviation from target. Moreover, the model establishes a framework for comparing combined and separate policies; when to test components and when to test assemblies; and for assessing trade-offs between important factors such as design and construction uncertainty, adjustment cost, and adjustment precision.

Because of the exploratory nature of our analysis many simplifying assumptions were made which raise several practical concerns:

1 The mean dimensions are assumed to be perfectly known at each adjustment stage. In practice they are estimated by measuring a (usually quite small) sample of parts. A partially observed MDP (POMDP) can be formulated to capture this source of uncertainty, and may be useful for analytically assessing the structure of the optimal policy. Approximation techniques could then be used to find the optimal policy.

2 The adjustment outcome is assumed to be random and Markovian. In some situations knowledge about previous observations and adjustments might provide substantial information about the outcome of future adjustments. For instance, a tool designer lacking knowledge about the effect of tool parameter settings may learn from experience and hence predict tool adjustment outcome more accurately over time. Or, the manufacturer may learn over time about the level of expertise of the tool designer or tool maker. When such learning occurs, an adaptive policy, where knowledge of model parameters is improved at each period, might be more appropriate.

3 The cost of pre-production process adjustment is balanced against quality-related costs during production. In practice, quality costs may be difficult to estimate or even define. A design specification approach may be more appropriate, with the performance measure being the expected amount of adjustment time needed to make assembled product dimensions conform to some given design specification.

4 The assembly function $g(\cdot)$ is often not known, in that the effect of component quality and assembly tooling settings on assembled product quality can only be assessed by constructing sample assemblies. In the context of high tool adjustment cost, factorial experiments are not feasible. However, a sequential estimation and improvement model could create a framework for evaluating heuristic policies that use sample assembly dimensions (built from sample components) for process adjustment decisions.

5 Two computational issues arise from the need to discretize the state space: (a) as the number of critical dimensions on component parts (and

the number of component parts) increases, the cardinality of the state space grows exponentially; (b) extending the action space beyond a binary action space increases the computational burden.

6 Adjusting component processes usually involves several agents – at the tool construction plant, the component production plant, and the assembly plant – each having their own incentives. Our approach identifies total potential savings only if all agents are motivated to achieve a common global objective. We suspect, however, that our model will provide insight into alternative incentive assumptions.

References

[1] Clark, K. B.; and Fujimoto, T. (1991). *Product Development Performance.* Harvard Business School Press, Boston, MA.

[2] Fine, C. H.; and Porteus, E. L. (1989). "Dynamic Process Improvement". *Operations Research* 37, pp. 580-91.

[3] Glenn, D.W.; and Pollock, S.M. (1998). "Process Centering With Uncertain Tool Adjustment". *IOE Technical Report 98-11* , University of Michigan Department of Industrial and Operations Engineering, Ann Arbor MI.

[4] Hammett, P. C. (1995). *Methods for Validating Stamping and Metal Assembly Processes During Automotive Body Development.* Ph.D. Dissertation. The University of Michigan, Ann Arbor, MI.

[5] Hu, S. J.; and Liu, S. C. (1994). "Joint Designs and Their Variation Characteristics". Lecture notes. The University of Michigan Mechanical Engineering Deptartment, Ann Arbor, MI.

[6] Hu, S. J.; and Liu, S. C. (1998). "Sheet Metal Joint Configurations and Their Variation Characteristics". *ASME Journal of Manufacturing Science and Engineering* 120, pp. 461-467.

[7] Kumar, P. R.; and Varaiya, P. (1986). *Stochastic Systems: estimation, identification, and adaptive control.* Prentice-Hall, Englewood Cliffs, NJ.

[8] Takezawa, N. (1980). "An Improved Method for Establishing the Process-Wise Quality Standard". *Reports of Statistical Application Research,* Union of Japanese Scientists and Engineers (JUSE), 27, pp. 63-76.

[9] Trietsch, D. (1998). "The Harmonic Rule for Process Setup Adjustment with Quadratic Loss". *Journal of Quality Technology* 30, pp. 75-84.

Chapter 4

EXACT ANALYSIS OF A CONTINUOUS MATERIAL MERGE SYSTEM WITH LIMITED BUFFER CAPACITY AND THREE STATIONS

Stefan Helber
Institut für Wirtschaftswissenschaft
Technische Universität Clausthal, Germany
stefan.helber@tu-clausthal.de

Nicole Mehrtens
Institut für Wirtschaftswissenschaft
Technische Universität Clausthal, Germany

Abstract We develop a Markov process model of a flow line consisting of three unreliable machines and one buffer of limited capacity. Two machines upstream of the buffer perform the same operation and feed the same buffer in such a way that one machine has priority over the other when the buffer is full. The third machine removes material from the buffer. Processing times are deterministic and machine-specific. Exponentially distributed times to failure and to repair are also machine-specific. We develop an exact procedure to compute the average production rate and inventory level. The procedure is intended to be used as a building block of decomposition methods for the analysis of flow lines with non-linear flow of material that may, for example, be due to rework loops in flow lines with quality inspections.

Keywords: Flow line, merge, buffer, priority, Markov process model

1. Introduction

1.1 The Problem of Merging Flow of Material

Complex flow lines with limited buffer capacity and stochastic disruptions in the flow of material can be analyzed using decomposition methods of the type described in [Gershwin, 1994]. In these methods, a flow line with multiple machines and buffers is decomposed into a set of virtual two-machine lines, one for each buffer. One tries to find parameters for these virtual machines such that the flow of material through the buffers of the virtual line mimics those through the buffers in the original system. The reason for this procedure is that two-machine lines are often analytically tractable whereas more complex systems are often intractable.

In practical situations, however, there may be situations where *more* than one machine performs one particular operation and where these machines feed the same buffer as depicted in Figure 1, for example because the machines upstream of the buffer are slower than the downstream machine.

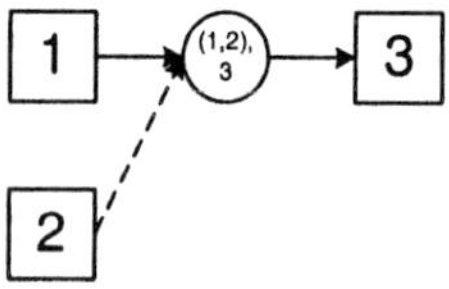

Figure 1. Flow line with a merging flow of material

In Figure 1, the squares indicate machines and the circle represents a buffer of limited capacity. The flow of material follows the directed arcs. Such a merge configuration may also serve as a building block of a decomposition approach for the analysis of a flow line with a rework loop like the one depicted in Figure 2.

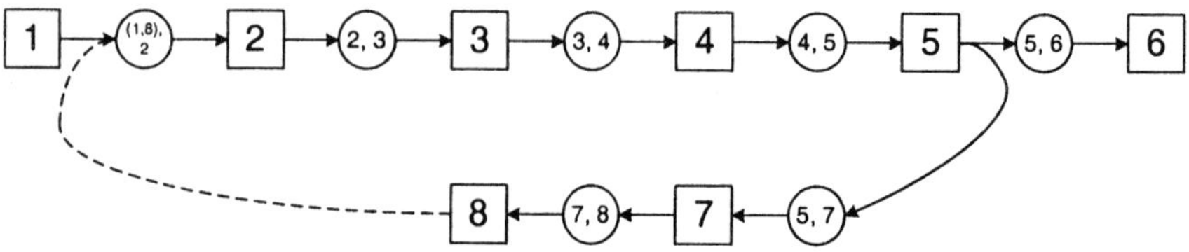

Figure 2. Flow line with a rework loop

In this system, there is a quality inspection at Machine M_5 and bad parts are reworked at Machines M_7 and M_8 before they are sent back to M_2. Assume that, from the point of view of M_2, there is no difference between the new and the reworked parts in the buffer upstream of M_2. In order to be able to analyze the flow of material in Figure 2 through a decomposition approach, one must be able to analyze a sub-system like the one in Figure 1 exactly, i.e. without falling back on decomposition.

Flow lines are often automated and therefore subject to random machine failures and repairs. These failures lead to random disruptions in the flow of material. Processing times, however, are often deterministic and machine-specific. These features should be considered in a useful model of a merge system.

Stochastic models of dynamic systems tend to be easiest to analyze whenever they possess the Markov property, i.e. whenever the future behavior of the system depends only on its current state, but not on its history. Deterministic processing times for the processing of discrete parts in flow lines, however, cannot easily be modeled in Markov process models. The reason is that in the case of deterministic processing times, the end of an operation can be forecasted, given the knowledge of when the operation started. In this case the Markov property is lost and the system is difficult to analyze. It is possible to escape this problem by defining the processing time as the time unit in systems where all processing times are identical (see [Buzacott, 1967, Gershwin and Schick, 1983]), but this approach excludes the general case of systems with machine-specific processing times. The standard approach to model deterministic *and* machine-specific processing times in a Markov process model is to assume that the machines process a *liquid* instead of discrete parts. It helps to imagine that the originally discrete parts have been molten and are processed continuously by machines that operate like pumps (see [Zimmern, 1956, Wijngaard, 1979, Gershwin and Schick, 1980]). In this case it is no longer possible to distinguish separate parts and therefore it is also not possible to predict the end of their processing, i.e. the future system behavior, based on the knowledge of its history. For this reason we now develop a continuous material Markov process of a merge system.

1.2 The Continuous Material Merge Model

The system in Figure 1 consists of three machines denoted as $M_i, i = 1, ..., 3$, and a buffer of capacity N. The supply of material upstream of M_1 and M_2 is unlimited and so is the available space downstream of M_3. Machines M_1 and M_2 send material into the buffer and M_3 removes material from the buffer. The inventory level in the buffer is denoted as x. Machines are unreliable and machine failures are operation dependent, i.e. machines can only fail while they operate. Failures and repairs of different machines are stochastically independent. The state of Machine M_i is denoted as α_i such that $\{\alpha_i = 1\}$ if M_i is operational (or up) and $\{\alpha_i = 0\}$ if M_i is under repair (or down). The complete description of the system state is therefore $(x, \alpha_1, \alpha_2, \alpha_3)$.

Times to failure of the machines $M_i, i = 1, ..., 3$, are exponentially distributed. The *rate* of failures depends on the instantaneous processing rate as we assume operation dependent failures. Repair times are also exponentially distributed, with rate r_i.

The instantaneous processing rate at time t of any machine that is operational depends on the buffer level x. If the buffer is neither full nor empty

$(0 < x < N)$, Machines $M_i, i = 1, ..., 3$, in Figure 1 process continuous material at a constant rate μ_i. In this case the instantaneous failure rate is p_i.

When the buffer is empty ($x = 0$) and both Machine M_1 and M_2 are down ($\alpha_1 = \alpha_2 = 0$), Machine M_3 cannot operate because it is (completely) starved. In this case it cannot fail as failures are operation dependent. However, if M_1 and M_2 are up *and* the combined processing speed $\mu_1 + \mu_2$ of Machines M_1 and M_2 is below the speed μ_3 of M_3 *and* the buffer is empty, then M_3 is slowed down by M_1 and M_2, i.e. it can operate only at the reduced speed $\mu_1 + \mu_2$ at which material flows through the buffer. If Machine M_3 is slowed down by M_1 and M_2, the buffer remains empty even though material is flowing through the buffer. In this case, M_3 is said to be *partially* starved. If M_3 is partially starved, it fails less often and the adjusted failure rate in this state is $\frac{\mu_1+\mu_2}{\mu_3}p_3$. This and other cases of partial starvation are discussed in more detail in the main section of the paper.

When the buffer is full ($x = N$) and Machine M_3 is down ($\alpha_3 = 0$), both M_1 and M_2 cannot operate because they are (completely) blocked. In this case, they both cannot fail. If the buffer is full while Machine M_3 is up, Machine M_1 has priority over Machine M_2 to fill the buffer. In this case it is possible that M_1 is partially blocked and M_2 can be partially or completely blocked. Assume, for example, that all machines are up, and that both M_1 and M_2 are (in isolation) slower than M_3, but their combined speed exceeds that of M_3, i.e. we have $\mu_1 < \mu_3 \wedge \mu_2 < \mu_3 \wedge (\mu_1 + \mu_2 > \mu_3)$. If the buffer is full ($x = N$) and all machines are up, Machine M_1 can operate at its normal speed μ_1 because it has priority over M_2 and M_3 is faster than M_1, i.e. M_3 cannot partially block M_1. The situation of Machine M_2, however, is different. Due to the priority of Machine M_1, Machine M_2 can only fill material into the buffer according to the *remaining* capacity of M_3, i.e. at rate $\mu_3 - \mu_1$. In this situation M_2 is partially blocked and fails at the reduced failure rate $\frac{\mu_3-\mu_1}{\mu_2}p_2$. The reasoning for other ranges of machine speeds is analogous and is treated in detail in the remainder of the paper.

1.3 Related Research

While there is a rich body of literature on the analysis of flow lines with linear flow of material (see the review in [Dallery and Gershwin, 1992] and the introduction in [Gershwin, 1994]), there is relatively little literature on non-linear patterns in the flow of material. Most of it focuses on assembly/disassembly systems, see, for example, [Di Mascolo et al., 1991, Gershwin, 1991]. A comprehensive review of this literature is given in [Helber, 1999] where a slightly different model of a merge system with *two* buffers is analyzed through decomposition. However, this led to serious numerical problems and motivated the work that led to this paper. One attempt to model a continuous material merge system is reported in [Dallery, 1999]. In that paper, the assumption is that material coming from Machines M_1 and M_2 is always processed in a *constant ratio* by Machine M_3. Due to this assumption, it is possible to reduce the problem to the (known) analysis of an assembly system. Instead of assuming constant ratios, we develop the merge system with a priority rule, and we do so for two reasons: Firstly, we want to have a variety of building blocks for the analysis of flow lines and networks in order to deal with the variety of real-world manufacturing systems. Secondly, if a merge system is operated in isolation, it is more attractive to use a priority rule as it leads to a higher production rate. If we demand fixed ratios, i.e. that for every part coming from M_1 we take any specific number of parts coming from M_2, a failure of any of the two machines leads to starvation of M_3. In our merge model, however, M_1 and M_2 can truly *replace* each other in cases of failures, and this is a useful and realistic feature. While the model in [Dallery, 1999] is analyzed through an approximate decomposition, we develop an exact solution by extending the two-machine model in [Gershwin, 1994, p.112-131]. A special case of our model was developed and analyzed simultaneously by Baris Tan [Tan, 2001]. In his model he assumes that the combined speed of the two machines M_1 and M_2 upstream of the buffer is less than the speed of the downstream machine, M_3, i.e $\mu_1 + \mu_2 < \mu_3$. This excludes the possibility of blocking of M_1 and/or M_2 and therefore there is no need for a priority rule in his model. In this paper, however, the general case is studied which allows for blocking and utilizes a priority mechanism. This is important not only as blocking of M_1 and/or M_2 changes some parts of the equations used to analyze the model. Our numerical results (both analytical and simulation) show that the three-machine merge system shows an interesting and possibly surprising behavior which is in fact due to the priority mechanism. We provide a complete characterization of all the internal as well as the boundary equations of the model for all possible cases. The approach to derive the boundary equations differs from the approach in [Tan, 2001] while we use essentially the same approach to solve the model numerically as in [Tan, 2001, Yeralan and Tan, 1997].

While the model and the solution approach is intended to be used as a building block of a decomposition method for larger systems with a merging flow of material to overcome the above-mentioned problems in [Helber, 1999], the focus of this paper is on the development, analysis, and documentation of the new building block. Due to the large number of cases that need to be analyzed, it appears to be appropriate to devote a complete paper to the new three-machine system. For this reason we cannot yet answer the question whether the new approach will eventually help to solve the problems we had with the decomposition of large merge systems. This is still work in process.

2. Derivation of Transition Equations

In order to determine performance measures such as the expected production rate or the expected inventory level for the model described above, we compute the probabilities of the possible system states. For system states $(x, \alpha_1, \alpha_2, \alpha_3)$ with full $(x = N)$ or empty $(x = 0)$ buffer, there are discrete probability masses $\mathbf{p}(x, \alpha_1, \alpha_2, \alpha_3)$. Some of these *boundary states* have zero probability, however.

System states with a buffer that is neither full nor empty $(0 < x < N)$ are called *internal states*. Since the buffer level is a continuous quantity, we use probability density functions $f(x, \alpha_1, \alpha_2, \alpha_3)$ to describe the random behavior of the system for this range of states. The analysis of these intermediate storage levels leads to a set of differential equations that are coupled.

The equations for boundary and internal states and the condition that all probabilities sum up to one lead to a system of equations that can be solved simultaneously to compute steady-state probabilities and density functions which in turn allow us to determine the above mentioned performance measures, i.e. average production rates and the inventory level.

2.1 Intermediate Storage Levels

The derivation of the transition equations for the intermediate storage levels is lengthy and the complete derivation of the first equation below is given in Appendix 1. Consider a situation where all machines are operational and the buffer is and remains neither empty nor full during a short time interval, i.e. we have $0 < x < N$ in the time interval $[t, t + \delta t]$. If we consider only events that have a probability that is of first order, this state can only be left if one of the machines fails during the time interval. The rate of failures of either of the machines is $(p_1 + p_2 + p_3)$. In steady state, the rate at which this state can be left is therefore proportional to the product of $(p_1 + p_2 + p_3)$ and the density $f(x, 1, 1, 1)$ of this particular state. This explains the left hand side of the following steady-state transition equation:

$$
\begin{aligned}
(p_1 + p_2 + p_3) f(x, 1, 1, 1) \quad = \quad & (\mu_3 - \mu_1 - \mu_2) \frac{\partial}{\partial x} f(x, 1, 1, 1) + \\
& r_1 f(x, 0, 1, 1) + r_2 f(x, 1, 0, 1) + \\
& r_3 f(x, 1, 1, 0). \qquad (1)
\end{aligned}
$$

The rate at which this state is entered must equal the rate at which it is left and it is given on the right hand side of (1). The state $(x, 1, 1, 1)$ can be reached either from another state $(\tilde{x}, 1, 1, 1)$ with $\tilde{x} \neq x$ due to a change in the buffer level, or from one of the states $(x, 0, 1, 1)$, $(x, 1, 0, 1)$ or $(x, 1, 1, 0)$ due to a repair of a broken machine (with rate r_1, r_2, or r_3, respectively).

The derivation for the remaining internal equations is completely analogous. If we use the notation $f'(x, \alpha_1, \alpha_2, \alpha_3) = \frac{\partial}{\partial x} f(x, \alpha_1, \alpha_2, \alpha_3)$ and rearrange the terms slightly, we can state a complete set of differential equations where (1) corresponds to (7):

$$
\begin{aligned}
\mu_2 f'(x,0,1,0) &= -(r_1 + p_2 + r_3) f(x,0,1,0) + p_1 f(x,1,1,0) + \\
&\quad r_2 f(x,0,0,0) + p_3 f(x,0,1,1) && (2)\\
(\mu_2 - \mu_3) f'(x,0,1,1) &= -(r_1 + p_2 + p_3) f(x,0,1,1) + p_1 f(x,1,1,1) + \\
&\quad r_2 f(x,0,0,1) + r_3 f(x,0,1,0) && (3)\\
\mu_1 f'(x,1,0,0) &= -(p_1 + r_2 + r_3) f(x,1,0,0) + r_1 f(x,0,0,0) + \\
&\quad p_2 f(x,1,1,0) + p_3 f(x,1,0,1) && (4)\\
(\mu_1 - \mu_3) f'(x,1,0,1) &= -(p_1 + r_2 + p_3) f(x,1,0,1) + r_1 f(x,0,0,1) + \\
&\quad p_2 f(x,1,1,1) + r_3 f(x,1,0,0) && (5)\\
(\mu_1 + \mu_2) f'(x,1,1,0) &= -(p_1 + p_2 + r_3) f(x,1,1,0) + r_1 f(x,0,1,0) + \\
&\quad r_2 f(x,1,0,0) + p_3 f(x,1,1,1) && (6)\\
(\mu_1 + \mu_2 - \mu_3) f'(x,1,1,1) &= -(p_1 + p_2 + p_3) f(x,1,1,1) + r_1 f(x,0,1,1) + \\
&\quad r_2 f(x,1,0,1) + r_3 f(x,1,1,0) && (7)\\
0 &= -(r_1 + r_2 + r_3) f(x,0,0,0) + p_1 f(x,1,0,0) + \\
&\quad p_2 f(x,0,1,0) + p_3 f(x,0,0,1) && (8)\\
-\mu_3 f'(x,0,0,1) &= -(r_1 + r_2 + p_3) f(x,0,0,1) + p_1 f(x,1,0,1) + \\
&\quad p_2 f(x,0,1,1) + r_3 f(x,0,0,0) && (9)
\end{aligned}
$$

Note that this is a set of coupled linear differential equations with constant coefficients. The same set of equations is reported in [Tan, 2001] as it also holds in the special case $\mu_1 + \mu_2 < \mu_3$ studied in the cited paper. If we have $\mu_1 \neq \mu_3 \wedge \mu_2 \neq \mu_3 \wedge \mu_1 + \mu_2 \neq \mu_3$, these are seven differential equations. With $\mu_1 = \mu_3 \wedge \mu_2 \neq \mu_3$, or $\mu_1 \neq \mu_3 \wedge \mu_2 = \mu_3$, or $\mu_1 + \mu_2 = \mu_3$, the number of differential equations reduces by one. In case of $\mu_1 = \mu_3 \wedge \mu_2 = \mu_3$, the number of differential equations even reduces by two.

Adding up all these eight equations leads to

$$
\begin{aligned}
&\mu_2 f'(x,0,1,0) + (\mu_2 - \mu_3) f'(x,0,1,1) + \mu_1 f'(x,1,0,0) + \\
&(\mu_1 - \mu_3) f'(x,1,0,1) + (\mu_1 + \mu_2) f'(x,1,1,0) + \\
&(\mu_1 + \mu_2 - \mu_3) f'(x,1,1,1) - \mu_3 f'(x,0,0,1) = 0 \qquad (10)
\end{aligned}
$$

Note that some parts on the left hand side of (10) vanish if we have $\mu_1 = \mu_3$, $\mu_2 = \mu_3$, or $\mu_1 + \mu_2 = \mu_3$.

In a case with $\mu_1 \neq \mu_3 \wedge \mu_2 \neq \mu_3 \wedge \mu_1 + \mu_2 \neq \mu_3$, integrating this equation with respect to x and rearranging the result yields:

$$\begin{aligned}
\mu_1\,(f(x,1,0,0) + f(x,1,0,1) + f(x,1,1,0) + f(x,1,1,1)) + \\
\mu_2\,(f(x,0,1,0) + f(x,0,1,1) + f(x,1,1,0) + f(x,1,1,1)) - \\
\mu_3\,(f(x,0,0,1) + f(x,0,1,1) + f(x,1,0,1) + f(x,1,1,1)) = K
\end{aligned} \tag{11}$$

The sum of the first two lines in Equation (11) is proportional to the rate at which the buffer level increases above a level x due to the production of Machines M_1 and M_2 while the remaining term on the left hand side term is proportional to the rate of decrease below x due to the production of Machine M_3. If the system reaches steady state, there must be an upward crossing for every downward crossing for every x, i.e. these two rates must be equal and we therefore have $K = 0$. We can use (11) to replace (9). If we divide the differential equations by the coefficients of the respective left hand side, we find in matrix form the equation

$$\begin{pmatrix} f'(x,0,1,0) \\ f'(x,0,1,1) \\ f'(x,1,0,0) \\ f'(x,1,0,1) \\ f'(x,1,1,0) \\ f'(x,1,1,1) \\ 0 \\ 0 \end{pmatrix} = \mathbf{A} \cdot \mathbf{F}(x) \tag{12}$$

where the matrix $\mathbf{A}$ is given in Figure 3 and where we have the vector of probability densities

$$\mathbf{F}(x) = \begin{pmatrix} f(x,0,1,0) \\ f(x,0,1,1) \\ f(x,1,0,0) \\ f(x,1,0,1) \\ f(x,1,1,0) \\ f(x,1,1,1) \\ f(x,0,0,0) \\ f(x,0,0,1) \end{pmatrix}. \tag{13}$$

If we have $\mu_2 - \mu_3 \neq 0$, $\mu_1 - \mu_3 \neq 0$, and $\mu_1 + \mu_2 - \mu_3 \neq 0$, this is a set of six coupled differential equations that describes the structure of the probability density functions of internal states. If one or two of these inequalities do not hold, the order of the coupled set of differential equations reduces to five or four, but the solution principle remains unaffected.

$$
\mathbf{A} = \begin{pmatrix}
-\frac{r_1+p_2+r_3}{\mu_2} & \frac{p_3}{\mu_2} & 0 & 0 & \frac{p_1}{\mu_2} & 0 & \frac{r_2}{\mu_2} & 0 \\
\frac{r_3}{\mu_2-\mu_3} & -\frac{r_1+p_2+p_3}{\mu_2-\mu_3} & 0 & 0 & 0 & \frac{p_1}{\mu_2-\mu_3} & 0 & \frac{r_2}{\mu_2-\mu_3} \\
0 & 0 & -\frac{p_1+r_2+r_3}{\mu_1} & \frac{p_3}{\mu_1} & \frac{p_2}{\mu_1} & 0 & \frac{r_1}{\mu_1} & 0 \\
0 & 0 & \frac{r_3}{\mu_1-\mu_3} & -\frac{p_1+r_2+p_3}{\mu_1-\mu_3} & 0 & \frac{p_2}{\mu_1-\mu_3} & 0 & \frac{r_1}{\mu_1-\mu_3} \\
\frac{r_1}{\mu_1+\mu_2} & 0 & \frac{r_2}{\mu_1+\mu_2} & 0 & -\frac{p_1+p_2+r_3}{\mu_1+\mu_2} & \frac{p_3}{\mu_1+\mu_2} & 0 & 0 \\
0 & \frac{r_1}{\mu_1+\mu_2-\mu_3} & 0 & \frac{r_2}{\mu_1+\mu_2-\mu_3} & \frac{r_3}{\mu_1+\mu_2-\mu_3} & -\frac{p_1+p_2+p_3}{\mu_1+\mu_2-\mu_3} & 0 & 0 \\
p_2 & 0 & p_1 & 0 & 0 & 0 & -(r_1+r_2+r_3) & p_3 \\
\mu_2 & \mu_2-\mu_3 & \mu_1 & \mu_1-\mu_3 & \mu_1+\mu_2 & \mu_1+\mu_2-\mu_3 & 0 & -\mu_3
\end{pmatrix}
$$

Figure 3. Content of matrix **A** for cases with $\mu_1 \neq \mu_3 \wedge \mu_2 \neq \mu_3 \wedge \mu_1 + \mu_2 \neq \mu_3$

In order to determine production rate and inventory level estimates, we will later determine closed-form expressions of this system of equations.

2.2 Boundary Storage Levels with Empty or Full Buffer

We use probability masses $\mathbf{p}(x, \alpha_1, \alpha_2, \alpha_3)$ with $x = 0$ or $x = N$ to denote the probability of seeing the system in a state with empty or full buffer. There are 16 of these boundary states that need to be analyzed. Some of the boundary states have zero probability. An example is state $(0, 1, 0, 0)$: If Machine M_1 is up, the buffer becomes non-empty immediately and therefore the probability of seeing the system in this state is zero. The reasoning for several other states is similar and it can easily be shown that

$$\begin{aligned} &\mathbf{p}(0,0,0,0) = \mathbf{p}(0,0,1,0) = \mathbf{p}(0,1,0,0) = \mathbf{p}(0,1,1,0) = 0, \\ &\mathbf{p}(N,0,0,0) = \mathbf{p}(N,0,0,1) = 0 \end{aligned} \tag{14}$$

Other states always have non-zero probabilities. Consider state (0,0,0,1). For all combinations of machine parameters it is perfectly possible that both Machine M_1 and M_2 are down, the buffer becomes empty, and the system remains in this state, at least until one machine is repaired. By a similar reasoning the following can be shown:

$$\mathbf{p}(0,0,0,1) > 0 \tag{15}$$
$$\mathbf{p}(N,0,1,0) > 0 \tag{16}$$
$$\mathbf{p}(N,1,0,0) > 0 \tag{17}$$
$$\mathbf{p}(N,1,1,0) > 0 \tag{18}$$

This accounts for 10 of the 16 boundary states. The situation for the remaining six boundary states is more complicated as we have to take the processing times into account. Assume, for example, that the combined processing rate of Machines M_1 and M_2 is not above the rate of M_3, i.e. $\mu_1 + \mu_2 \leq \mu_3$. In this case it is possible that all three machines are up while the buffer is empty. The only reason for the buffer to become non-empty is a failure of Machine M_3 and we therefore have $\mathbf{p}(0, 1, 1, 1) \neq 0$ in this case. If, however, M_3 is slower, i.e. we have $\mu_1 + \mu_2 > \mu_3$, then the buffer becomes non-empty immediately and we have $\mathbf{p}(0, 1, 1, 1) = 0$. Table 1 shows which states have zero or non-zero probability, depending on the processing rates.

In order to determine the probabilities of the boundary states, we again relate probabilities at time $t + \delta t$ to probabilities of possible previous states at time t and the probabilities of the respective transition. Consider, for example, the

Table 1. States that can have zero (z) or non-zero (n) probability

Case	Processing rates	p(0,0,1,1)	p(0,1,0,1)	p(0,1,1,1)	p(N,0,1,1)	p(N,1,0,1)	p(N,1,1,1)
1	$\mu_1 + \mu_2 < \mu_3$	n	n	n	z	z	z
2	$\mu_1 + \mu_2 = \mu_3$	n	n	n	z	z	n
	$\mu_1 + \mu_2 > \mu_3$,						
3	$\mu_1 < \mu_3, \mu_2 < \mu_3$	n	n	z	z	z	n
4	$\mu_1 < \mu_3, \mu_2 = \mu_3$	n	n	z	n	z	n
5	$\mu_1 = \mu_3, \mu_2 < \mu_3$	n	n	z	z	n	n
6	$\mu_1 = \mu_3, \mu_2 = \mu_3$	n	n	z	n	n	n
7	$\mu_1 > \mu_3, \mu_2 < \mu_3$	n	z	z	z	n	n
8	$\mu_1 > \mu_3, \mu_2 = \mu_3$	n	z	z	n	n	n
9	$\mu_1 < \mu_3, \mu_2 > \mu_3$	z	n	z	n	z	n
10	$\mu_1 = \mu_3, \mu_2 > \mu_3$	z	n	z	n	n	n
11	$\mu_1 > \mu_3, \mu_2 > \mu_3$	z	z	z	n	n	n

probability $\mathbf{p}(0,0,0,1)$ of Machines M_1 and M_2 being down and M_3 being up while the buffer is empty. If this state is observed at time $t+\delta t$, the system may have been in four possible states at time t:

- Machines M_1 and M_2 were down and the buffer was empty and Machine M_3 was up and idle such that it cannot fail. The probability of this state at time t is $\mathbf{p}(0,0,0,1,t)$. The probability that the system remains in this state is essentially the probability that neither M_1 nor M_2 is repaired:

$$(1 - r_1\delta t)(1 - r_2\delta t) + o(\delta t) = 1 - (r_1 + r_2)\delta t + o(\delta t) \qquad (19)$$

- The buffer was empty and Machines M_1 and M_3 were up and M_2 was down at time t, with probability $\mathbf{p}(0,1,0,1,t)$. Note that $\mathbf{p}(0,1,0,1,t) > 0$ only if $\mu_1 \leq \mu_3$. Otherwise the system cannot remain in this states as the buffer fills immediately. The probability of the transition is essentially the probability of a failure of Machine M_1:

$$p_1\delta t(1 - r_2\delta t)(1 - p_3\delta t) + o(\delta t) = p_1\delta t + o(\delta t) \qquad (20)$$

- The buffer was empty and Machines M_2 and M_3 were up and M_1 was down at time t, with probability $\mathbf{p}(0,0,1,1,t)$. Note that $\mathbf{p}(0,0,1,1,t) > 0$ only if $\mu_2 \leq \mu_3$. The transition probability is essentially the probability of a failure of Machine M_2:

$$(1 - r_1\delta t)p_2\delta t(1 - p_3\delta t) + o(\delta t) = p_2\delta t + o(\delta t) \qquad (21)$$

- The buffer level was between 0 and $\mu_3\delta t$, Machines M_1 and M_2 were down and Machine M_3 was up at time t. The probability is

$$\int_0^{\mu_3\delta t} f(x,0,0,1,t)dx = F(\mu_3\delta t,0,0,1,t) - F(0,0,0,1,t) \qquad (22)$$

if we define $F(x,0,0,1,t)$ such that $\frac{\partial}{\partial x}F(x,0,0,1,t) = f(x,0,0,1,t)$. A Taylor series expansion of $F(x,0,0,1,t)$ at $\tilde{x} = 0$ yields

$$\begin{aligned} F(\mu_3\delta t,0,0,1,t) &= F(0,0,0,1,t) + \frac{\partial}{\partial x}\frac{F(x,0,0,1,t)}{1!}(\mu_3\delta t - 0)^1 + o(\delta t) \\ &= F(0,0,0,1,t) + \mu_3\delta t f(x,0,0,1,t) + o(\delta t) \end{aligned} \qquad (23)$$

and inserting this result in (22) we find

$$\int_0^{\mu_3\delta t} f(x,0,0,1,t)dx = \mu_3\delta t f(0,0,0,1,t) + o(\delta t). \qquad (24)$$

Between times t and $t + \delta t$ no machine fails or gets repaired. The transition probability is

$$(1 - r_1\delta t)(1 - r_2\delta t)(1 - p_3\delta t) + o(\delta t) = 1 - (r_1 + r_2 + p_3)\delta t + o(\delta t) \quad (25)$$

If we assemble all these results, we get:

$$\begin{aligned} \mathbf{p}(0,0,0,1,t+\delta t) &= (1 - (r_1 + r_2)\delta t)\mathbf{p}(0,0,0,1,t) \\ &+ p_1\delta t\mathbf{p}(0,1,0,1,t) + p_2\delta t\mathbf{p}(0,0,1,1,t) \\ &+ \mu_3\delta t f(0,0,0,1,t) + o(\delta t). \end{aligned} \quad (26)$$

Rearranging this equation and dividing by δt we find:

$$\begin{aligned} &\frac{\mathbf{p}(0,0,0,1,t+\delta t) - \mathbf{p}(0,0,0,1,t)}{\delta t} = \\ &\quad -(r_1 + r_2)\mathbf{p}(0,0,0,1,t) + p_1\mathbf{p}(0,1,0,1,t) + p_2\mathbf{p}(0,0,1,1,t) \\ &\quad +\mu_3 f(0,0,0,1,t) + \frac{o(\delta t)}{\delta t} \end{aligned} \quad (27)$$

If we take the limit $\delta t \to 0$ and assume that the system is in steady state, the time derivative vanishes and we can omit the time index to conclude:

$$\begin{aligned} &(r_1 + r_2)\mathbf{p}(0,0,0,1) = \\ &\quad p_1\mathbf{p}(0,1,0,1) + p_2\mathbf{p}(0,0,1,1) + \mu_3 f(0,0,0,1) \end{aligned} \quad (28)$$

The left hand side of (28) is the rate at which the system leaves state (0,0,0,1) and the right hand side is the rate at which this state is entered.

Note that an Equation like (28) relates probability masses like $\mathbf{p}(0,0,0,1)$ to probability density functions like $f(x,0,0,0,1)$. These relations are needed as *boundary conditions* when we solve the system of differential equations developed in Section 2.1 that describe probability density functions of intermediate storage levels.

Equation (28) holds for all 11 cases in Table 1. However, in several of these cases $\mathbf{p}(0,0,1,1)$ and/or $\mathbf{p}(0,1,0,1)$ are zero as the respective state is left immediately because Machines M_1 and/or M_2 are faster then M_3.

In order to state the remaining equations in a compact form, we first define an indicator function

$$I(x) = \begin{cases} 1, & x > 0 \\ 0, & x \leq 0 \end{cases} \tag{29}$$

Using this indicator function, the complete set of boundary equations is given in Table 2. Since some of the states can have both zero and non-zero probability, depending on the processing rates, not all of the equations apply to all of the cases in Table 1. Table 3 shows which of the boundary equations are used to solve which case.

2.3 Boundary Storage Levels with *Almost* Empty or Full Buffer

A second set of equations that relates probability masses of boundary states to probability density functions of internal states can be derived if we analyze states where the buffer is *almost* empty or full at time $t + \delta t$ and the processing rates are such that it must have been *completely* empty or full at time t. Consider, for example, in Case 1 with $\mu_1 + \mu_2 < \mu_3$ the situation that Machine M_1 is up while Machines M_2 and M_3 are down at time $t + \delta t$. We now ask for the probability of having a buffer level x between 0 and $\mu_1 \delta t$, the amount of material Machine M_1 can have filled into the buffer between time t and $t + \delta t$. This probability is given by

$$\int_0^{\mu_1 \delta t} f(x, 1, 0, 0, t + \delta t) dx. \tag{40}$$

The only possible (previous) boundary state is $(0, 1, 0, 1, t)$, with probability $\mathbf{p}(0, 1, 0, 1, t)$, and there must have been a failure of Machine M_3. The probability of this failure during a time interval of length δt is

$$\frac{\mu_1}{\mu_3} p_3 \delta t. \tag{41}$$

Note that the probability of state $(0, 1, 0, 1, t)$ is non-zero if and only if $\mu_1 \leq \mu_3$, i.e. if Machine M_3 can be partially starved and the failure rate in this state has to be reduced by a factor $\frac{\mu_1}{\mu_3}$. There is no other possible state at time t with non-zero probability or a transition of at most first order. Furthermore, we do *not* have to take internal states with $\mu_1 \delta t < x$ into account. These would only add second or higher order terms. For this reason, we have

$$\int_0^{\mu_1 \delta t} f(x, 1, 0, 0, t + \delta t) dx = \frac{\mu_1}{\mu_3} p_3 \delta t \mathbf{p}(0, 1, 0, 1, t). \tag{42}$$

Applying the same reasoning as in the derivation of Equation (24), we find

$$\mu_1 \delta t f(0, 1, 0, 0, t + \delta t) = \frac{\mu_1}{\mu_3} p_3 \delta t \mathbf{p}(0, 1, 0, 1, t). \tag{43}$$

Table 2. Boundary equations

$$(r_1 + r_2)\mathbf{p}(0,0,0,1) = p_1\mathbf{p}(0,1,0,1) + p_2\mathbf{p}(0,0,1,1) + \mu_3 f(0,0,0,1) \tag{30}$$

$$(r_1 + p_2 + \frac{\mu_2}{\mu_3}p_3)\mathbf{p}(0,0,1,1) = p_1\mathbf{p}(0,1,1,1) + r_2\mathbf{p}(0,0,0,1) + (\mu_3 - \mu_2)f(0,0,1,1)I(\mu_3 - \mu_2) \tag{31}$$

$$(p_1 + r_2 + \frac{\mu_1}{\mu_3}p_3)\mathbf{p}(0,1,0,1) = r_1\mathbf{p}(0,0,0,1) + p_2\mathbf{p}(0,1,1,1) + (\mu_3 - \mu_1)f(0,1,0,1)I(\mu_3 - \mu_1) \tag{32}$$

$$\begin{aligned}(p_1 + p_2 + \frac{\mu_1 + \mu_2}{\mu_3}p_3)\mathbf{p}(0,1,1,1) &= r_1\mathbf{p}(0,0,1,1) + r_2\mathbf{p}(0,1,0,1) + \\ &\quad (\mu_3 - \mu_1 - \mu_2)f(0,1,1,1)I(\mu_3 - \mu_1 - \mu_2)\end{aligned} \tag{33}$$

$$(r_1 + r_3)\mathbf{p}(N,0,1,0) = p_3\mathbf{p}(N,0,1,1) + \mu_2 f(N,0,1,0) \tag{34}$$

$$\begin{aligned}(r_1 + \frac{\mu_3}{\mu_2}p_2 + p_3)\mathbf{p}(N,0,1,1) &= \min\{\frac{\mu_3}{\mu_1}, 1\}p_1\mathbf{p}(N,1,1,1) + r_3\mathbf{p}(N,0,1,0) + \\ &\quad (\mu_2 - \mu_3)f(N,0,1,1)I(\mu_2 - \mu_3)\end{aligned} \tag{35}$$

$$(r_2 + r_3)\mathbf{p}(N,1,0,0) = p_3\mathbf{p}(N,1,0,1) + \mu_1 f(N,1,0,0) \tag{36}$$

$$(\frac{\mu_3}{\mu_1}p_1 + r_2 + p_3)\mathbf{p}(N,1,0,1) = r_3\mathbf{p}(N,1,0,0) + (\mu_1 - \mu_3)f(N,1,0,1)I(\mu_1 - \mu_3) \tag{37}$$

$$\begin{aligned}r_3\mathbf{p}(N,1,1,0) &= r_1\mathbf{p}(N,0,1,0) + r_2\mathbf{p}(N,1,0,0) + p_3\mathbf{p}(N,1,1,1) + \\ &\quad (\mu_1 + \mu_2)f(N,1,1,0)\end{aligned} \tag{38}$$

$$\begin{aligned}(\min\{\frac{\mu_3}{\mu_1}, 1\}p_1 &+ \max\{\frac{\mu_3 - \mu_1}{\mu_2}, 0\}p_2 + p_3)\mathbf{p}(N,1,1,1) \\ &= r_1\mathbf{p}(N,0,1,1) + r_2\mathbf{p}(N,1,0,1) + r_3\mathbf{p}(N,1,1,0) + \\ &\quad (\mu_1 + \mu_2 - \mu_3)f(N,1,1,1)I(\mu_1 + \mu_2 - \mu_3)\end{aligned} \tag{39}$$

Table 3. Cases and boundary equations

Cases	Valid Equations
1	(30), (31), (32), (33), (34), (36), (38)
2	(30), (31), (32), (33), (34), (36), (38), (39)
3	(30), (31), (32), (34), (36), (38), (39)
4	(30), (31), (32), (34), (35), (36), (38), (39)
5	(30), (31), (32), (34), (36), (37), (38), (39)
6	(30), (31), (32), (34), (35), (36), (37), (38), (39)
7	(30), (31), (34), (36), (37), (38), (39)
8	(30), (31), (34), (35), (36), (37), (38), (39)
9	(30), (32), (34), (35), (36), (38), (39)
10	(30), (32), (34), (35), (36), (37), (38), (39)
11	(30), (34), (35), (36), (37), (38), (39)

If we assume that the system is in steady state and hence omit the time indicator t and if we multiply by μ_3, we eventually get:

$$\mu_3 f(0, 1, 0, 0) = p_3 \mathbf{p}(0, 1, 0, 1) \tag{44}$$

for all cases with $\mu_1 \leq \mu_3$. Now consider cases where Machine M_1 is faster that M_3, i.e. $\mu_1 > \mu_3$. There is *no* possible state at time t that requires a transition of at most first order. It is, for example, possible that the previous state was $(0, 0, 0, 1, t)$. In order for the buffer level to increase, we need a repair of M_1, with probability $r_1 \delta t$. However, we also need a failure of M_3, with probability $p_3 \delta t$. This leads to second order terms like $r_1 p_3 (\delta t)^2$ that can be omitted. In all these cases with $\mu_1 > \mu_3$ we therefore have

$$\int_0^{\mu_1 \delta t} f(x, 1, 0, 0, t + \delta t) dx = 0 \tag{45}$$

and this is included in (44) as $\mathbf{p}(0, 1, 0, 1) = 0$ if $\mu_1 > \mu_3$. We conclude that (44) holds in all cases. Proceeding in the same way, we can derive the following second set of equations for near-boundary states:

$$\begin{aligned}
\mu_3 f(0,0,1,0) &= p_3 \mathbf{p}(0,0,1,1) && (46)\\
\mu_3 f(0,1,0,0) &= p_3 \mathbf{p}(0,1,0,1) && (47)\\
\mu_3 f(0,1,1,0) &= p_3 \mathbf{p}(0,1,1,1) && (48)\\
(\mu_1 - \mu_3) f(0,1,0,1) &= r_1 \mathbf{p}(0,0,0,1) && (49)\\
(\mu_2 - \mu_3) f(0,0,1,1) &= r_2 \mathbf{p}(0,0,0,1) && (50)\\
(\mu_1 + \mu_2 - \mu_3) f(0,1,1,1) &= r_1 \mathbf{p}(0,0,1,1) + r_2 \mathbf{p}(0,1,0,1) && (51)\\
\mu_3 f(N,0,0,1) &= \frac{\mu_3}{\mu_1} p_1 \mathbf{p}(N,1,0,1) &&\\
&+ \frac{\mu_3}{\mu_2} p_2 \mathbf{p}(N,0,1,1) && (52)\\
(\mu_3 - \mu_2) f(N,0,1,1) &= \min\{\frac{\mu_3}{\mu_1}, 1\} p_1 \mathbf{p}(N,1,1,1) &&\\
&+ r_3 \mathbf{p}(N,0,1,0) && (53)\\
(\mu_3 - \mu_1) f(N,1,0,1) &= \frac{\mu_3 - \mu_1}{\mu_2} p_2 \mathbf{p}(N,1,1,1) &&\\
&+ r_3 \mathbf{p}(N,1,0,0) && (54)\\
(\mu_3 - \mu_1 - \mu_2) f(N,1,1,1) &= r_3 \mathbf{p}(N,1,1,0) && (55)
\end{aligned}$$

Table 4 shows which of the equations hold in which case. (The equations that we actually used in our algorithm—to be described below—are underlined.)

Table 4. Cases and near-boundary equations

Cases	Valid Equations
1	(46), (47), (48), (53), (54), (55)
2	(46), (47), (48), (53), (54)
3	(46), (47), (48), (51), (52), (53), (54)
4	(46), (47), (48), (51), (52), (54)
5	(46), (47), (48), (51), (52), (53)
6	(46), (47), (48), (51), (52)
7	(46), (47), (49), (48), (51), (52), (53)
8	(46), (47), (49), (48), (51), (52)
9	(46), (50), (47), (48), (51), (52), (54)
10	(46), (50), (47), (48), (51), (52)
11	(46), (47), (50), (49), (48), (51), (52)

2.4 Solution of the Transition Equations

To determine production rate and inventory level estimates, we first have to determine probability masses $\mathbf{p}(x, \alpha_1, \alpha_2, \alpha_3)$ and densities $f(x, \alpha_1, \alpha_2, \alpha_3)$. In this section we show how to derive general analytic expressions for these quantities. (We do not state these expressions explicitly as they are so lengthy that we were only able to handle them using a program like *Mathematica* or *Maple*.) These general symbolic expressions can then be plugged into a computer program to quickly determine numerical results for a *given* set of system parameters. The procedure consists essentially of four steps that are briefly outlined as follows:

1 Determine a general solution of the set of up to six differential equations for probability density functions $f(x, \alpha_1, \alpha_2, \alpha_3)$ of internal states. This solution is general as it contains up to six coefficients c_i that need to determined later on.

2 Determine expressions for the probabilities masses $\mathbf{p}(x, \alpha_1, \alpha_2, \alpha_3), x \in \{0, N\}$, of the boundary states in terms of the probability density functions by applying the equations for boundary states.

3 Solve a set of up to six equations in order to determine the coefficients c_i by applying the near-boundary equations and the condition that all probabilities sum up to one. This leads to unique expressions for the probabilities.

4 Use the probabilities to eventually determine the performance measures.

The remainder of this subsection describes these steps in detail for the relatively complicated Case 3 in Table 1. Several of the other cases are slightly easier to deal with as the number of differential equations for internal states decreases to five or even four. Consider, for example, Case 2 in Table 1 where we have $\mu_1 + \mu_2 = \mu_3$. In this case the left side of Equation (7) vanishes and the second to last row of the matrix $\mathbf{A}$ in Figure 3 has to be adjusted accordingly.

2.4.1 Solution of the Internal Equations. In this section we describe how to solve the set of differential equations (12) for Case 3. Define

$$\mathbf{F}_1(x) = \begin{pmatrix} f(x,0,1,0) \\ f(x,0,1,1) \\ f(x,1,0,0) \\ f(x,1,0,1) \\ f(x,1,1,0) \\ f(x,1,1,1) \end{pmatrix} \qquad \mathbf{F}_2(x) = \begin{pmatrix} f(x,0,0,0) \\ f(x,0,0,1) \end{pmatrix}$$

and the matrices

$$\mathbf{A}_{11} = \begin{pmatrix} -\frac{r_1+p_2+r_3}{\mu_2} & \frac{p_3}{\mu_2} & 0 & 0 & \frac{p_1}{\mu_2} & 0 \\ \frac{r_3}{\mu_2-\mu_3} & -\frac{r_1+p_2+p_3}{\mu_2-\mu_3} & 0 & 0 & 0 & \frac{p_1}{\mu_2-\mu_3} \\ 0 & 0 & -\frac{p_1+r_2+r_3}{\mu_1} & \frac{p_3}{\mu_1} & \frac{p_2}{\mu_1} & 0 \\ 0 & 0 & \frac{r_3}{\mu_1-\mu_3} & -\frac{p_1+r_2+p_3}{\mu_1-\mu_3} & 0 & \frac{p_2}{\mu_1-\mu_3} \\ \frac{r_1}{\mu_1+\mu_2} & 0 & \frac{r_2}{\mu_1+\mu_2} & 0 & -\frac{p_1+p_2+r_3}{\mu_1+\mu_2} & \frac{p_3}{\mu_1+\mu_2} \\ 0 & \frac{r_1}{\mu_1+\mu_2-\mu_3} & 0 & \frac{r_2}{\mu_1+\mu_2-\mu_3} & \frac{r_3}{\mu_1+\mu_2-\mu_3} & -\frac{p_1+p_2+p_3}{\mu_1+\mu_2-\mu_3} \end{pmatrix}$$

$$\mathbf{A}_{12} = \begin{pmatrix} \frac{r_2}{\mu_2} & 0 \\ 0 & \frac{r_2}{\mu_2-\mu_3} \\ \frac{r_1}{\mu_1} & 0 \\ 0 & \frac{r_1}{\mu_1-\mu_3} \\ 0 & 0 \\ 0 & 0 \end{pmatrix} \qquad \mathbf{A}_{22} = \begin{pmatrix} -(r_1 + r_2 + r_3) & p_3 \\ 0 & -\mu_3 \end{pmatrix}$$

as well as

$$\mathbf{A}_{21} = \begin{pmatrix} p_2 & 0 & p_1 & 0 & 0 & 0 \\ \mu_2 & \mu_2 - \mu_3 & \mu_1 & \mu_1 - \mu_3 & \mu_1 + \mu_2 & \mu_1 + \mu_2 - \mu_3 \end{pmatrix}$$

in order to write system (12) as:

$$\begin{pmatrix} \mathbf{F}_1'(x) \\ 0 \end{pmatrix} = \begin{pmatrix} \mathbf{A}_{11} & \mathbf{A}_{12} \\ \mathbf{A}_{21} & \mathbf{A}_{22} \end{pmatrix} \cdot \begin{pmatrix} \mathbf{F}_1(x) \\ \mathbf{F}_2(x) \end{pmatrix} \tag{56}$$

A similar partitioning is possible in all the other cases as well if we adjust the matrices accordingly. If we solve the second equation of (56) for $\mathbf{F}_2(x)$, we find the linear matrix equation

$$\mathbf{F}_2(x) = -\mathbf{A}_{22}^{-1}\mathbf{A}_{21}\mathbf{F}_1(x) = \mathbf{\Omega}\mathbf{F}_1(x). \tag{57}$$

and, using (57) in the first equation of (56), the linear matrix differential equation

$$\mathbf{F}_1'(x) = (\mathbf{A}_{11} - \mathbf{A}_{12}\mathbf{A}_{22}^{-1}\mathbf{A}_{21})\mathbf{F}_1(x) = \mathbf{\Lambda}\mathbf{F}_1(x). \tag{58}$$

with

$$\Omega = -\mathbf{A}_{22}^{-1}\mathbf{A}_{21} \tag{59}$$

$$\Lambda = \mathbf{A}_{11} - \mathbf{A}_{12}\mathbf{A}_{22}^{-1}\mathbf{A}_{21} \tag{60}$$

In order to solve the set of equations $\mathbf{F}_1'(x) = \Lambda\mathbf{F}_1(x)$ in (58), we have to determine the eigenvalues s_i and corresponding eigenvectors $\mathbf{y}^{(i)}$ of Λ that are defined as the solution to equation $(\Lambda - s\mathbf{I})\mathbf{y} = 0$. These eigenvalues and eigenvectors have to be computed numerically using standard procedures like those in [Englin-Müllges and Uhlig, 1996]. In Cases 1, 3, 7, 9, and 11, there are up to six different eigenvalues. Assuming that we find six independent eigenvectors, we get the fundamental matrix for Cases 1, 3, 7, 9, and 11:

$$\Psi(x) = \left(\begin{pmatrix} y_1^{(1)} \\ y_2^{(1)} \\ y_3^{(1)} \\ y_4^{(1)} \\ y_5^{(1)} \\ y_6^{(1)} \end{pmatrix} e^{s_1 x} \quad \begin{pmatrix} y_1^{(2)} \\ y_2^{(2)} \\ y_3^{(2)} \\ y_4^{(2)} \\ y_5^{(2)} \\ y_6^{(2)} \end{pmatrix} e^{s_2 x} \quad \cdots \quad \begin{pmatrix} y_1^{(6)} \\ y_2^{(6)} \\ y_3^{(6)} \\ y_4^{(6)} \\ y_5^{(6)} \\ y_6^{(6)} \end{pmatrix} e^{s_6 x} \right) \tag{61}$$

In Cases 2, 4, 5, 8, and 10, we assume to find only five independent eigenvectors as the number of differential equations reduces by one. In Case 6 we assume to find only four independent eigenvectors as the number of differential equations reduces by two. Note that the relevant question is whether we find the suitable number of independent eigenvectors. It is not harmful to have a multiplicity of eigenvalues as long as we find for each eigenvalue of multiplicity k the same number k of independent eigenvectors. This is solution methodology is very similar to the one used in [Yeralan and Tan, 1997].

Defining a coefficient vector $\mathbf{c}^T = (c_1, c_2, \ldots, c_6)$, we can write the general solution to (58) as

$$\mathbf{F}_1(x) = \Psi(x)\mathbf{c}. \tag{62}$$

In Cases 1, 3, 7, 9, and 11, we find six equations:

$$
\begin{aligned}
f(x,0,1,0) &= c_1 y_1^{(1)} \exp(s_1 x) + c_2 y_1^{(2)} \exp(s_2 x) + \ldots + c_6 y_1^{(6)} \exp(s_6 x) && (63)\\
f(x,0,1,1) &= c_1 y_2^{(1)} \exp(s_1 x) + c_2 y_2^{(2)} \exp(s_2 x) + \ldots + c_6 y_2^{(6)} \exp(s_6 x) && (64)\\
f(x,1,0,0) &= c_1 y_3^{(1)} \exp(s_1 x) + c_2 y_3^{(2)} \exp(s_2 x) + \ldots + c_6 y_3^{(6)} \exp(s_6 x) && (65)\\
f(x,1,0,1) &= c_1 y_4^{(1)} \exp(s_1 x) + c_2 y_4^{(2)} \exp(s_2 x) + \ldots + c_6 y_4^{(6)} \exp(s_6 x) && (66)\\
f(x,1,1,0) &= c_1 y_5^{(1)} \exp(s_1 x) + c_2 y_5^{(2)} \exp(s_2 x) + \ldots + c_6 y_5^{(6)} \exp(s_6 x) && (67)\\
f(x,1,1,1) &= c_1 y_6^{(1)} \exp(s_1 x) + c_2 y_6^{(2)} \exp(s_2 x) + \ldots + c_6 y_6^{(6)} \exp(s_6 x) && (68)
\end{aligned}
$$

In these cases, Equation (57) boils down to the following system:

$$
\begin{aligned}
f(x,0,0,0) &= \omega_{11} f(x,0,1,0) + \ldots + \omega_{16} f(x,1,1,1) && (69)\\
f(x,0,0,1) &= \omega_{21} f(x,0,1,0) + \ldots + \omega_{26} f(x,1,1,1) && (70)
\end{aligned}
$$

which indirectly relates density functions to eigenvalues, eigenvectors and unknowns c_i.

For any given matrix $\mathbf{\Lambda}$ and corresponding eigenvalues and eigenvectors, these are six functions in x with unknowns c_i. These equations are used in two ways: Firstly, they are applied to the boundary and near-boundary equations to determine numerical values for the coefficients c_i for any given set of system parameters. Secondly, for any given set of both system parameters *and* coefficients c_i, they are used to compute production rate and inventory level estimates.

2.4.2 Determination of the Boundary Probabilities.

Given the expressions for the probability density functions in (62), we now use equations (30) to (39) in order to determine expressions for the probabilities $\mathbf{p}(x,\alpha_1,\alpha_2,\alpha_3)$ of boundary states. Note that these equations relate probability masses to densities $f(x,\alpha_1,\alpha_2,\alpha_3)$ for specific numerical values x of 0 or N. In order to solve Case 3, we first use from Table 3 Equations (30), (31), and (32) for the empty buffer case and write

$$
\begin{pmatrix} p(0,0,0,1) \\ p(0,0,1,1) \\ p(0,1,0,1) \end{pmatrix} = \mathbf{H}_{3,0}^{-1} \cdot \begin{pmatrix} \mu_3 f(0,0,0,1) \\ (\mu_3-\mu_2) f(0,0,1,1) \\ (\mu_3-\mu_1) f(0,1,0,1) \end{pmatrix} \tag{71}
$$

where we define

$$
\mathbf{H}_{3,0} = \begin{pmatrix} r_1+r_2 & -p_2 & -p_1 \\ -r_2 & r_1+p_2+\frac{\mu_2}{\mu_3}p_3 & 0 \\ -r_1 & 0 & p_1+r_2+\frac{\mu_1}{\mu_3}p_3 \end{pmatrix}. \tag{72}
$$

Equations (34), (36), (38), and (39) for the full buffer case yield

$$\begin{pmatrix} p(N,0,1,0) \\ p(N,1,0,0) \\ p(N,1,1,0) \\ p(N,1,1,1) \end{pmatrix} = \mathbf{H}_{3,N}^{-1} \cdot \begin{pmatrix} \mu_2 f(N,0,1,0) \\ \mu_1 f(N,1,0,0) \\ (\mu_1+\mu_2) f(N,1,1,0) \\ (\mu_1+\mu_2-\mu_3) f(N,1,1,1) \end{pmatrix} \tag{73}$$

where we define

$$\mathbf{H}_{3,N} = \begin{pmatrix} r_1+r_3 & 0 & 0 & 0 \\ 0 & r_2+r_3 & 0 & 0 \\ -r_1 & -r_2 & r_3 & -p_3 \\ 0 & 0 & -r_3 & p_1+\frac{\mu_3-\mu_1}{\mu_2}p_2+p_3 \end{pmatrix} \tag{74}$$

We now have a way to compute expressions for the probabilities of boundary states. The densities $f(x,\alpha_1,\alpha_2,\alpha_3)$ on the right hand side of (71) and (73) are solutions to the set of (differential) equations in (56). If the coefficient vector $\mathbf{c}^T$ in (62) has not yet been numerically determined, they define equations relating probability masses on the left hand side of (71) and (73) to the general solutions to the set of (differential) equations in (56). However, once the coefficient vector $\mathbf{c}^T$ is numerically determined, these equations can be used to determine the numerical values of the probability masses.

2.4.3 Determination of Coefficients c_i. Even for a given set of system parameters and the corresponding eigenvalues and eigenvectors, all we have so far are equations for the probability masses and densities with the unknowns c_i. What needs to be done now is to determine numerical values for these unknown coefficients. If we have, as in Case 3, six coefficients, we need six independent equations. The equations we derive in this section are *linear* in the six unknowns c_i and it is therefore straightforward to solve them.

One equation is used in all cases. It states that all the probabilities sum up to one, or formally:

$$\sum_{\alpha_1=0}^{1}\sum_{\alpha_2=0}^{1}\sum_{\alpha_3=0}^{1}\left[\int_0^N f(x,\alpha_1,\alpha_2,\alpha_3)\,dx + p(0,\alpha_1,\alpha_2,\alpha_3) + p(N,\alpha_1,\alpha_2,\alpha_3)\right] = 1 \tag{75}$$

Since all expressions for probability masses $\mathbf{p}(x,\alpha_1,\alpha_2,\alpha_3)$ and all densities $f(x,\alpha_1,\alpha_2,\alpha_3)$ are linear in the unknowns c_i, and since summation and integration are linear operations, Equation (75) is also linear in c_i. In order to determine six unknowns c_i in Case 3, we need five additional equations. For this purpose we use the *near*-boundary equations (46) to (55) to establish the remaining conditions. From Table 4 we know that seven equations hold in Case 3. Out of these we chose (46), (47), (51), (53), and (54), i.e. we demand that

$$0 = \mu_3 f(0,0,1,0) - p_3 \mathbf{p}(0,0,1,1) \tag{76}$$
$$0 = \mu_3 f(0,1,0,0) - p_3 \mathbf{p}(0,1,0,1) \tag{77}$$
$$0 = (\mu_1 + \mu_2 - \mu_3) f(0,1,1,1) - r_1 \mathbf{p}(0,0,1,1) - r_2 \mathbf{p}(0,1,0,1) \tag{78}$$
$$0 = (\mu_3 - \mu_2) f(N,0,1,1) - p_1 \mathbf{p}(N,1,1,1) - r_3 \mathbf{p}(N,0,1,0) \tag{79}$$
$$0 = (\mu_3 - \mu_1) f(N,1,0,1) - \frac{\mu_3 - \mu_1}{\mu_2} p_2 \mathbf{p}(N,1,1,1) - r_3 \mathbf{p}(N,1,0,0) \tag{80}$$

Note that these equations are also linear in the unknowns c_i, i.e. we can write equations (75) to (80) as

$$\mathbf{B} \cdot \begin{pmatrix} c_1 \\ c_2 \\ \vdots \\ c_6 \end{pmatrix} = \begin{pmatrix} 1 \\ 0 \\ \vdots \\ 0 \end{pmatrix}. \tag{81}$$

where the matrix **B** contains symbolic expressions for the coefficients of the unknowns c_i that we get by inserting the symbolic expressions for probability masses and densities. If we solve this linear system of equations for given numerical values of the system parameters and eigenvalues and eigenvectors of $\mathbf{\Lambda}$, we get the desired numerical values of the probability masses and densities. These can then be used to determine performance measures as described in the next section. The underlined entries in Table 4 indicate those equations that we used in each of the cases to establish the required set of equations. Note that the *number* of required equations is case-specific. In Case 6, for example, the set of differential equations for internal states has only four degrees of freedom and we need only four equations (including (75)) to determine the four unknowns $c_i, i = 1, \ldots, 4$.

2.4.4 Determination of Performance Measures. For given probability masses and density functions, we define the average inventory level $\overline{x}$ as

$$\bar{x} = \sum_{\alpha_1=0}^{1} \sum_{\alpha_2=0}^{1} \sum_{\alpha_3=0}^{1} \left[\int_0^N x \cdot f(x, \alpha_1, \alpha_2, \alpha_3)\, dx + N \cdot \mathbf{p}(N, \alpha_1, \alpha_2, \alpha_3) \right]. \tag{82}$$

The production rate can be calculated along several paths. Firstly, we can ask for the he rate at which material leaves the system in the possible states and evaluate the expression

$$\begin{aligned} PR &= \mu_3 \int_0^N (f(x,0,0,1) + f(x,0,1,1) + f(x,1,0,1) + f(x,1,1,1))dx \\ &+ \mu_3 (\mathbf{p}(N,0,1,1) + \mathbf{p}(N,1,0,1) + \mathbf{p}(N,1,1,1)) \\ &+ \mu_1 \mathbf{p}(0,1,0,1) + \mu_2 \mathbf{p}(0,0,1,1) + (\mu_1 + \mu_2)\mathbf{p}(0,1,1,1) \end{aligned} \tag{83}$$

However, it is also possible to determine the production rates related to the separate machines. If p_i and r_i are the rates of failures and repairs, respectively, Machine M_i is operational a fraction of time that equals

$$e_i = \frac{r_i}{r_i + p_i}. \tag{84}$$

If Machine M_i operates in isolation, its production rate is $e_i \cdot \mu_i$. However, if the machines are part of a system, we have to take blocking and starvation into account. For Machines M_1 and M_2 we denote by p_{b_1} and p_{b_2} the probability of being blocked and find the following production rates:

$$PR_1 = \mu_1 e_1 (1 - p_{b_1}) \tag{85}$$

$$PR_2 = \mu_2 e_2 (1 - p_{b_2}) \tag{86}$$

With p_{s_3} as starvation probability of Machine M_3 we also find:

$$PR_3 = \mu_3 e_3 (1 - p_{s_3}) \tag{87}$$

Since Machine M_3 processes all the material coming from M_1 and M_2, we must have

$$PR_1 + PR_2 = PR_3. \tag{88}$$

Furthermore, PR_3 must equal PR from Equation (83). It is useful to determine the production rate estimates along several paths in order to detect possible modeling or programming bugs.

In order to determine the probabilities of blocking and starvation, we have to take into account that M_2 tends to be blocked more often that M_1 as the latter machine has priority to fill the buffer.

The blocking probabilities are

$$
\begin{aligned}
p_{b_1} &= \mathbf{p}(N,1,1,0) + \mathbf{p}(N,1,0,0) + \frac{\mu_1 - \mu_3}{\mu_1}\mathbf{p}(N,1,0,1) \\
&+ \frac{\max\{\mu_1 - \mu_3, 0\}}{\mu_1}\mathbf{p}(N,1,1,1) \qquad (89)
\end{aligned}
$$

$$
\begin{aligned}
p_{b_2} &= \mathbf{p}(N,1,1,0) + \mathbf{p}(N,0,1,0) + \frac{\mu_2 - \mu_3}{\mu_2}\mathbf{p}(N,0,1,1) \\
&+ \frac{\mu_2 - \max\{\mu_3 - \mu_1, 0\}}{\mu_2}\mathbf{p}(N,1,1,1) \qquad (90)
\end{aligned}
$$

and the probability of starvation is

$$
\begin{aligned}
p_{s_3} &= \mathbf{p}(0,0,0,1) + (1 - \frac{\mu_2}{\mu_3})\mathbf{p}(0,0,1,1) \\
&+ (1 - \frac{\mu_1}{\mu_3})\mathbf{p}(0,1,0,1) + (1 - \frac{\mu_1 + \mu_2}{\mu_3})\mathbf{p}(0,1,1,1) \qquad (91)
\end{aligned}
$$

where some of the probabilities $\mathbf{p}(x, \alpha_1, \alpha_2, \alpha_3)$ can be zero (see Table 1).

The algorithm used to solve the model is summarized in Appendix 2 to facilitate a possible re-implementation.

3. Numerical Results

The purpose of this section is to show that the algorithm yields numerical results whose structural properties agree with our knowledge of how the system should behave.

In a first set of experiments we assumed the following failure and repair rates:

- $p_1 = 0,03$, $r_1 = 0,07$ (M_1 is available $e_1 = \frac{0{,}07}{0{,}07+0{,}03}$ =70% of the time)
- $p_2 = 0,04$, $r_2 = 0,06$ (M_2 is available $e_2 = \frac{0{,}06}{0{,}06+0{,}04}$ =60% of the time)
- $p_3 = 0,01$, $r_3 = 0,09$ (M_3 is available $e_3 = \frac{0{,}09}{0{,}09+0{,}01}$ =90% of the time)

Processing rates for the four experiments are given in Table 5, the buffer size was $N = 10$ in all cases.

Average production rates and inventory levels as a function of varying processing rates are depicted in Figures 4 to 8. We report the production rate estimates related to Machines M_1, M_2, and M_3 (see Equations (85) to (87)).

Figure 4 shows that in the first experiment PR_1 increases almost linearly with μ_1 for low values of μ_1. As μ_1 increases further, Machine M_2 is blocked more often and PR_2 decreases, but the total production rate of the system still

Table 5. Processing rates in the numerical experiments

Experiment	μ_1	μ_2	μ_3
1	0.1,...,20	2	5
2	0.1,...,20	7.5	5
3	2	0.1,...,20	5
4	7.5	4	0.1,...,20

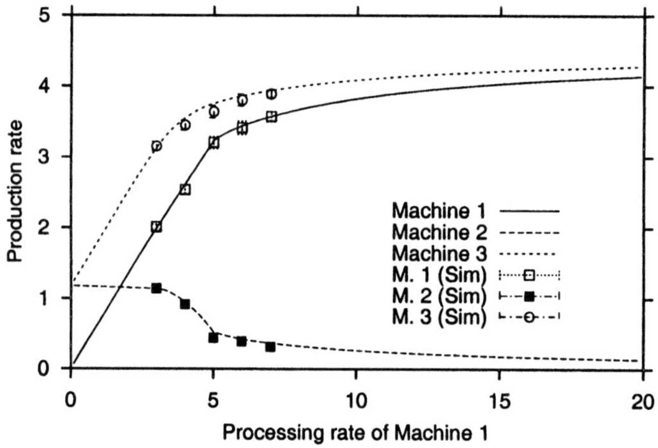

Figure 4. Production rates in Experiment 1 ($\mu_1 = 0.1, \ldots, 20, \mu_2 = 2, \mu_3 = 5$)

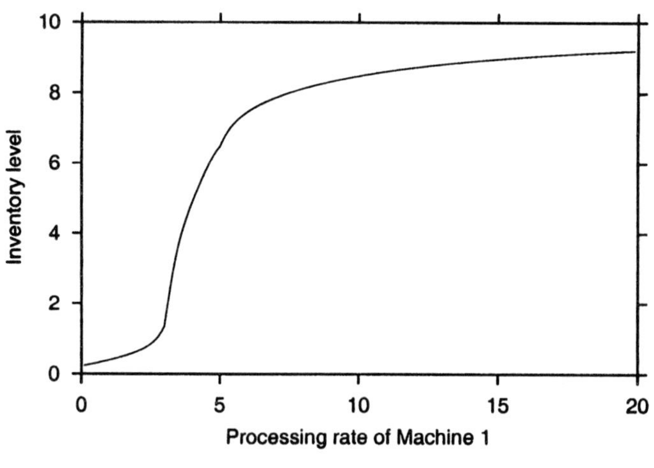

Figure 5. Inventory level in Experiment 1

increases. As μ_1 approaches 20, M_2 is almost always blocked and PR_2 approaches zero while M_3 is almost never starved and PR_3 approaches $\mu_3 \frac{r_3}{r_3+p_3} =$ 4.5. Note the kink of the production rate of Machine M_3 at $\mu_1 = 3$ as well as the kink of the production rate of Machine M_1 at $\mu_1 = 5$. The first kink is due to the fact that as μ_1 approaches 3.0, the combined rate $\mu_1 + \mu_2$ begins to exceed the rate $\mu_3 = 5$ of Machine M_3. Now the priority rule leads to a sharp decrease of M_2's production while M_1's production keeps increasing almost linearly. As μ_1 approaches $\mu_3 = 5$, M_1 is now also blocked more often which explains the second kink. Note also that the total production rate (which equals the production rate of Machine M_3) does *not* exhibit any such kink. That is, the kinks are only related to the change in the allocation of the capacity of Machine M_3 between M_1 and M_2 which is due to the priority mechanism. Figure 5 indicates a strong increase of the average buffer level as μ_1 increases.

To convince the reader that this system behavior is actually due to the modeling assumptions and not the result of an erroneous analysis, Figure 4 also reports the results of a discrete-event simulation of a three-machine merge system. We report the averages as well as the 95% confidence intervals for the production rates of the three machines for values of μ_1 of 3.0, 4.0, 5.0, 6.0, and 7.0. The simulation model was developed in ARENA. In the simulation model, we analyzed the production of *discrete* material assuming deterministic processing times. To implement a strict priority of Machine M_1 over M_2, we used blocking *after* service at M_1 and blocking *before* service at M_2. We set the buffer size in the simulation model to 9 as there is a 10th space for the parts at Machine M_3. In each simulation experiment we performed 10 independent replications over 110,000 time units each where the first 10,000 time units were truncated as a warm-up phase. This led to very tight confidence intervals as depicted in Figure 4. Note that the simulation results exhibit *all* the kinks the numerical analysis predicts. We conjecture that the slight deviation between simulation and analytical results is due to the fact that the simulation reports the result of a random experiment and that in the simulation model we worked with discrete material as opposed the fluid studied in the analytical model. The revelation of this surprising system behavior is considered to be one of the contributions of the research reported in this paper. It was only possible to detect this system behavior—with kinks in the production of M_1 and M_2—by analyzing the general case which is not restricted to $\mu_1 + \mu_2 < \mu 3$ (as opposed to the work reported in [Tan, 2001]).

The second experiment in Figure 6 shows that both PR_1 and PR_3 increase with an increasing value of μ_1. Note, however, that Machine M_1's production almost completely displaces those of M_2, i.e. the increase of PR_1 is much higher than those of PR_3.

In the third experiment we increased the speed of Machine M_2 in a situation where the upstream machines are initially the bottleneck. Note that both PR_2

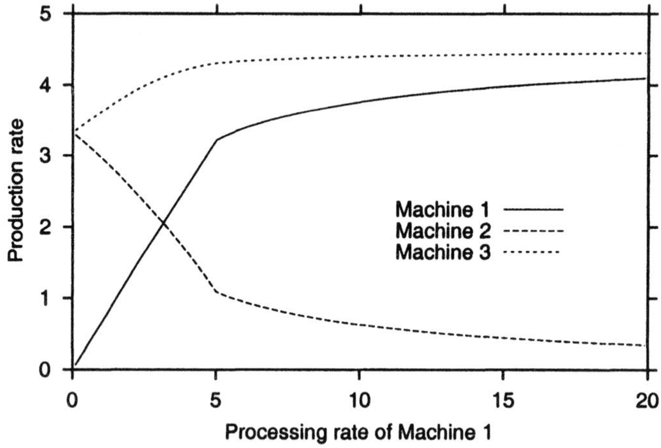

Figure 6. Production rates in Experiment 2 ($\mu_1 = 0.1, \ldots, 20$, $\mu_2 = 7.5$, $\mu_3 = 5$)

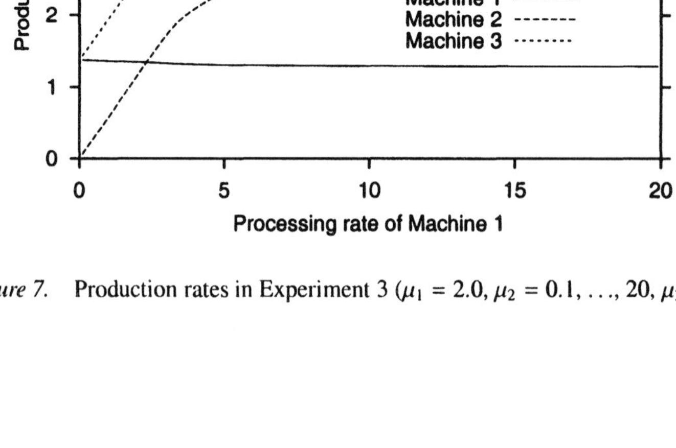

Figure 7. Production rates in Experiment 3 ($\mu_1 = 2.0$, $\mu_2 = 0.1, \ldots, 20$, $\mu_3 = 5$)

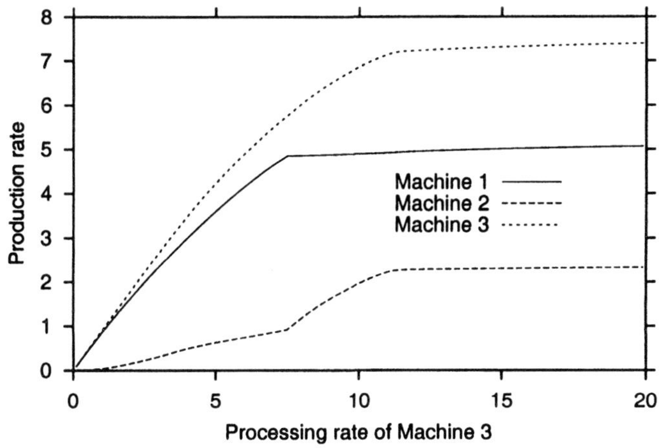

Figure 8. Production rates in Experiment 4 ($\mu_1 = 7.5, \mu_2 = 4, \mu_3 = 0.1, \ldots, 20$)

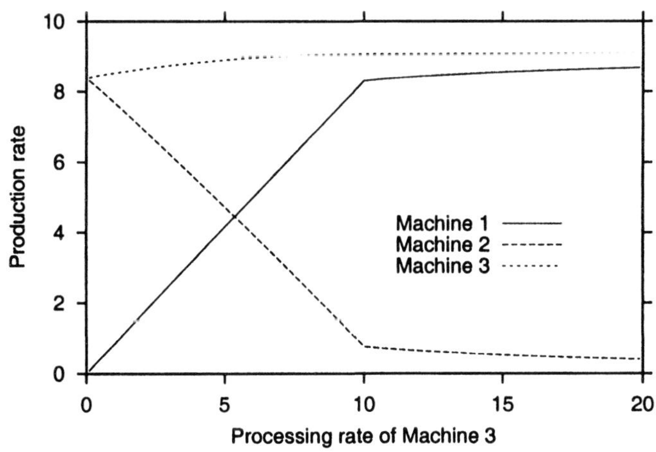

Figure 9. Production rates in Experiment 5 ($\mu_1 = 0.1, \ldots, 20, \mu_2 = 10, \mu_3 = 10$)

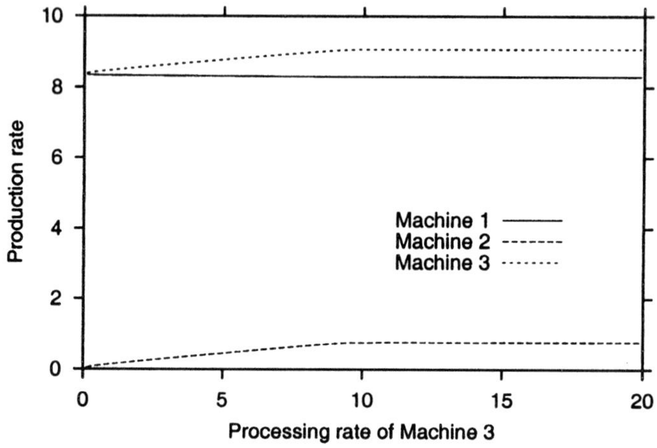

Figure 10. Production rates in Experiment 6 ($\mu_1 = 10, \mu_2 = 0.1, \ldots, 20, \mu_3 = 10$)

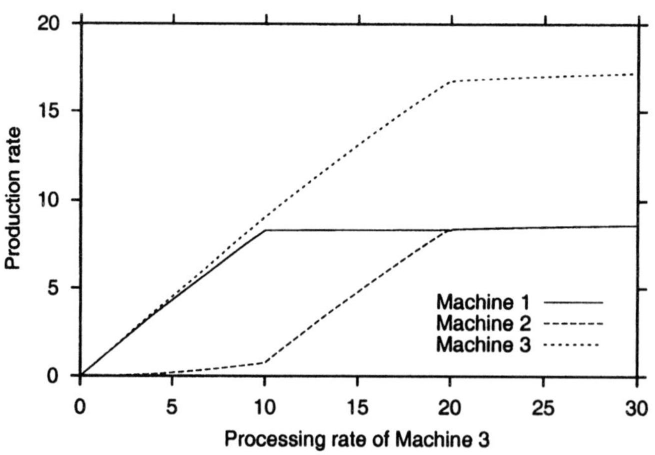

Figure 11. Production rates in Experiment 7 ($\mu_1 = 10, \mu_2 = 10, \mu_3 = 0.1, \ldots, 30$)

and PR_3 increase (as we would expect). What is interesting is a slight *decrease* of the production rate of the priority one machine M_1 in Figure 7. The reason for this decrease of PR_1 is that as μ_2 increases, the operation-dependent failures of M_3 occur more frequently and M_1 is therefore blocked more often.

The experiment in Figure 8 shows what happens as the speed of the last machine increases. Initially, almost all of the increase is dedicated to the priority-one machine M_1. However, as M_1 is eventually saturated, Machine M_2 shows a stronger increase of its production rate and PR_3 keeps increasing until M_2 is saturated as well.

In a second set of three experiments we studied systems with machines that are identical with respect to the failure rate $p_1 = p_2 = p_3 = 0.01$, the repair rate $r_1 = r_2 = r_3 = 0.1$ as well as the buffer size $N = 10$. We also set the processing rates of two of the three machines equal and varied the rate of the third machine to study the effect of the priority rule. The processing rates in the three experiments are given in Table 6.

Table 6. Processing rates in experiments 5 to 7

Experiment	μ_1	μ_2	μ_3
5	0.1,...,20	10	
6	10	0.1,...,20	10
7	10	10	0.1,...,30

In the Figure 9, we varied the speed of M_1 to see how the production of M_1 replaces that of M_2 as μ_1 increases. We observe essentially the same behavior as in Figure 6. However, we now find even more symmetry as now the failure and repair rates are all identical.

The experiment in Figure 10 indicates that a speed $\mu_2 = 10$ of Machine M_2 is sufficient to make sure that M_3 is almost never starved, its production now approaches $\frac{r_3}{p_3+r_3}\mu 3 = \frac{0.1}{0.01+0.1}10 \approx 9.1$. In the final experiment in Figure 11, we varied μ_1 between 0.1 and 30. Initially, almost all of the capacity of M_3 is devoted to M_1. As μ_3 reaches $\mu_1 = 10$, the capacity of M_1 is exhausted. Now the production of M_2 increases substantially until μ_3 reaches $\mu_1 + \mu_2 = 20$. A further increase of μ_3 has almost no additional effect on the production of M_3.

4. Conclusion

We have developed a Markov process model of a three-machine merge system with limited buffer capacity. In order to cope with machine-specific deterministic processing times, we used a continuous material model. The transition equations for the model were formulated and we described a procedure to solve them numerically in a specialized computer program. In order to handle the large symbolic expressions required in this program, we explained how to develop these expressions in *Mathematica* or *Maple*. Numerical experiments show that the procedures predicts the behavior of the merge system in a way that agrees with our intuition. Simulation experiments agree with the analytical results. The developed model is a new building block for the analysis of larger networks that contain merge structures, for example due to rework loops.

Appendix: Details of the Derivation and the Algorithm

1. Detailed Derivation of an Internal Transition Equation

Assume that all machines are up at time $t + \delta t$ and that the buffer level is between x and $x + \delta x$. The probability of seeing the system in this state is

$$f(x, 1, 1, 1, t + \delta t)\delta x.$$

It is equal to the sum of the probabilities of all previous states at time t times the probabilities of the respective transitions. In a continuous time Markov process model, we need to consider explicitly only those transitions with at most *one* change of a machine state since all other transitions of second and higher order have zero probability as δt approaches zero.

The four relevant (previous) states at time t, their probabilities, and the respective transition probabilities are:

- Machines M_1, M_2, and M_3 are up and the buffer level is between $x + (\mu_3 - \mu_1 - \mu_2)\delta t$ and $x + (\mu_3 - \mu_1 - \mu_2)\delta t + \delta x$. The probability of this state is

  $$f(x + (\mu_3 - \mu_1 - \mu_2)\delta t, 1, 1, 1, t)\delta x.$$

  In this case, none of the machines failed between time t and $t + \delta t$. During this time interval, the buffer level decreased by the amount $\mu_3 \delta t$ that Machine M_3 processed and increased by the amount processed by M_1 and M_2. The net decrease between time t and $t + \delta t$ is therefore $(\mu_3 - \mu_1 - \mu_2)\delta t$. The transition probability that all machines are up at time $t + \delta t$, given that all are up at time t, is

  $$(1 - p_1\delta t)(1 - p_2\delta t)(1 - p_3\delta t) + o(\delta t) = 1 - (p_1 + p_2 + p_3)\delta t + o(\delta t)$$

  since machine failures and repairs are assumed to be independent. This is the probability that none of the machines fails. All second and higher order terms are collected in $o(\delta t)$. Therefore

  $$(1 - (p_1 + p_2 + p_3)\delta t + o(\delta t)) \cdot f(x + (\mu_3 - \mu_1 - \mu_2)\delta t, 1, 1, 1, t)\delta x$$

  is the joint probability that all machines are up *both* at time t and at time $t + \delta t$.

- Machines M_2 and M_3 are up, M_1 is down, and the buffer level is between $x + (\mu_3 - \mu_2)\delta t$ and $x + (\mu_3 - \mu_2)\delta t + \delta x$. The probability of this state is

  $$f(x + (\mu_3 - \mu_2)\delta t, 0, 1, 1, t)\delta x.$$

  Machine M_1 must have been repaired between time t and $t + \delta t$ and the other machines must not have failed. The probability of this transition is

  $$r_1\delta t(1 - p_2\delta t)(1 - p_3\delta t) + o(\delta t) = r_1\delta t + o(\delta t)$$

  and the joint probability is therefore:

  $$(r_1\delta t + o(\delta t)) \cdot f(x + (\mu_3 - \mu_2)\delta t, 0, 1, 1, t)\delta x$$

- Machines M_1 and M_3 are up, M_2 is down, and the buffer level is between $x + (\mu_3 - \mu_1)\delta t$ and $x + (\mu_3 - \mu_1)\delta t + \delta x$. The probability of this state is

$$f(x + (\mu_3 - \mu_1)\delta t, 1, 0, 1, t)\delta x.$$

Machine M_2 must have been repaired between time t and $t + \delta t$ and the other machines must not have failed. The probability of this transition is

$$(1 - p_1\delta t)r_2\delta t(1 - p_3\delta t) + o(\delta t) = r_2\delta t + o(\delta t)$$

and the joint probability is therefore:

$$(r_2\delta t + o(\delta t)) \cdot f(x + (\mu_3 - \mu_1)\delta t, 1, 0, 1, t)\delta x$$

- Machines M_1 and M_2 are up, M_3 is down, and the buffer level is between $x - (\mu_1 + \mu_2)\delta t$ and $x - (\mu_1 + \mu_2)\delta t + \delta x$. The probability of this state is

$$f(x - (\mu_1 + \mu_2)\delta t, 1, 1, 0, t)\delta x.$$

Machine M_3 must have been repaired between time t and $t + \delta t$ and the other machines must not have failed. The probability of this transition is

$$(1 - p_1\delta t)(1 - p_2\delta t)r_3\delta t + o(\delta t) = r_3\delta t + o(\delta t)$$

and the joint probability is therefore:

$$(r_3\delta t + o(\delta t)) \cdot f(x - (\mu_1 + \mu_2)\delta t, 1, 1, 0, t)\delta x$$

Assembling all these expressions, we have:

$$\begin{aligned}
&f(x, 1, 1, 1, t + \delta t)\delta x = \\
&\quad (1 - (p_1 + p_2 + p_3)\delta t) \cdot f(x + (\mu_3 - \mu_1 - \mu_2)\delta t, 1, 1, 1, t)\delta x + \\
&\quad r_1\delta t\, f(x + (\mu_3 - \mu_2)\delta t, 0, 1, 1, t)\delta x + \\
&\quad r_2\delta t\, f(x + (\mu_3 - \mu_1)\delta t, 1, 0, 1, t)\delta x + \\
&\quad r_3\delta t\, f(x - (\mu_1 + \mu_2)\delta t, 1, 1, 0, t)\delta x + o(\delta t) \qquad (1)
\end{aligned}$$

Using a Taylor series expansion of the probability density function $f(x, 1, 1, 1, t)$ at x, we can write

$$\begin{aligned}
&f(x + (\mu_3 - \mu_1 - \mu_2)\delta t, 1, 1, 1, t) \\
= \ &f(x, 1, 1, 1, t) + (\mu_3 - \mu_1 - \mu_2)\delta t\frac{\partial}{\partial x}f(x, 1, 1, 1, t) + o(\delta t). \qquad (2)
\end{aligned}$$

Using the same type of expansion for the other density functions on the right hand side of (1) and collecting all second and higher order terms in $o(\delta t)$, we find

$$\begin{aligned}
f(x, 1, 1, 1, t + \delta t)\delta x \ = \ &(1 - (p_1 + p_2 + p_3)\delta t)f(x, 1, 1, 1, t)\delta x + \\
&(\mu_3 - \mu_1 - \mu_2)\delta t\frac{\partial}{\partial x}f(x, 1, 1, 1, t)\delta x + \\
&r_1\delta t f(x, 0, 1, 1, t)\delta x + r_2\delta t f(x, 1, 0, 1, t)\delta x + \\
&r_3\delta t f(x, 1, 1, 0, t)\delta x + o(\delta t). \qquad (3)
\end{aligned}$$

Subtracting $f(x,1,1,1,t)\delta x$ and dividing by δx and δt, we find:

$$\frac{f(x,1,1,1,t+\delta t)-f(x,1,1,1,t)}{\delta t} =$$
$$-(p_1+p_2+p_3)f(x,1,1,1,t)+(\mu_3-\mu_1-\mu_2)\frac{\partial}{\partial x}f(x,1,1,1,t)+$$
$$r_1 f(x,0,1,1,t)+r_2 f(x,1,0,1,t)+r_3 f(x,1,1,0,t)+\frac{o(\delta t)}{\delta t}. \tag{4}$$

If we now take the limit for $\delta t \to 0$, this leads to a partial differential equation in time t and buffer level x. If we assume that the system reaches steady state where the probabilities of the states remain constant, we have

$$\lim_{\delta t\to 0}\frac{f(x,1,1,1,t+\delta t)-f(x,1,1,1,t)}{\delta t}=0$$

and we can omit the time variable t in (4) to find:

$$\begin{aligned}(p_1+p_2+p_3)f(x,1,1,1) &= (\mu_3-\mu_1-\mu_2)\frac{\partial}{\partial x}f(x,1,1,1)+\\ &\quad r_1 f(x,0,1,1)+r_2 f(x,1,0,1)+\\ &\quad r_3 f(x,1,1,0).\end{aligned} \tag{5}$$

Equation (5) is a differential equation in the buffer level x. The right side of the equation is related to the rate at which state $(x,1,1,1)$ is reached and the left side is related to the rate at which it is left. In steady state, both rates have to be equal.

2. The Algorithm

It is possible to use a program supporting the symbolic manipulation of algebraic expressions to evaluate the equations derived above numerically to determine the performance measures. However, the same type of program can also be used to determine *analytic* expressions symbolically that can be transferred to, for example, a C-program for the analysis of the merge system. This is necessary if one wants to incorporate this merge system in larger decomposition methods for systems with rework loops.

In this section we firstly describe step by step how to *develop* the required symbolic expressions and to incorporate them in a specialized program. We do *not* state the expressions explicitly.

1 State density functions (62), i.e. (63) to (68) for Case 3, *symbolically* (using a program for symbolic manipulation of algebraic expressions) for undetermined values c_i, s_i, and $\mathbf{y}^i$.

2 Use these symbolic density functions and symbolic expressions for the undetermined matrices $\mathbf{H}_{3,0}^{-1}$ and $\mathbf{H}_{3,N}^{-1}$ (in Case 3) in Equations (71) and (73) to determine symbolic expressions for the boundary state probabilities.

3 Plug these symbolic expressions for probabilities and density functions into (75) and the corresponding underlined equations from Table 4. Determine symbolically the matrix $\mathbf{B}$ in Equation (81), i.e. the coefficients of the unknowns c_i. These coefficients are symbolic expressions that can be transferred into a C-program where it is easy to evaluate them numerically for given numerical values of the system parameters.

4 Plug the same symbolic expressions for probabilities and density functions into (82), (83), (85), (86), and (87) and evaluate these expressions symbolically to determine general symbolic expressions for the inventory level and production rate estimates.

We now describe the steps of the program that uses the symbolic expressions for the matrix **B**, the probability masses and densities, and the performance measures.

1 Determine the matrices $\mathbf{A}_{11}$, $\mathbf{A}_{12}$, $\mathbf{A}_{21}$, and $\mathbf{A}_{22}$ from the given parameters (μ_i, p_i, and $r_i, i = 1, ..., 3$ and the buffer capacity N). Determine matrices $\mathbf{\Omega}$ and $\mathbf{\Lambda}$ via Equations (59) and (60), respectively.

2 Determine the eigenvalues s_i and eigenvectors $\mathbf{y}^{(i)}$ of matrix $\mathbf{\Lambda}$. Determine numerically the matrices that are needed to compute upper and lower boundary probability masses. In Case 3, for example, compute numerically matrices $\mathbf{H}_{3,0}^{-1}$ and $\mathbf{H}_{3,N}^{-1}$, starting with (72) and (74). Plug all these numerical values into the symbolic expressions of matrix **B**. Compute numerically the inverse of **B** to solve (81) for the numerical values of the coefficients c_i.

3 Plug the all the numerical values computed so far into the symbolic expressions for buffer level and production rate estimates (82) and (83) or (85) to (87). This gives the numerical values of the sought performance measures.

The computer program we wrote essentially consists of 11 very similar parts for the 11 cases. The code is relatively lengthy due to the clumsy symbolic expressions, but its execution times are negligible.

References

[Buzacott, 1967] Buzacott, J. A. (1967). Automatic transfer lines with buffer stocks. *International Journal of Production Research*, 5(3):183–200.

[Dallery, 1999] Dallery, Y. (1999). Extending the scope of analytical methods for performance evaluation of manufacturing flow systems. In *Second Aegean International Conference on "Analysis and Modeling of Manufacturing Systems", May 16-20, 1999, Tinos Island, Greece*, pages 1–16. University of the Aegean, Department of Business Administration.

[Dallery and Gershwin, 1992] Dallery, Y. and Gershwin, S. B. (1992). Manufacturing flow line systems: A review of models and analytical results. *Queuing Systems Theory and Applications*, 12(1-2):3–94. Special issue on queuing models of manufacturing systems.

[Di Mascolo et al., 1991] Di Mascolo, M., David, R., and Dallery, Y. (1991). Modeling and analysis of assembly systems with unreliable machines and finite buffers. *IIE Transactions*, 23(4):315–330.

[Englin-Müllges and Uhlig, 1996] Englin-Müllges, G. and Uhlig, F. (1996). *Numerical Algorithms with C.* Springer, Berlin et al.

[Gershwin, 1991] Gershwin, S. B. (1991). Assembly/disassembly systems: An efficient decomposition algorithm for tree-structured networks. *IIE Transactions*, 23(4):302–314.

[Gershwin and Schick, 1980] Gershwin, S. and Schick, I. (1980). Continuous model of an unreliable two-stage material flow system with a finite interstage buffer. Technical Report LIDS-R-1039, Massachusetts Institute of Technology, Cambridge, Massachusetts.

[Gershwin and Schick, 1983] Gershwin, S. and Schick, I. (1983). Modeling and analysis of three-stage transfer lines with unreliable machines and finite buffers. *Operations Research*, 31(2):354–380.

[Gershwin, 1994] Gershwin, S. B. (1994). *Manufacturing Systems Engineering*. PTR Prentice Hall, Englewood Cliffs, New Jersey.

[Helber, 1999] Helber, S. (1999). *Performance Analysis of Flow Lines with Non-Linear Flow of Material*, Volume 473 of *Lecture Notes in Economics and Mathematical Systems*. Springer, Berlin et al.

[Tan, 2001] Tan, B. (2001). A three-station merge system with unreliable stations and a shared buffer. *Mathematical and Computer Modelling*, 33:1011–1026.

[Wijngaard, 1979] Wijngaard, J. (1979). The effect of interstage buffer storage on the output of two unreliable production units in series, with different production rates. *AIIE Transactions*, 11(1):42–47.

[Yeralan and Tan, 1997] Yeralan, S. and Tan, B. (1997). A station model for continuous materials flow production systems. *International Journal of Production Research*, 35(9):
2525–2542.

[Zimmern, 1956] Zimmern, B. (1956). Etudes de la propagation des arrêts aleatoires dans les chaines de production. *Review Statististical Applications*, 4:85–104.

Chapter 5

OPTIMAL SCHEDULING FOR PIECEWISE DETERMINISTIC MULTI-ARMED BANDIT PROBLEM

Max-Olivier Hongler

Département de Microtechnique, (DMT)

Institut de Production Microtechnique, (IPM)

E.P.F.L.

CH-1015 LAUSANNE.

max.hongler@epfl.ch

Fabrice Dusonchet*

Département de Microtechnique, (DMT)

Institut de Production Microtechnique, (IPM)

E.P.F.L.

CH-1015 LAUSANNE.

fabrice.dusonchet@epfl.ch

Abstract We derive explicit expressions for the priority indices (Gittins' indices), associated with a class of multi-armed Bandit processes. The underlying dynamic governing the arms belong to piecewise deterministic random evolutions. We then use this class of model to discuss a simple version of the scheduling problem of a flexible manufacturing resource, with limited capacity, operating in a random environment.

Keywords: Multi-armed Bandit processes, Gittins' indices, Optimal stopping problems, Dynamic scheduling of jobs.

*In part supported by the "Fonds National Suisse pour la Recherche Scientifique"

1. Introduction

The scheduling of a multiclass make-to-stock facility with a limited resource, operating in a fluctuating environment, is a favorite subject in the ongoing research devoted to manufacturing systems, [Wei 92], [Vea 96], [Ha 97], [Yao 97], [Pen 97], [Sriv 98], [Fed 99], [Ver 99]. Due to the problem complexity it is, in general, impossible to analytically derive exact optimal production policies. Moreover, even when numerical results are available, the optimal solutions are often too complex to be implemented in actual situations. Therefore, one heavily relies on approximation schemes yielding efficient though, in general, only sub-optimal policies. When the set-up costs and/or times required to change from one production class to another are negligible, efficient approximation schemes can be obtained by using priority indices $M_j(X_j(t))$, $j = 1, 2, \ldots, N$, with N denoting the numbers of items that can be processed by the machine. These indices $M_j(X_j(t))$ indicate the relative urgency to produce the particular item j when its surplus level is at position $X_j(t)$. Using such priority indices, the resulting scheduling policy reads for example as: *at each time produce the item having the largest index value* $M_j(X_j(t))$.

For a rather large class of systems, scheduling policies based on priority indices may actually be optimal. *For which class of dynamics is this realized ?* This fundamental question has been addressed and answered in a recent series of papers by [Sha 92], [Ber 96] and [Yao 97]. As an illustration, the scheduling problems which can be modeled by Multi-armed Bandit processes introduced by J.G. Gittins and D.M. Jones [Git 74], [Git 89] belong to the above mentioned class of dynamics [Ber 96]. Gittins' work has stimulated a wealth of studies devoted to both discrete and continuous-time evolutions for the arms of the Bandits. For the continuous-time case, which will be studied in section 3, let us mention the contributions: [Whi 82], [Kar 84], [Epl 86], [Man 87], [ElK 93], [Kas 95], [Dal 96], [ElK 97] and [Hon 01]. In these contributions, the indices $M_j(x_j)$ are characterized and calculated for several types of random evolutions. In particular, in an elegant paper, I. Karatzas [Kar 84] explicitly calculates the priority indices (also named Gittins' indices), for arms obeying to diffusive dynamics. More recently, H. Kaspi and A. Mandelbaum [Kas 95] discussed the situation where Levy's processes govern the arms' dynamics.

In the present paper, we shall consider another class of random evolutions given by piecewise deterministic processes (PD) for which explicit Gittins indices can be calculated. A complete characterization of the PD processes can be found in [Pin 91]. One motivation to study the PD processes originates from the fact that they arise naturally in the fluid modeling of production flows delivered by failure prone machines, [Bie 88], [Ger 94]. In this manufacturing context, a simple illustration of these processes can be given by considering a flexible, failure prone machine able to produce N different types of items. The machine can produce only a single item at the same time. We assume the set-up costs and time to change from one production type to another to be negligible. The underlying dynamic $X_k(t)$ when production k is engaged is described by the stochastic differential equation:

$$\frac{d}{dt}X_k(t) = U_k(t) - D_k, \quad k = 1, 2, \dots N$$

where $U_k(t) \in \{0, U_k\}$ is an alternating renewal process [Ros 83]. The quantities $U_k(t)$ and D_k represent the production respectively the demand rates for the type k. The sojourn times in the $\{0\}$ and $\{U_k\}$ states have to be specified and will be chosen as exponential distributions in the following. In general, one can define a reward function $h_k(X_k)$ which describes the instantaneous advantage gained when engaging the production of type k at time t. The dynamic scheduling problem is therefore to select which type of production, among the N possibilities, to engage as time t increases, in order to maximize, for instance, the discounted reward that can be expected with this production process.

The organization of the paper is as follows. In section 2, the reader will find a short overview of Multi-armed Bandit processes (MABP). In particular, the calculation of the Gittins' indices via the solution of an associated optimal stopping problem is briefly reviewed in section 2.1. In section 3, we apply the formal theory of section 2.1 to the MABP with arms evolving as piecewise deterministic processes. Explicit expressions for the Gittins' indices are derived and the associated optimal scheduling is given in section 3.3. In section 4, we show how the mathematical modeling developed in section 3 can be used in the manufacturing context. We end by conclusions and perspectives.

2. Classical Multi-armed Bandit problems. A brief overview.

The Multi-armed Bandit process (MABP) is an idealized, mathematical description of the conflict existing between taking those actions which yield immediate reward and those whose benefit will only come later. This paradigmatic situation is represented by the possibility to choose one among N parallel projects (also named arms from now on). At each decision time t_i ($\{t_i\}$, $i = 1, 2, 3, \dots$ is a discrete set in this section), one is allowed to engage one project at most. The decision to engage (or disengage) a project, as a function of time, is called a scheduling policy π. Besides other technicalities, the fundamental property of an admissible policy is its non anticipating character [Whit 88] i.e.

the decisions are taken only on the past information and no premonition is allowed. If the decision is to engage the arm j in state $X_j(t_i) = x_j \in \mathcal{Y}_j$, where $\mathcal{Y}_j$ is the set of possible states for the arm j, then one gets an instant reward $h_j(x_j)$ and the project moves into the state $X_j(t_{i+1})$, where $X_j(t)$ is a stochastic process. We assume here the independence of $X_j(t)$ and $X_k(t)$ for $j \neq k$. Moreover $X_j(t)$ will be here chosen to be markovian. The states of the $N-1$ disengaged projects remain frozen (i.e. $X_k(t_{i+1}) = X_k(t_i)$, $k \neq j$). We write $\vec{X}(t) = (X_1(t), \ldots, X_N(t))$ for the state of the system at time t. Let us define the indicator $\mathcal{I}_j^\pi(t_i)$:

$$\mathcal{I}_j^\pi(t_i) = \begin{cases} 1 & \text{if arm } j \text{ is played at time } t_i, \\ 0 & \text{elsewhere.} \end{cases}$$

Given a policy π, we can now write the total discounted reward as:

$$J^\pi(\vec{X}(t_0)) = E_\pi \left\{ \sum_{i=0}^{\infty} \beta^{t_i} \sum_{j=1}^{N} h_j(X_j(t_i)) \mathcal{I}_j^\pi(t_i) \;\middle|\; \vec{X}(t_0) \right\}, \tag{1}$$

where $E_\pi\{\cdot \mid \vec{X}(t_0)\}$ is the mathematical expectation under policy π when the initial condition is $\vec{X}(t_0)$ and $0 < \beta < 1$ is a discounting factor.

The objective is to find an optimal allocation policy π^* which maximizes the global reward:

$$J^{\pi^*}(\vec{X}(t_0)) = \max_\pi J^\pi(\vec{X}(t_0)).$$

Define the one-step operator $L_j(\cdot)$ for arm j as:

$$L_j J(\vec{X}(t_i)) = h_j(X_j(t_i)) + \beta E[J(\vec{X}(t_{i+1})) \mid I_j(t_i) = 1)].$$

Then $J^{\pi^*}(\vec{X}(t_0))$ is the unique solution of [Whit 82]:

$$J^{\pi^*}(\vec{X}(t_0)) = \max_j L_j J^{\pi^*}(\vec{X}(t_0)).$$

2.1 Optimal scheduling policy for MABP.

Let us now recall the form of the optimal policy first discovered by J.C. Gittins [Git 74]. To this end, one first introduces the concept of priority indices (Gittins'indices).

Definition 1, (Priority index policy): An index policy is a scheduling rule, based on the existence of indices $M_j(x_j)$, depending only on j and x_j. In terms of the $M_j(x_j)$, the priority index policy specifies: "*At each decision time en-*

gage the project exhibiting the largest index value".

Gittins showed that with the appropriate indices which will be defined below, the MABP is optimally solved by the priority index policy. The calculation of the priority indices is done by splitting the initial N-armed Bandit problem into N stopping subproblems which will be denoted as problems $\mathcal{SP}_j$, $j = 1, \ldots, N$. The Problem $\mathcal{SP}_j$ reads as follows: Given a terminal reward $\frac{M}{1-\beta}$ (remember that $0 < \beta < 1$ is a discounting factor), find an optimal stopping time τ^* which maximizes the reward $J_j^*(X_j(t_0), M)$ gained by engaging arm j until time τ^*, then stop and collect the reward $\frac{\beta^{\tau^*} M}{1-\beta}$. This maximal expected reward is the unique bounded solution, of the equation:

$$J_j^*(X_j(t_0), M) = \max\left[M \,;\, L_j J_j^*(X_j(t_0), M)\right].$$

Alternatively, we can view the problem $\mathcal{SP}_j$ as being equivalent to the following two-armed Bandit problem which will be denoted as $\mathcal{P}_j$, $j = 1, \ldots, N$. The first arm of problem $\mathcal{P}_j$ (arm 1) corresponds to the former arm j, with dynamics $X_j(t)$ and yielding a reward $h_j(X_j(t))$ as before. The arm 2 evolves trivially as $X_2(t) \equiv \xi$, $\forall t$ and yields a systematic reward M when engaged. We assume that initially arm 1 is engaged. Observe that once the arm 2 is engaged, it is never optimal to reengage arm 1. Indeed, if at time t_i, it is optimal to engage arm 2, the states of the system at time t_{i+1} remain identical with those given at time t_i. It is then optimal to carry on playing arm 2 for ever. Then, problem $\mathcal{SP}_j$ and problem $\mathcal{P}_j$ are equivalent. In accord with intuition, it is proved that $J_j^*(X_j(t_0), M)$ is a non decreasing function of M which equals M for M large enough (see [Whit 82]). The Gittins index $M_j(X_j(t_0))$ for arm j, is now defined as the smallest value of M for which $J_j^*(X_j(t_0), M) = M$. Then the index $M_j(X_j(t_0))$ is exactly the critical value for which one has:

$$M_j(X_j(t_0)) = L_j J_j^*\big(X_j(t_0), M_j(X_j(t_0))\big).$$

It is important to understand that the index value $M_j(X_j(t_0))$ is not the expected reward received for a time infinite operation of arm j, when the initial state is $X_j(t_0)$. Rather, $M_j(X_j(t_0))$ can be regarded as the minimal value of M which makes the following options equivalent:

a) disengage project j immediately and collect a reward $\frac{M}{1-\beta}$,

b) engage project j until τ^*, then disengage it and collect the reward $\frac{\beta^{\tau^*} M}{1-\beta}$.

Lemma: The optimal stopping time τ^* of problem $\mathcal{SP}_j$ is:

$$\tau^* = \inf\{t \in \mathbb{R}^+ \mid M_j(X_j(t_i)) \leq M\}$$

Proof: Problem $\mathcal{SP}_j$ is equivalent to problem $\mathcal{P}_j$. As the Bandit problems $\mathcal{P}_j$ is solved optimally with a Gittins' index policy, the solution of the stopping problem $\mathcal{SP}_j$ is as follows: engage arm j as long as $M_j(X_j(t_i)) > M_2(X_2(t_i))$ otherwise stop and collect $\frac{M}{1-\beta}$. Here $M_2(X_2(t_i))$ is the Gittins' index for arm 2 of problem $\mathcal{P}_j$ and by construction we have:

$$M_2(X_2(t_i)) = M.$$

□

3. Optimal stopping for piecewise deterministic processes.

In this section we calculate the priority index $M_j(x_j)$ associated with a class of piecewise deterministic processes, by solving a stopping problem of the type discussed in section 2.1. The random processes used below have been studied in detail by [Pin 91], under the name of random evolution models. Here we focus our attention on the two-states random velocity model.

3.1 A simple class of dynamics

Let us consider the scalar time evolution for $t \geq 0$ given by:

$$\frac{d}{dt}X(t) = I(t), \quad X(0) = x, \quad I(0) = i, \quad \text{and} \quad X(t) \in \mathbb{R} \tag{2}$$

and where $I(t)$ is an alternating renewal process ([Ros 83], [Pin 91]) taking the alternate constant values U, (up) and D, (down). Assume that the sojourn times in the U and D phases are exponentially distributed with parameters λ and μ respectively. We further impose:

$$D\lambda + U\mu > 0 \quad \text{with} \quad D < 0 \quad \text{and} \quad D + U > 0.$$

so that the asymptotic average of $X(t)$ increases with time. Note that the realization of the stochastic differential equations Eq.(2) are continuous in contrary to the class of dynamics discussed in [Dav 84] where jumps in the realization are present. A typical realization of the solution of Eq.(2) is sketched in Figure 1. It is important to note that the process $X(t)$ solution of Eq.(2) is not markovian. However the pair process defined by $\zeta(t) := (X(t), I(t)) \in \mathbb{R} \times \{D, U\}$ is itself a markov process.

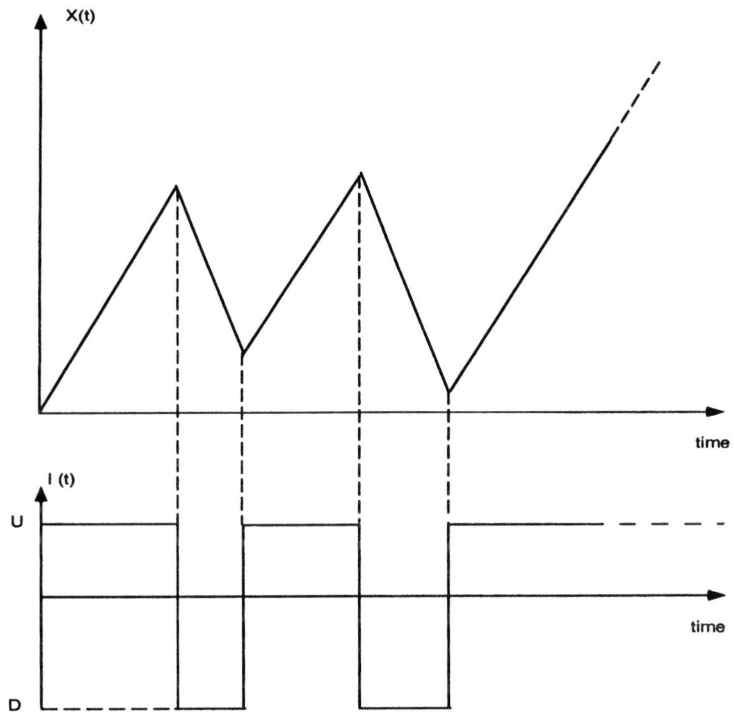

Figure 1. Qualitative behavior of the solution of Eq.(2) as a function of a realization the noise $I(t)$.

3.2 The associated optimal stopping problem

Now we introduce a simple optimal stopping problem defined by the cumulate reward function $J(x, i, M)$:

$$J(x, i, M) = E_{(x,i)} \int_0^{\tau} h(X(s))\, e^{-\beta s} ds + M e^{-\beta \tau}, \quad M \in \mathbb{R}^+, \tag{3}$$

where τ is a stopping time, M a terminal reward and $\beta > 0$ a discount factor. The instant reward function $h(x)$ is a twice differentiable function, which is strictly increasing. In addition, we follow [Whi 82] and [Kar 84] and impose the following asymptotic properties:

$$\lim_{x\to\infty} h(x) = \beta K, \quad \lim_{x\to-\infty} h(x) = \beta k, \quad 0 \le k < K, \tag{4}$$

with $\beta > 0$ and:

$$\lim_{|x|\to\infty} \frac{d}{dx} h(x) = \lim_{|x|\to\infty} h'(x) = 0.$$

In Eq. (3), the notation $E_{(x,i)}$ stands for the mathematical expectation when the initial conditions are $X(t = 0) = x$ and $I(t = 0) = i$.
The problem is to find an optimal stopping time τ^* which maximizes the expected reward $J(x, i, M)$.

To solve the optimal stopping problem, we proceed, as usual, by observing that when the process $(X(t), I(t))$ defined by Eq.(2) is in a given state, we can either wait a certain period of time before stopping or stop immediately. The resulting rewards are calculated by assuming that the subsequent decisions are optimal. Accordingly, we can write:

i) If at time t, the decision is to wait for an (infinitesimal) time ξ before stopping, the optimal reward $J(X(t), I(t), M)$ is greater than the running reward from t to $t + \xi$ and the optimal reward from time $t + \xi$ onwards. This yields the inequality:

$$\begin{aligned} &J(X(t), I(t), M) \geq \\ &\int_t^{t+\xi} e^{-\beta(s-t)} h(X(s))\, ds + e^{-\beta\xi} J(X(t+\xi), I(t+\xi), M). \end{aligned} \tag{5}$$

ii) If the decision at time t is to stop immediately, the reward is precisely: $J(X(t), I(t), M) = M$.

Let us focus on case i) for which we consider two possible rewards depending on the realization of the process $I(t)$, namely: $I = U$ or $I = D$. To this two possibilities, we associate the rewards functions $J(x, U, M)$ and $J(x, D, M)$ respectively. With this notation and the markov character of the alternating process $I(t)$, the first order time expansion of Eq.(5) yields:

$$0 \geq h(x) - (\beta + \lambda)J(x, U, M) + U\frac{d}{dx}J(x, U, M) + \lambda J(x, D, M) \tag{6}$$

and

$$0 \geq h(x) - (\beta + \mu)J(x, D, M) + D\frac{d}{dx}J(x, D, M) + \mu J(x, U, M). \tag{7}$$

The strictly increasing nature of $h(x)$ suggests to distinguish between three regimes holding in three adjacent intervals O_j, $j = 1, 2, 3$ defined by two thresholds $b_u < b_d$. Accordingly, we shall write:

a) $O_1 = [b_d, \infty)$,

b) $O_2 = [b_u, b_d]$,

c) $O_3 = (-\infty, b_u]$.

Now we construct the reward functions $J^{(k)}(x, D, M)$ and $J^{(k)}(x, U, M)$ for $k = 1, 2, 3$ holding in the intervals O_1, O_2, and O_3 respectively. The qualitative behavior of the reward functions is sketched in Figure 2.

a) *Solution in the interval* O_1. This is a continuation region for both $J^{(1)}(x, D, M)$ and $J^{(1)}(x, U, M)$. In this region Eqs.(6) and (7) hold simultaneously with equality. We can therefore combine Eqs.(6) and (7) to get two second order differential equations having the form:

$$\mathcal{A}\left(J^{(1)}(x, U, M)\right) + h_D(x) = 0 \tag{8}$$

and

$$\mathcal{A}\left(J^{(1)}(x, D, M)\right) + h_U(x) = 0, \tag{9}$$

where $\mathcal{A}$ stands for the differential operator:

$$\mathcal{A}(\cdot) = -UD\frac{d^2}{dx^2}(\cdot) + \delta\frac{d}{dx}(\cdot) - \beta\rho(\cdot)$$

with

$$\delta = D(\beta + \lambda) + U(\beta + \mu) > 0, \quad \text{and} \quad \rho = (\beta + \mu + \lambda)$$

and

$$h_D(x) = \rho h(x) - Dh'(x) \quad \text{and} \quad h_U(x) = \rho h(x) - Uh'(x).$$

For later use, we shall restrict our attention to the class $\mathcal{H}$ of strictly increasing functions $h_I(x)$ where I is the state of the noise $I \in \{U, D\}$,

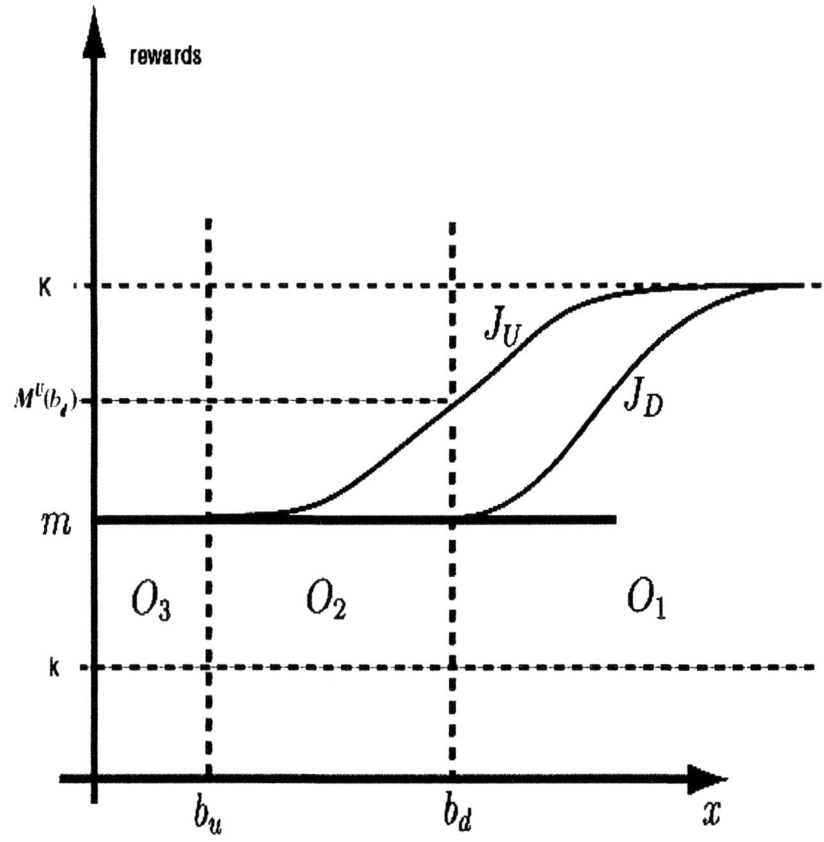

Figure 2. Qualitative behavior of the rewards functions $J(x, U, M)$ and $J(x, D, M)$ in the intervals O_k, $k = 1, 2, 3$.

namely:

$$h_I(x) \in \mathcal{H} \quad \Leftrightarrow \quad h_I'(x) = \rho h'(x) - Ih''(x) > 0. \tag{10}$$

The solutions of the linear differential Eqs.(8) and (9) are easy to derive. Identical equations arise in [Kar 84] and using these results, we can directly write:

$$J^{(1)}(x, D, M) = A_{d,1} \exp\{-\gamma_+ x\} + p_d(x), \quad \text{for} \quad x > b_d \tag{11}$$

and

$$J^{(1)}(x, U, M) = A_{u,1} \exp\{-\gamma_+ x\} + p_u(x), \quad \text{for} \quad x > b_d \tag{12}$$

with $A_{d,1}$ and $A_{u,1}$ being integration constants and:

$$\gamma_+ = \frac{-1}{2DU}\left[\sqrt{\delta^2 - 4DU\rho\beta} + \delta\right] > 0,$$

The functions $p_d(x)$ and $p_u(x)$ are particular solutions of Eqs.(8) and (9). They represent the rewards when running forever, (i.e. when no stopping occurs). Hence, we have $p_u(x) > k$ and $p_d(x) > k$. In our case, two particular solutions of the inhomogeneous Eqs.(8) and (9), respectively read as:

$$p_u(x) = E_{(x,U)} \int_0^\infty e^{-\beta t} h_D(X(t))\, dt$$

$$= \mathcal{N}\left[e^{-\gamma_+ x}\int_{-\infty}^{x} h_D(y)e^{\gamma_+ y}dy + e^{\gamma_- x}\int_x^{\infty} h_D(y)e^{-\gamma_- y}dy\right]$$

and

$$p_d(x) = E_{(x,D)} \int_0^\infty e^{-\beta t} h_U(X(t))\, dt$$

$$= \mathcal{N}\left[e^{-\gamma_+ x}\int_{-\infty}^{x} h_U(y)e^{\gamma_+ y}dy + e^{\gamma_- x}\int_x^{\infty} h_U(y)e^{-\gamma_- y}dy\right],$$

with:

$$\mathcal{N} = [-DU(\gamma_+ + \gamma_-)]^{-1}$$

and

$$\gamma_- = \gamma_+ + \frac{\delta}{UD} \leq 0.$$

b) *Solution in the interval* O_2. This is a stopping region when $I(t) = D$ and hence $J(x, D, M) = M$ and a continuation region when $I(t) = U$. Hence the use of Eq.(6) with $J(x, D, M) = M$ immediately yields:

$$J^{(2)}(x, U, M) = e^{\omega x}\left[A_{u,2} - \int_0^x \left[\frac{h(\xi) + \lambda M}{U}\right] e^{-\omega\xi} d\xi\right], \qquad (13)$$

with $A_{u,2}$ being an integration constant and $\omega = \frac{\beta+\lambda}{U}$.

c) *Solution in the interval* O_3. This is a global stopping region and we therefore have $J^{(3)}(x, U, M) = J^{(3)}(x, D, M) = M$

From Eqs.(11), (12), (13), we see that we have to determine three constants of integration $A_{u,1}$, $A_{u,2}$ and $A_{d,1}$. Moreover the thresholds b_u and b_d are yet unknown. Hence, we need five relations to determine these five unknowns. These relations are, see figure (2):

i) $J^{(1)}(b_d, D, M) = M$.

ii) $\frac{d}{dx}J^{(1)}(x, D, M)\mid_{(x=b_d)}= 0$, smooth fitting in b_d.

iii) $J^{(2)}(b_u, U, M) = M$.

iv) $\frac{d}{dx}J^{(2)}(x, U, M)\mid_{(x=b_u)}= 0$, smooth fitting in b_u.

v) $J^{(1)}(b_d, U, M) = J^{(2)}(b_d, U, M)$.

The relations i) and ii) determine alone the constants b_d and $A_{d,1}$ and the details of the resulting algebra is identical with the those yielding the Eqs.(3.14) and (3.20) of the Karatzas' paper [Kar 84]). Accordingly, we obtain:

$$\begin{cases} M(b_d) & = & M \\ A_{d,1} & = & \frac{p_d'(b_d)}{\gamma_+}\exp(\gamma_+ b_d) \end{cases} \qquad (14)$$

with:

$$M(x) = \frac{1}{\beta}\int_0^\infty h_U(x + \frac{z}{\gamma_-})e^{-z}dz. \qquad (15)$$

Finally, the relations iii) and iv) imply:

$$\begin{cases} \frac{h(b_u)}{\beta} & = & M \\ A_{u,2} & = & M\,e^{-\omega b_u} + \int_0^{b_u}\left[\frac{h(\xi)+\lambda M}{U}\right] e^{-\omega\xi}d\xi \end{cases} \qquad (16)$$

which, together with the relation v), gives:

$$A_{u,1} = e^{\gamma^+ b_d}\left[M e^{\omega(b_d - b_u)} - \int_{b_u}^{b_d}\left[\frac{h(\xi) + \lambda M}{U}\right] e^{-\omega\xi}d\xi - p_u(b_d)\right].$$

As we imposed in Eq.(10) $h_U(x)$ to be strictly increasing, the function $M(x)$ defined in Eq.(15) is itself strictly increasing, (see Eq. (3.19) in [Kar 84] for the direct demonstration). Hence, the parameters b_d and b_u are the unique solutions of Eqs.(14) and (16) respectively.

In addition to the property given in Eq.(4), note that the following properties consistently hold:

$$\lim_{x\to\infty} M(x) = K \quad \text{and} \quad \lim_{x\to-\infty} M(x) = k. \tag{17}$$

Theorem: Assume that the instant reward function $h_U(x)$ is strictly increasing. Then for the initial conditions $X(t=0) = x$ and $I(0) = D$ respectively $I(0) = U$, the optimal stopping times are $\tau^*(x, D, M)$ and $\tau^*(x, U, M)$ read:

a) $k < M < K$.

$$\tau^*(x, D, M) = \inf\{t \geq 0; X(t) \leq b_d \text{ and } I(t) = D\} \tag{18}$$

and

$$\tau^*(x, U, M) = \begin{cases} 0 & \text{if } \mathrm{x} < \mathrm{b_u} \\ \inf\{t \geq 0; X(t) \leq b_d \text{ and } I(t) = D\} & \text{if } \mathrm{x} \geq \mathrm{b_u} \end{cases} \tag{19}$$

where the thresholds b_d and b_u are implicitly given in Eqs.(14) and (16). Remark that if $x \geq b_u$, we never stop when $I(t) = U$.

b) $M > K$.

In this case we have: $b_d = b_u = \infty$ and therefore $\tau^*(x, D, M) = \tau^*(x, U, M) = 0$.

In this case, the associated reward functions read $J(x, U, M) \equiv J(x, D, M) \equiv J^{(3)}(x, U, M) \equiv J^{(3)}(x, D, M) = M$.

c) $M < k$. In this cases we have: $b_d = b_u = -\infty$ and hence $\tau^*(x, D, M) = \tau^*(x, U, M) = \infty$.

In this case, the associated reward functions read:

$$J(x, D, M) \equiv J^{(1)}(x, D, M) = p_d(x) > k$$

and

$$J(x, U, M) \equiv J^{(1)}(x, U, M) = p_u(x) > k.$$

Remark: When $0 \leq k < M < K$, Eqs.(18) and (19) implies that at the stopping time $\tau^*(x, D, M)$ or $\tau^*(x, U, M)$, the process $I(t)$ will always be in the D phase, namely:

$$I(\tau^*(x, D, M)) = D \quad \text{or} \quad I(\tau^*(x, U, M)) = D.$$

3.3 Multi-armed Bandit with piecewise deterministic arms.

Let us now return to the MABP introduced in section 2 and consider N arms with piecewise deterministic dynamics:

$$\frac{d}{dt}X_j(t) = \mathbf{1}_{[\chi(t)=j]}I_j(t), \quad j = 1, 2, ..., N; \;\; X_j(0) = x_j, \; I_j(0) = i_j \tag{20}$$

where $\mathbf{1}_{[\chi(t)=j]}$ is the indicator function with $\chi(t) \in \{1, 2, \ldots, N\}$ and $I_j(t)$, $j \in \{1, 2, \ldots, N\}$ are independent alternating renewal processes similar to the one introduced in Eq.(2) with states U_j and D_j. The process $\chi(t)$ indicates which project is engaged at time t. Remember that, when a project is not engaged its dynamic remains "frozen". Note that with the previous assumptions the processes $\zeta_k(t) := (X_k(t), I_k(t)) \in \mathbb{R} \times \{D_k, U_k\}$ for $k = 1, 2, \ldots N$ are independent and markovian.

As in section 2, the problem is to find an optimal allocation policy $\pi^* = \{\chi^*(t)\}$ which maximizes the expected reward (see Eq.(1). More specifically, we have here:

$$J(\vec{x}, \vec{i}; \pi^*) = E_{\vec{x},\vec{i}} \int_0^\infty e^{-\beta t} h\left(\chi^*(t), X_{\chi^*(t)}(t)\right) dt, \tag{21}$$

with the notation:

$$\vec{x} = (x_1, x_2, ..., x_N) \quad \text{and} \quad \vec{i} = (i_1, i_2, ..., i_N),$$

where:

$$X_k(t = 0) = x_k \quad \text{and} \quad i_k(t = 0) = i_k; \quad \text{for} \quad k = 1, 2, \ldots, N.$$

The reward functions $h(i, x)$, for $i = 1, 2, ..., N$, are assumed to belong to the class $\mathcal{H}$, defined by Eq.(10). As before in Eq.(4), we impose:

$$\lim_{x\to\infty} h(j, x) = \beta K, \quad \lim_{x\to-\infty} h(j, x) = \beta k, \quad \forall\, j = 1, 2, \ldots N; \quad 0 \le k < K.$$

In [Men 90], (see also the section 3.9 of [Man 87]), it is shown that the optimal reward function Eq.(21) associated with the MABP for general Markov processes has a simple product form in term of individual stopping problems without any smoothness properties of the optimal reward function neither for the global problem nor for the individual stopping problems. The processes $\zeta_k(t) := (X_k(t), I_k(t))$ being markovian, the MABP defined by Eqs. (20) and

(21) belongs to the class discussed in [Men 90]. For each individual process $X_k(t)$, given by Eq.(20), we then consider the associated stopping problem. We can define, for a given position x_k, two Gittins' indices $M_k^{D_k}(x_k)$ and $M_k^{U_k}(x_k)$ depending on the respective states D_k and U_k of the process i_k. Remember that, by definition, the Gittins' indices are the smallest value of the terminal reward M_k which make immediate stopping profitable when the state of the process is respectively $\zeta_k = (x_k, D_k)$ or $\zeta_k = (x_k, U_k)$. In view of the results obtained in section 3.1, the index $M_k^{D_k}(x_k)$ coincides with the expression Eq.(15) which was first derived by I. Karatzas. To identify the index $M_k^{U_k}(x_k)$, we can, from Figure 2, conclude:

$$M_k^{U_k}(x_k) = J(x_k, U_k, M_k^{D_k}(x_k))$$

Note in particular that the index $M_k^{U_k}(x_k)$ is monotonously increasing and that it obviously satisfies the relation Eq.(17).
The expressions $M_k^{D_k}(x_k)$ and $M_k^{U_k}(x_k)$ can be used as priority indices to characterize the optimal allocation policy π^*. Accordingly, the optimal strategy follows the rule "*play the project with the leading Gittins' index*". Formally, we shall then write, (see [Kar 84] for the notation):

$$\chi^*(t) = \min\left\{\{1 \leq q \leq N\}, \mathcal{M}_q\left(X_q(t)\right)\right\},$$

with

$$\mathcal{M}_q\left(X_q(t)\right) = \begin{cases} \max_{\{1\leq j\leq d\}} M_j^{D_j}\left(X_j(t)\right) & \text{if } i_j(t) = D_j; \\ \max_{\{1\leq j\leq d\}} M_j^{U_j}\left(X_j(t)\right) & \text{if } i_j(t) = U_j. \end{cases}$$

Observe that for identical arms occupying a given identical position x_k, the fact that $M_k^{D_k}(x_k) \leq M_k^{U_k}(x_k)$ implies that one should optimally engage the arm exhibiting the U_k state of the noise. This is obviously what is intuitively expected.

4. Illustration for the manufacturing context

Consider a flexible failure prone machine able to produce N different types of items. The machine has a limited capacity which implies that it is able to produce only a single type of item at the same time. We consider here that the set-up costs and/or times are negligible. Let us introduce a time process $T_k(t)$ per item's class, defined as the total amount of time allocated to the production of item k over the interval $[0, t]$. We then have a stochastic process

$$T(t) = (T_1(t), \ldots, T_N(t))$$

with the following properties: i) $T(t)$ is nondecreasing in $t \geq 0$, with $T(0) = 0$ and ii) $T_1(t) + \ldots, T_N(t) = t$. As in [Bie 88] and in [Ger 94], we assume that the cumulate production $Y_k(t)$ of the class k at time t, $k = 1, 2, \ldots, N$ can be described by a fluid equation of the form:

$$\frac{dY_k(t)}{dt} = U_k(T_k(t))\mathbf{1}_k(t), \quad Y_k(t) \geq 0, \quad \forall\ t, \tag{22}$$

where $\mathbf{1}_k(t) \in \{0, 1\}$ is the indicator which has the value 1 when the production of the item k is engaged and 0 otherwise and $U_k(t) \in \{0, U_k\}$ is an alternating, markovian, stochastic renewal process [Ros 83]. In particular, the sojourn time in the "on" and "failed" phases are exponentially distributed with parameters λ_k and μ_k respectively. Let us now introduce a set of constants D_k, $k = 1, 2, \ldots, N$ which are the target production rates for the class k. Using the D_k's, we can write the cumulate production balance equations $X_k(t)$ at time t in the form:

$$\frac{dX_k(t)}{dt} = [U_k(t_k) - D_k]\, \mathbf{1}_k(t), \quad k = 1, 2, \ldots, N. \tag{23}$$

We can interpret $X_k(t)$ as a performance measure of the production balance for item k. Indeed from Eq.(23), we observe that when $X_k(t) \geq 0$, we fulfill or exceed the production target and, conversely when $X_k(t) < 0$ we are below our objective. Here the D_k are chosen so that $D_k\lambda_k + U_k\mu_k > 0$ as it is imposed in section 3.1. Note from Eq.(22), that when the class k production is disengaged, both the time evolution of $X_k(t)$ and the state of the process $U_k(T_k(t))$ are "frozen". As long as the production process k is engaged, we assume that an instantaneous performance gain $h(k, X_k(t))$ is achieved. The gain $h(k, X_k(t)) = h_0(k) + h_{extra}(k, X_k(t))$ admits two contributions, namely a systematic contribution due to the intrinsic value of the item produced say $h_0(k)$ and an extra contribution $h_{extra}(k, X_k(t))$ which depends on the production balance $X_k(t)$. We shall assume that:

$$h_{extra}(k, X_k(t)) = 0 \quad for \quad X_k(t) \geq 0,$$

$$h_{extra}(k, X_k(t)) < 0 \quad \text{for} \quad X_k(t) < 0. \tag{24}$$

The contribution $h_{extra}(k, X_k(t))$ reflects that extra costs are incurred when the production is below the target rate. We assume that these costs increase monotonously with the distance to the target production and that:

$$\lim_{x \to -\infty} h_{extra}(k, x) = -h_0(k) \tag{25}$$

From Eqs.(24) and (25), we directly have:

$$0 \leq h(k, X_k(t)) \leq h_0(k) \tag{26}$$

and therefore $h(k, X_k(t))$ is monotonously increasing as required by Eq.(4). Moreover the technical condition given by Eq.(10) implies that the variations of $h_{extra}(k, X_k(t))$ should not be too drastic.

Remark: Note that for a single class of item, (i.e. when $N = 1$), Eq.(23) coincides (see [Bie 88] and [Ger 94]), with the time evolution of a surplus $X_k(t)$ with D_k being an external demand rate ($\mathbf{1}_k(t) = 1$ for all t in this case). However when $N > 1$, the model defined by Eq.(23) differs fundamentally from the make-to-stock multiclass production context. Remember that for a make-to-stock multiclass production problem of the type described in [Vea 96], $X_k(t)$ represents the surplus of type k items and Eq.(23) should be replaced by:

$$\frac{dX_k(t)}{dt} = U_k(T_k(t))\mathbf{1}_k(t) - D_k(t), \quad k = 1, 2, \ldots, N. \tag{27}$$

In view of Eq.(23) and Eq.(27) we observe that, for disengaged items, the demand continue to accrue in the make-to-stock context (Eq.(27)), while $X_k(t)$ given by Eq.(23) remains frozen. Clearly, in Eq.(23), $X_k(t)$ is a performance measure which describes the operating characteristics of the machine when it delivers items of class k and this independently from the external demands. The scheduling problem defined by the dynamics Eq.(27) with relevant costs functions, belongs to the class of "restless Bandit" which can only be suboptimally solved by priority index policies [Nino 00] and [Ber 00]. This problem is explicitly addressed in [Dus 01].

5. Conclusion and perspectives

The conflict existing between taking an action which yields an immediate reward (i.e. "myopic" type of decision) or wait and take the higher potential benefits only later is the crucial difficulty inherent to the dynamic scheduling problems. In the flexible manufacturing devices for which the set-up costs and/or delays can be neglected, the scheduling of the production can be naturally formulated in terms of generalized multi-armed Bandit problems (MABP), each arm representing the production dynamics of one specific type of item. The MABP can be divided into two subclasses, the so called "classical" MABP characterized by the fact that disengaged arms remind in "frozen" states and "restless" MABP for which this constraint is relaxed. It is established that for the classical MABP, the optimal scheduling rule reads as a priority index policy which is particularly appealing due to its simplicity of implementation, i.e. "*"engage the project exhibiting the largest priority index*". The remaining work is to actually compute the indices, a task which we explicitly perform here for a class of piecewise deterministic processes. The calculation of the priority indices is performed via the resolution of appropriate stopping problems. As noted in [Kas 95], it is rather exceptional that these optimal stopping prob-

lems can be solved explicitly. Besides the diffusion processes and the class of Lévy's processes solved respectively in [Kar 84] and [Kas 95], we have shown that the random velocity models do also yield explicit results. The analytic form of the priority indices enables to obtain the optimal scheduling rule for a simple class of manufacturing processes. In these models, the gain is connected to the operating characteristics of the production device when it delivers one particular item and this independently from the external demand. Hence, disengaged arms remain "frozen" as required for the classical MABP.
Unfortunately, for many manufacturing scheduling problems, the classical MABP's are not suitable and rather the restless MABP's are mandatory. For example, when the scheduling problem does depend on the external demands (as typically for the inventory levels), disengaged arms do evolve with time. Indeed, the demands for the different types of items steadily accrue independently of the fact that a particular production is engaged or not. Recent research efforts devoted to the "restless" MABP's have explicitly shown that scheduling rules given by Gittins' indices are not optimal and the optimal policy is not known in general. Moreover, the complexity of the "restless" MABP's often precludes numerical investigations, even in relatively simple situations. Therefore, we may consider that a sub-optimal policy given by a set of suitably defined priority indices constitutes a simple and reliable algorithm. Finally, the introduction of set-up costs and/or times which is a generalization necessary for several flexible systems, makes the scheduling problems even more complex. Set-up costs generate policies exhibiting "hysteretic" characteristics, (i.e. the history of the production affects the production scheduling see [Dus 02]). The "restless" MABP's and the scheduling problems involving set-up costs and/or times are the core of an ongoing research activity.

Acknowledgments. We would like to thank the excellent referee's work which enables to improve a first version of this paper.

References

[1] [Ber 96] D. Bertsimas and J. Niño-Mora. "Conservation laws, extended polymatroids and multi-armed Bandit problems ; A polyhedral approach to indexable systems." Math. Oper. Res. **21** (2), (1996), 257-306.

[2] [Ber 00] D. Bertsimas and J. Nino-Mora. "Restless Bandits, linear programming relaxations and primal-dual index heuristic." Oper. Res. **48**, (2000), 80-90.

[3] [Bie 88] T. Bielecki and T. R. Kumar. "Optimality of zero inventory policies for unreliable manufacturing systems." Oper. Res. **36**, (1988) 532-541.

[4] [Dav 84] M. H. A. Davis. "Piecewise-deterministic Markov processes: a general class of non-diffusion stochastic models". J. Roy. Statist. Soc. **B** **46**, (1984), 353–388.

[5] [Dus 01] F. Dusonchet and M.O. Hongler. "Continuous time Restless Bandits and Dynamic scheduling for make-to-stock production." Working Paper, EPFL - DMT/IPM, (2001).

[6] [Dus 01] F. Dusonchet and M.O. Hongler. "Optimal Hysteresis for Deterministic Deteriorating Two-armed Bandit Problem with Switching Costs." Working Paper, EPFL - DMT/IPM, (2001).

[7] [ElK 94] N. El Karoui and I. Karatzas. "Dynamic allocation problems in continuous time." The Annals of Prob. **2**, (1994), 255-286.

[8] [ElK 97] N. El Karoui and I. Karatzas. "Synchronization and optimality for multi-armed Bandit problems in continuous time." Special invited paper in Computational and Applied Math. **16**(2), (1997), 117-152.

[9] [Epl 86] W.J.R. Eplett. "Continuous- time allocation indices and their discrete time approximation". Adv. Appl. Probab. **18**, (1986), 724-746.

[10] [Fed 99] A. Federgruen and Z. Katalan. "The impact of adding a make-to-order item to a make-to-stock production system." Management Sci. **45**, (1999), 980-994.

[11] [Ger 94] S. B. Gershwin. *Manufacturing Systems Engineering*. Prentice Hall, (1994).

[12] [Git 74] J. C. Gittins and D. M. Jones."A dynamic allocation index for the sequential design of experiments". In Progress in Statistics, (J. Gani Ed.), (1974), 241-266, North Holland.

[13] [Git 89] J. C. Gittins. *Multi-Armed Bandits Allocation Indices*. J. Wiley, (1989).

[14] [Ha 97] A. Ha. "Optimal dynamic scheduling policy for a make-to-stock production system." Operation Res.**45**, (1997), 42-54.

[15] [Hon 01] M.O. Hongler and F. Dusonchet. "Optimal stopping and Gittins' indices for piecewise deterministic evolution process." To appear in Discrete Event Dyn. Sys. **11**, no. 3, (2001)

[16] [Kar 84] I. Karatzas. "Gittins indices in the dynamic allocation problem for diffusion processes." Ann. of Probab. **12**, (1984), 173-192.

[17] [Kas 95] H. Kaspi and A. Mandelbaum. "Levy Bandits ; multi-armed Bandits driven by Levy processes." Ann. Appl. Probab. **5**, (1995), 541-565.

[18] [Man 86] A. Mandelbaum. "Discrete multi-armed Bandits and multi-parameter processes." Probab. Theory Rel. Fields **71**, (1986), 129-147.

[19] [Man 87] A. Mandelbaum. "Continuous multi-armed Bandits and multi-parameter processes." Annals of Probab. **15**, (1987), 1527-1556.

[20] [Men 90] J. L. Menaldi and M. Robin. "On the optimal reward function of the continuous time multiarmed Bandit problem." SIAM J. Optim. and Control **28**, (1990), 97-112.

[21] [Mil 63] R. G. Miller. "Continuous time stochastic storage processes with random linear inputs and outputs." J. of Math. and Mech. **12**, (1963), 275-291.

[22] [Nino 00] J. Nino-Mora "On certain greedoid polyhedra, partially indexable scheduling problems, and extended restless Bandit allocation indices" Submitted to *mathematical programming*

[23] [Pin 91]. M. A Pinsky. *Lectures on Random Evolution.* World Scientific, (1991).

[24] [Pen 97] A. Peña and P. Zipkin. "Dynamic scheduling rules for a multi-product make-to-stock queue." Operation Res. **45**, (1996), 919-930.

[25] [Ros 83] S.M. Ross *Stochastic Processes.* Wiley (1983).

[26] [Sha 92] J. G. Shanthikumar and D. D. Yao. Multiclass queuing systems: Polymatroid structures and optimal scheduling control. Oper. Res. **40**, (1992), S293-S299.

[27] [Sriv 98] N. Srivatsan and Y. Dallery "Partial characterization of Optimal Hedging Point policies in Unreliable Two-Part-Type Systems." Oper. Res. **46**, (1998), 36-45.

[28] [Vea 96] M.H. Veatch and L.M. Wein. "Scheduling a make to stock queue: Index policies and hedging points." Oper. Res. **44**, (1996), 634-647.

[29] [Ver 99] F. de Véricourt, F. Karaesmen and Y. Dallery. "Dynamic scheduling in a make-to-stock system: A partial Characterization of optimal policies." Oper. Res. **48** (5), (2000), 811-819.

[30] [Wei 92] L. Wein. "Dynamic scheduling of a multi class make-to-stock queue." Oper. Res.**40**, (1992), 724-735.

[31] [Whi 82] P. Whittle. *Optimization over Time. Dynamic Programming and Stochastic Control.* J. Wiley, New-York, (1982).

[32] [Whi 88] P. Whittle "Restless Bandits: Activity in a changing world." In *A Celebration of Applied Probability*, J. Gani (Ed.), J. of Appl. Probab. **25A**, (1988), 287-298.

[33] [Yao 97] D. D. Yao and Li Zhang. "Stochastic Scheduling via polymatroid optimization." In Lecture in Applied Mathematics (Eds. G.G. Yin and Q. Zhang), Vol **33**, (1997), 333-364.

Chapter 6

PRODUCTION PLANNING FOR SHORT LIFE-CYCLE PRODUCTS IN CONSIDERATION OF CLEARANCE SALE

H. Ishikura
Faculty of Corporate Intelligence, Osaka Gakuin University, Japan

Abstract: *A method that reduces economical risks in product planning for short life-products is proposed in this paper. In this method, it is assumed that production is decided before sales and at the end of the selling period, the products are sold at clearance prices and the optimum production level is calculated using the knowledge of experts. The effectiveness and convenience of the method is described with a numerical example.*

Key words: *production planning, short life-cycle product, clearance sale, optimum production.*

1. INTRODUCTION

It is well known that continuous new product development is a necessary task for manufacturers, due to customer demand the keen competition. Such developments have reduced product life-cycles and production planning should include closely monitoring the sale of products to reduce economic risk. However, recently manufacturers who produce short life-cycle products do not have enough time to revise their production planning according to the demand of the consumers, because of the shortened life-cycles. Many products are in such situations. Apparel goods, computers, and mobile phones to mention only a few, are all very short life-cycle products. Therefore, the manufacturers need to decide the quantities of their products to be manufactured in advance of sales. In addition to this demanding situation, when the manufacturers make their production planning they have to consider how to anticipate the clearance sales. This is because the value of the short life-cycle products usually reduces quickly with time and the products are sold with discount prices at the end of the product life-cycle. In addition, complete clearance sales are becoming a good business practice for manufacturers. Some researches have been done for uncertain demand [1], [2]. However few studies have been under-taken in to this interesting problem that is including clearance sales [3]. This paper proposes a method of production planning which copes with this present-day situation whereby the quantities of products are decided in advanced of sales and clearance sales are planned for the end of the product life-cycle. The effectiveness of the method is demonstrated though a numerical example.

2. NUMERICAL FORMULAE

In this study the following assumptions for the short life-cycle products are given:

(1) The selling prices are classified into a list price and a clearance price for sale. Unsold products are scrapped.
(2) The demand at a clearance price is in proportion to the demand at the list price.
(3) The probability distribution of demand on a product model is given by an expert. It means that the mean value and standard deviation are known.

If a discrete distribution x defined as the total demand of a new product model that includes demand at a list price and demand at a clearance price both takes x_0, x_1, $\cdots$, then

$$g_i = \Pr\{x = x_i\} \quad (i = 0,1,2,\cdots) \tag{1}$$

$$g_i \geq 0 \tag{2}$$

and

$$\sum_{i=0}^{\infty} g_i = 1 \tag{3}$$

is obtained. Where g_i is the probability at i products is demanded by consumers. In this case, if the manufacturer produces p products, the possibility that the manufacturer sold i products is given by the following equation.

$$f_i = \begin{cases} g_i & 0 \leq i \leq p-1 \\ \sum_{i=p}^{\infty} g_i & i = p \\ 0 & i \geq p+1 \end{cases} \tag{4}$$

The expectation of total demand E and variance V are expressed as follows:

$$E(x) = \sum_{i=0}^{\infty} x_i\, f(x_i) = \sum_{i=0}^{p-1} x_i\, g(x_i) + p\sum_{i=p}^{\infty} g(x_i) \tag{5}$$

$$V(x) = \sum_{i=0}^{p-1} [x_i - E(x)]^2 g(x_i) + [p - E(x)]^2 \cdot \sum_{i=p}^{\infty} g(x_i) \tag{6}$$

Thc function for expected profit is given as follows by assuming the profit per products at the list price a, the profit per product at thc clearance price b, and the cost c, where α is the ratio of demand at the clearance price to the demand at the list price.

$$h(x) = \begin{cases} \dfrac{ax_i + \alpha x_i b}{1+\alpha} - (p - x_i)c & x_i < p \\ \dfrac{ax_i + \alpha x_i b}{1+\alpha} & p \le x_i < p(1+\alpha) \\ apg & p(1+\alpha) \le x_i \end{cases} \tag{7}$$

The expected profit, variance, and standard deviation are given by the following equations:

$$E(h(x)) = \sum_{i=0}^{p-1}\left[\frac{ax_i + \alpha x_i b}{1+\alpha} - (p - x_i)c\right]g(x_i) + \sum_{i=p}^{p(1+\alpha)-1}\left(\frac{ax_i + \alpha xb}{1+\alpha}\right)g(x_i) + ap \sum_{i=p(1+\alpha)}^{\infty} g(x_i)ap \tag{8}$$

$$V(h(x)) = \sum_{i=0}^{p-1}\left[\{\frac{ax_i + \alpha x_i b}{1+\alpha} - (p - x_i)c\} - h(x)\right]g(x_i) + \sum_{i=p}^{p(1+\alpha)-1}\left(\frac{ax_i + \alpha xb}{1+\alpha} - h(x)\right)g(x_i) + \sum_{i=p(1+\alpha)}^{\infty}[ap - h(x)]^2 g(x_i) \tag{9}$$

$$s(h(x)) = \sqrt{V} \tag{10}$$

Based on the above calculation it is possible to find a reasonable production plan for maximum expected profits and the variances of one product for high profit and low risk. When the manufacturer has m new product models, the optimization for the maximum profit and a low risk could be also formulated by using portfolio selection as follows [4]:

Maximize

$$\sum_{j=1}^{m} E(h_j(x_j)) \tag{11}$$

subject to

$$\left[\sum_{j=1}^{m} V(h_j(x_j))\right]^{0.5} \leq \gamma \tag{12}$$

$$\sum_{j=1}^{m} x_j \leq T \tag{13}$$

Where γ is some number which depends on the expectation of profit (risk), and T is based on the productive capacity of the manufacturer. In this analysis it was assumed that the demand of each model is independent of other models. Even if prices have a number of incremental reductions through the end of the lifecycle, the model can correspond with adding new terms in equation (7).

3. NUMERICAL EXAMPLE

Let us consider a situation where a manufacturer produces product models A, B, and C, of which the probability density of demands, the cost, and the profit of each model, which is given by an expert, is shown in Figure 1 and Table 1. It is obvious that the production of model A is steady but the manufacturer cannot expect high profits by producing the product model. On the other hand, models B and C are risky, but the possibility of realizing good profits could be higher than models A. If the manufacturer has the ability to produce models B and C then which of the models should be selected from the 3 predicated models? How many products should be produced for each model? It may be not easy to answer these questions by intuition even if the manufacturer has enough experience.

Table 1. Samples

	Model A	Model B	Model C
Expectation of demand	150	80	100
Standard deviation of demand	10	42	20
Profit per product at list price	5000	10000	8000
Profit per product at clearance price	1000	2000	1500
Cost per product	4000	5000	4500

Figures.2 and 3 show the relation between the expectations and standard deviations of the profit and the production of each model as calculated by equations (8) and (10). It is possible to find a suitable production plan using figures 2 and 3 when the manufacturer has a boundless capacity and does not care about the total variance. The results for the total maximum profit are shown in Table 2. It is however impossible to obtain an optimum selection from these figures. Therefore, the optimal selection is made using the equations (11), (12), and (13), in the case that the manufacturer has conditions on the capacity and/or risk. The following equations give the conditions of optimization:

Table 2. Production for maximum profit

	Model A	Model B	Model C	Total
Production for maximum profit	142	63	70	275
Expectation of profit	376002	359512	490675	1226189
Standard deviation of profit	29119	256584	49025	262843

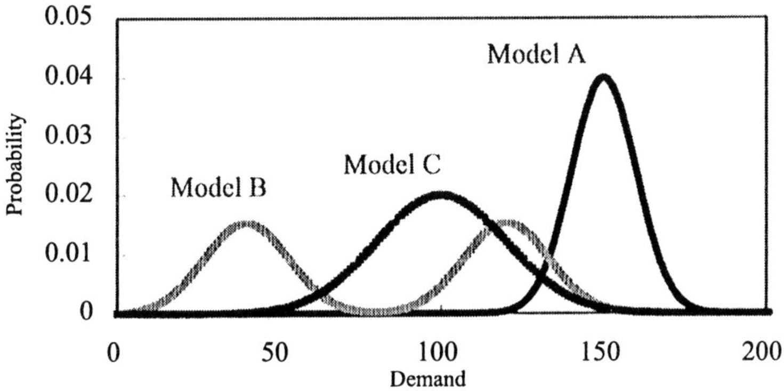

Figure 1. The anticipated probability densities of product models for sale

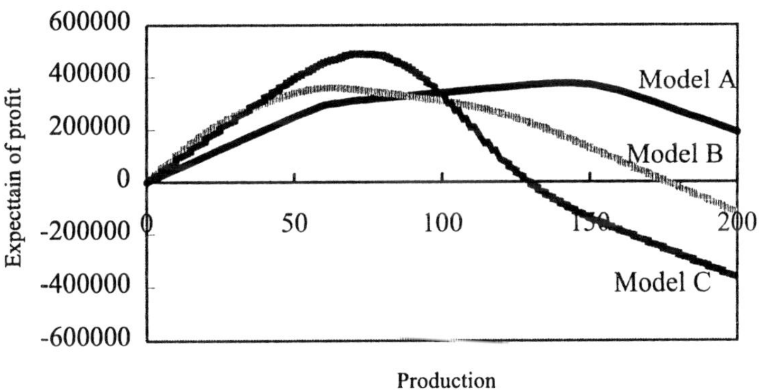

Figure 2. The relation between expectation of profits and production

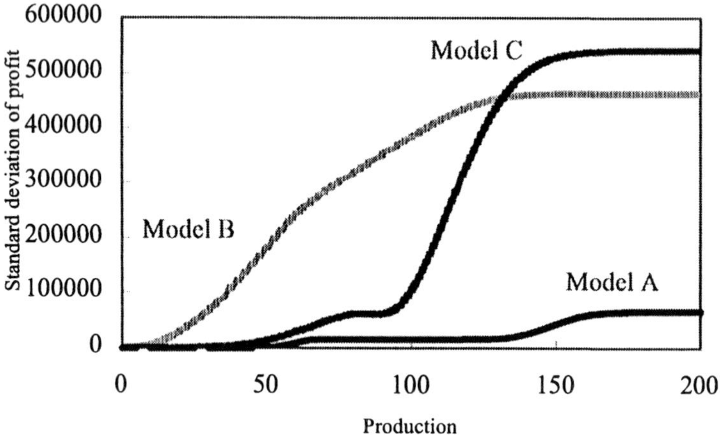

Figure 3. The relation between standard deviation and production

Maximize

$$\sum_{j=1}^{m} E(h_j(x_j)) \tag{14}$$

subject to

$$\left[\sum_{j=1}^{m} V(h_j(x_j))\right]^{0.5} \leq 20000 \tag{15}$$

$$\sum_{j=1}^{m} x_j \leq 250 \tag{16}$$

Table 3 shows the optimal production, the maximum expectation of profit, and the standard deviation and total products under equations (15) and (16). In equation (15), 20000 is given for getting the minimal profit, and 250 in equation (16) is taken just as an example. If 3 σ is assumed an aim, the minimal profit is about 600000.

Table 3. Production for maximal profit under conditions on total production and standard deviation of profit

	Model A	Model B	Model C	Total
Production for maximal profit	129	51	70	250
Expectation of profit	368678	341621	490675	1200974
Standard deviation of profit	17177	188713	49025	195732

4. CONCLUSIONS

A method that reduces economical risks in product planning for short life-products is proposed. In this method it was assumed that the production is decided in advance of sales and at the end of the sale the products are sold at clearance sales, and the optimum production is calculated using the knowledge of experts. The effectiveness and convenience of the method has been elucidated with the help of a numerical example.

REFERENCES

F.W.Ciarallo and R.Akella (1993) Planning Production of Multiple Items with Uncertain Capacity and Uncertain Demand, *Working paper*, University of Arizona.

F.W.Ciarallo, R.Akella, and T.E.Morton (1994) A Periodic review, Production Planning Model with Uncertain Capacity and Uncertain Demand — Optimality of Extended Myopic Policies, *Management Science* 40(3).

H. Ishikura (1994) Study on Production Planning of Apparel Products: Determining Optimal Production Times Quantities, *Computers ind. Engng.* Vol.27, Nos 1-4.

H.Markowitz (1959) Portfolio Selection: Efficient Diversification of Investments, John Wiley

Chapter 7

ANALYSIS OF AUTOMATED FLOW LINE SYSTEMS WITH REPAIR CREW INTERFERENCE

Heinrich Kuhn
Catholic University of Eichstaett-Ingolstadt
Chair of Production and Operations Management
heinrich.kuhn@ku-eichstaett.de

Abstract The interrelations between production and maintenance are mostly neglected during the design phase of automated production systems. Thus, the relevant performance measures of a planned production system like throughput time, production rate, work in process etc. are often estimated inaccurately. The paper presents an analytical approach for performance evaluation of an automated flow line system (AFLS) which takes into account the dependency between the production and the repair system. The suggested model and solution approach are particularly helpful in the initial design phase as well as during a redesign process in order to evaluate alternative configurations of the planned production and repair system.

Keywords: transfer line system, operator/workstation interference, performance evaluation

Introduction

An automated flow line system (AFLS) is a production system in which a series of machining stations, such as milling, drilling, and turning machines, or assembly stations are interconnected along a line or around a circle. The workpieces are predominantly mounted on specialized pallets. Then they are automatically transferred from one workstation to the next. After finishing the work process pallets and workpieces are divorced and the pallets are transferred back to the setup station. Human operators and sometimes inspection stations may also be incorporated along the line (Figure 1).

In automated flow line systems the impact of machine breakdown and subsequent repair has a dominant effect on the performance of these kind of systems. Thus, a dedicated repair crew usually exists at the line to clear station

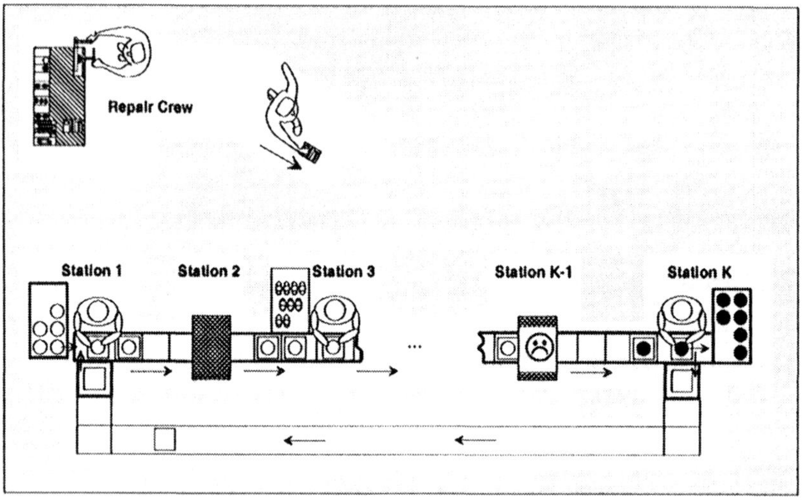

Figure 1. Flow line system with automated and manual stations and a dedicated repair crew.

jams or machine failures. One important design question of those production systems is how to specify the necessary repair crew. The objective is to repair a machine failure or to clear a station jam immediately after its occurrence, without having the repair system acting as a bottleneck on the performance of the whole system. Thus, the system should be designed in such a way that the machine tools or automated assembly stations need not have to wait after a machine failure because the required repairman or operator is not yet available. On the other hand, the repair crew should be designed as small as possible to save operation cost of the system. Thus, the number of repairmen or operators assigned to an AFLS might be much smaller than the number of stations in the systems, e.g., one or two persons for a 20-station line. When a machine failure or a station jam occurs and all operators are busy repairing or clearing other stations, the station must wait to be repaired or cleared. This form of machine idleness is known in the literature as *interference*. Interference or waiting for operator service is undesirable because it decreases the productivity of the AFLS. Obviously, the amount of interference primarily depends on the number of operators assigned. An insufficient number can give rise to a large amount of interference and hence a substantial drop in the system's production rate. An AFLS model that can handle operator/workstation interference can provide useful sensitivity information such as the increase (decrease) in the production rate of the system by adding or removing one operator. Such information is crucial for economic justification purposes.

The described problem is known in the literature as the machine-repairman or machine interference problem ([Stecke and Aronson, 1985]). Queueing theory is often used to solve this problem ([Buzacott and Shanthikumar, 1993], Chapter 2; [Gross and Harris, 1998]). The queueing models, however, assume that information such as machine breakdown time or the time to the next request for operator service is known. If the machines or automatic assembly stations are part of a production network, then these times have to be calculated from other system parameters. For example, for a particular station in an AFLS the mean time between machine failures and the probability that a station is jammed at any given moment are functions of the production rate of the system.

The production rate of an AFLS can be obtained by several specialized algorithms (e.g., [Dallery and Gershwin, 1992]; [Dallery and Frein, 1993] or the monographs by [Papadopoulos et al., 1993]; [Buzacott and Shanthikumar, 1993]; [Gershwin, 1994]). The fact that the system's production rate depends on the waiting time for the repair crew complicates the solution procedure. Hence, the problem at hand calls for a simultaneous solution to the machine-repairman problem and the problem of finding the production rate in an AFLS. In other words, a simultaneous solution to two interdependent queueing models is needed because the solution of one model depends on that of the other.

Only a few models for performance evaluation of automated production or flow line systems consider the impact of a limited repair crew. If this is the case the authors consider mainly two-station systems ([Dudick, 1979]; [Elsayed and Turley, 1980]; [Savsar and Biles, 1984]; [Li, 1987]; [Biles and Usher, 1997]).

[Kamath and Sanders, 1991], describe an analytic modeling approach for performance evaluation of an asynchronous automatic assembly system with a closed-loop structure and several stations. The assembly system is modeled by a closed queueing network with general service times and unlimited waiting room at any queue. The interference time due to the limited number of operators are determined by the help of a probabilistic machine-repairman model. The model of Kamath and Sanders, however, has a major drawback. In the machine-repairman model they assume that the probabilities of the machines being down are mutually independent, which means that the machines act independently of each other. In a machine-repairman model, however, the machines interact among themselves as they share a common repair crew. In addition, they assume an infinite buffer in front of each assembly station. In real automatic flow line systems, however, the space available for storage of workpieces waiting to be processed by the stations is limited.

Thus, it might be concluded that previous research neglects important design parameters involved in automatic flow line systems, e.g., more than two stations per line, finite buffers or limited repair personnel. The analytical model presented in this paper, however, considers the influence of all these design variables on system's performance. The procedure may be helpful for analyzing the effects of the finite storage space and the limited number of available repairmen or operators on the performance of an AFLS.

The suggested approach is based on the decomposition technique which is frequently used to evaluate the relevant performance measures of multi-stage manufacturing systems (e.g., [Dallery and Gershwin, 1992]; [Dallery and Frein, 1993]; [Di Mascolo et al., 1996]; [Helber, 1998]; [Helber, 2000]; [Bonvik et al., 2000]; [Tempelmeier and Bürger, 2001]; and the textbooks by [Papadopoulos et al., 1993]; [Buzacott and Shanthikumar, 1993]; [Gershwin, 1994]). The new issue integrated is the fact that the personnel interference is taken into account in the building blocks of the decomposition approach so that the phenomenon is represented as accurately as possible.

In the next section the model that depicts both systems, the production and the repair system, are explained. Section 2 and Section 3 present two alternative solution procedures for analyzing the considered model. Finally, in Section 4 two numerical examples are provided.

1. Model of the production and repair system

The underlying assumptions of the considered model are the following:

(a) The flow line system consists of K stations.

(b) The processing times at the stations are random and assumed to be exponentially distributed with known mean, $1/\mu_k$, $k = 1, 2, ..., K$. This might be either because a single operation at the workpiece is performed by workers or the line is a flexible transfer line with the processing requirements of each operation varying from workpiece to workpiece.

(c) All workpieces produced are assumed to be perfect. Thus, no inspection procedure, no rework, and no rejects are considered.

(d) The time to failure of station k is assumed to be exponentially distributed with known mean, $1/\lambda_k$, $k = 1, 2, ..., K$.

(e) A station may fail only when processing a workpiece. Thus, operation-dependent failures are assumed.

(f) No parts are scrapped, that is, if a station fails while processing a part, the part remains at the station until the station is repaired. Then the station continues processing the parts.

(g) Station repair times (without the waiting times for a repairman) are assumed to be exponentially distributed with known and identical means $1/\beta_k = 1/\beta$, $k = 1, 2, ..., K$.

(h) Waiting times for a repairman, $w_k = 1/\omega_k$, $k = 1, 2, ..., K$, are **not** included in the repair times and assumed to be exponentially distributed.

(i) The number of repair personnel is limited to c persons.

(j) If the number of stations that are down is greater than c, then a repair queue is formed and the failed stations are forced to wait until a repair person is available. The failed stations are served in FCFS order.

(k) Between station k and $k + 1$ exists a finite intermediate buffer with Z_k, $k = 1, 2, ..., K - 1$, places for workpieces.

(l) A station is blocked if its downstream buffer is full. Thus, the blocking model "blocking after service" is considered ([Dallery and Gershwin, 1992]).

(m) A station is starved if its upstream buffer is empty.

(n) An inexhaustible supply of unprocessed workpieces is available upstream the first station in the line. Thus, the first station is never starved.

(o) An unlimited storage area exists downstream of the last station in the line. Thus, the last station is never blocked.

(p) The time for transporting the workpieces between the stations is negligibly small compared to the processing times at the stations. Thus, transportation times are not considered.

The performance measures of the described automated flow line model can be determined by the help of two coupled queueing systems (Figure 2), one queueing system consisting of $(K-1)$ single-stage queueing systems with finite storage and failure-prone servers which are arranged in-line and one queueing system with finite customer population representing the repair process.

Analyzing these queueing systems is rather complex because the flow of customers through each system depends on that of the other. For example, a failure of station 2 can occur only if a customer is being served. This induces an arrival of a customer at the repair system. The end of a failure situation at station 2, however, depends on the flow time of the "failure customer 2" through the second queueing system. Thus, a simultaneous analytical evaluation of the considered queueing network model is a task with an extremely high degree of difficulty.

2. Iterative solution procedure

One way to analyze the described queueing network model is to decompose it into two independent queueing systems. One system for modeling the

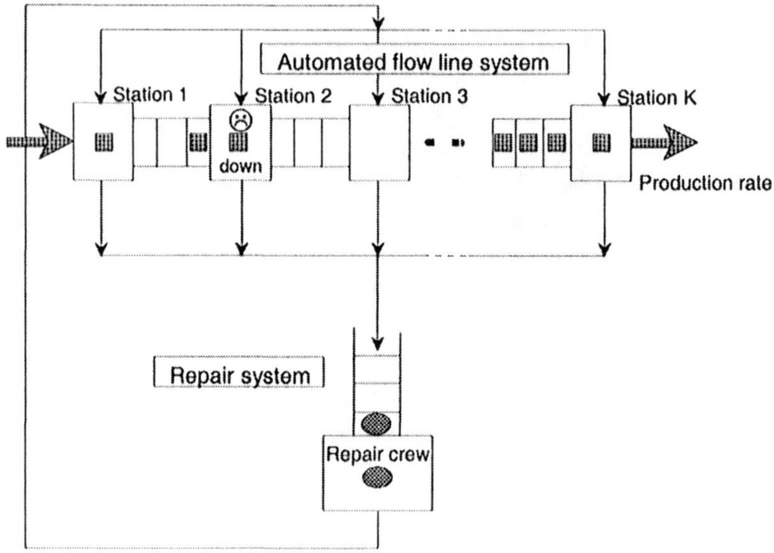

Figure 2. Two coupled queueing systems.

automatic flow line system with limited buffers and one for modeling the interference problem. Having analyzed each queueing model separately the results of the interference model are used to modify the input parameters of the flow line model. On the other hand, the results of the flow line model are used to modify the input parameters of the interference model. This is being done iteratively until a convergence criterion is met ([Kuhn, 1999]). [Tetzlaff, 1995], describes a similar approach, but for a different application.

For the considered model, the interference model is used to approximate the expectation of the station's repair times including the waiting times for the repairman, $E[R_k]$, $k = 1, 2, ..., K$ (Figure 3).

Next, by the help of the completion time approach suggested by [Gaver, 1962], a modified processing time, T_k, $k = 1, 2, ..., K$, for each station k is evaluated (s. also [Tempelmeier and Kuhn, 1993], Chapter 3.1.3). Underlying the modified processing times, an analytical model for performance evaluation of transfer lines ([Dallery et al., 1988]; [Gershwin, 1994], Chapter 4.3) is used to approximate the utilization of each station k, U_k, $k = 1, 2, ...K$, of the AFLS. Using the utilization of each station k we can compute a modified time to failure, F_k^*, $k = 1, 2, ..., K$ which considers the fact, that station k cannot fail during a starving or blocking period (Figure 3). Since we assume operation-dependent failures the given mean times to failure of the stations, $1/\lambda_k$, $k = 1, 2, ..., K$, have to be modified.

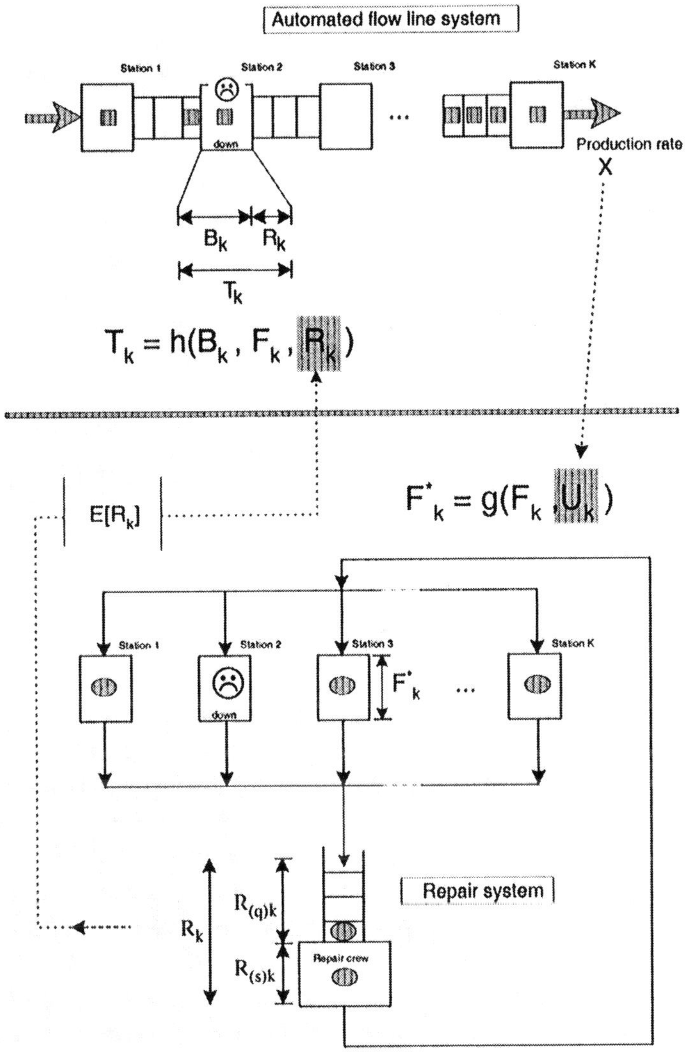

Figure 3. Iterative solution procedure.

A major drawback of this approach is that the achieved results are sometimes not very accurate, especially if the buffer areas are small or if the repair system acts as a bottleneck of the total system ([Kuhn, 1998]).

3. Integrative solution approach

An alternative to the described approach of iterative parameter adjustment is to integrate the repair system into the decomposition approach for analyzing the flow line system. This approach is called "integrative solution approach." This solution approach considers the repair system explicitly during the evaluation process of the flow line system.

3.1 Decomposition of the production and repair system

The decomposition procedure decomposes the original K-station flow line system into $(K-1)$ subsystems. Each subsystem consists of two production stations with its intermediate buffer and one repair station. Let us consider the four-station line shown in Figure 4 which has been decomposed into three subsystems. The upstream station of subsystem $L(2)$, $M_u(2)$, for example, models the behavior of all stations upstream of the intermediate buffer between station two and three. The downstream station $M_d(2)$ models the behavior of all stations downstream of the considered buffer. Station $I(2)$ represents the repair process of station $M_u(2)$ and $M_d(2)$ as well as all other stations of the flow line system.

3.2 Analyzing the subsystems

The subsystems are modeled as a continuous-time Markov process with discrete states. Thus, the subsystems can be analyzed by the help of standard procedures and exact analytical results can be achieved. The components of the subsystems (Figure 5) are defined as follows: The stations $M_u(k)$ and $M_d(k)$ are defined by their mean service rates $[\mu_u(k), \mu_d(k)]$, failure rates $[\lambda_u(k), \lambda_d(k)]$, starving and blocking rates $[\lambda s_u(k), \lambda b_d(k)]$, and resumption of flow rates $[\beta s_u(k), \beta b_d(k)]$. Station $I(k)$, however, is defined by the mean repair rates $[\beta_u(k), \beta_d(k)]$ and the mean waiting rates $[\omega_u(k), \omega_d(k)]$ considering the fact that all repairmen may be busy repairing other stations of the systems when a machine failure of one of the subsystem stations occurs.

The decomposition methods seeks to determine the parameters of each subsystem so that the behavior of the material flow in the buffer closely matches that of the flow in the equivalent buffer of the original K-station flow line. This procedure is very similar to the decomposition approach suggested by Gershwin ([Gershwin, 1987, Gershwin, 1989, Gershwin, 1994]) for analyzing transfer lines (s. also [Dallery et al., 1988]; [Dallery and Frein, 1993]).

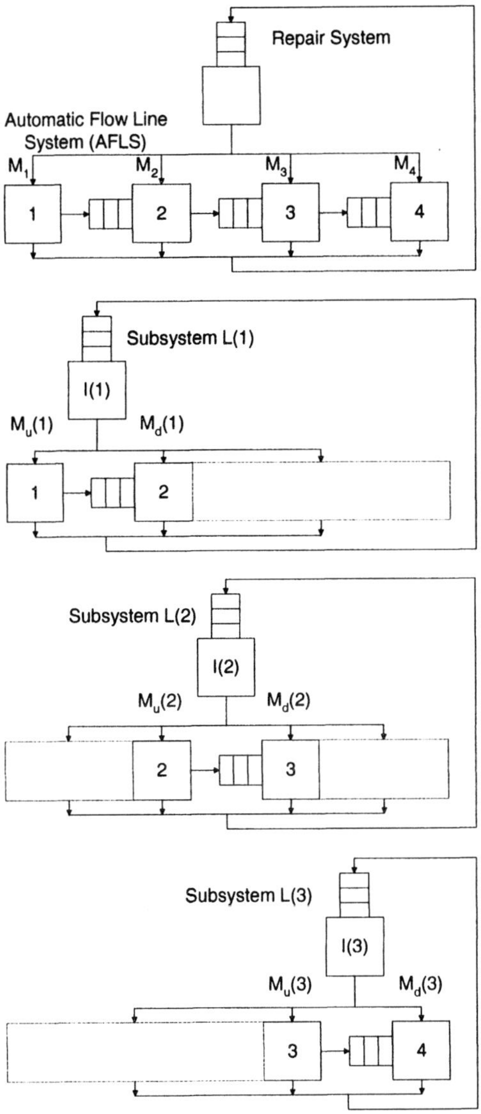

Figure 4. Decomposition of a four-station line with a repair crew into three two-station lines each having its own repair system.

3.3 Determining the parameters of the subsystems

The parameters of the subsystems can be determined by the help of six sets of equations. During a forward and backward procedure these equations are used to evaluate the parameters of the subsystems.

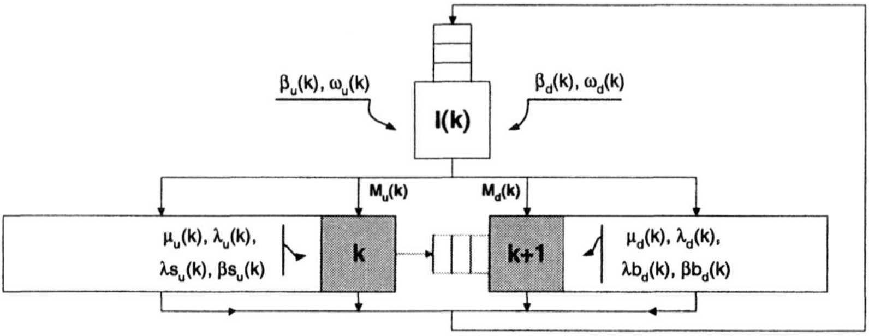

Figure 5. Subsystem $L(k)$, two-station line with its own repair system.

Conservation of flow condition. Because there is no mechanism for creating or destroying workpieces in the line, the conservation of flow condition holds for the considered model. Thus, the mean production rates of the decomposed two-station lines are all equal and they are also equal to the mean production rate of the original line:

$$X = X_k = X_K = X(k) = X_u(k) = X_d(k) \qquad k = 1, 2, ..., K - 1 \tag{1}$$

The mean production rate of the original line and the production rate of station k, $k = 1, 2, ..., K$, are denoted by X and X_k, respectively. The mean production rates of the decomposed two-station lines $L(k)$, $k = 1, 2, ..., K - 1$, and the mean production rates of the pseudo-machines $M_u(k)$ and $M_d(k)$, $k = 1, 2, ..., K - 1$, however, are denoted by $X(k)$, $X_u(k)$, and $X_d(k)$, respectively.

Flow-rate-idle time relationship. The production rate of station k can be approximated as follows:

$$X_k = \mu_k \cdot e_k \cdot (1 - \Pr[n_{k-1} = 0] - \Pr[n_k = N_k]) \tag{2}$$

The term $\Pr[n_{k-1} = 0]$ stands for the probability of station k being starved, the term $\Pr[n_k = N_k]$ stands for the probability of station k being blocked and e_k is the isolated efficiency of station k:

$$e_k = \frac{MTTF_k}{MTTF_k + MTTR_k + w_k} \tag{3}$$

Since the downstream station of subsystem $L(k - 1)$, $M_d(k - 1)$, is never blocked and the upstream station of subsystem $L(k)$, $M_u(k)$, is never starved, the flow rate-idle time relationship for station k can be expressed by the following two equations:

$$\frac{X_d(k-1)}{\mu_d(k-1)} = e_d(k-1) \cdot [1 - p_s(k-1)] \tag{4}$$

and

$$\frac{X_u(k)}{\mu_u(k)} = e_u(k) \cdot [1 - p_b(k)]; \tag{5}$$

where $e_d(k-1)$ and $e_u(k)$ are the isolated efficiencies of the pseudo-machines $M_d(k-1)$ and $M_u(k)$, respectively.

The conservation of flow $[X_k = X_u(k) = X_d(k)]$ and the flow rate-idle time relationship (production rate, see above) yield in the **second set** of equations:

$$\frac{1}{e_k \cdot \mu_k} + \frac{1}{X_k} = \frac{1}{e_d(k-1) \cdot \mu_d(k-1)} + \frac{1}{e_u(k) \cdot \mu_u(k)} \qquad k = 2, 3, ..., K-1 \tag{6}$$

Considering the possible states of upstream station $M_u(k)$ and downstream station $M_d(k)$ the efficiencies $e_u(k)$ and $e_d(k)$ can be determined by the help of the continuous-time, discrete-state Markov chain shown in Figure 6.

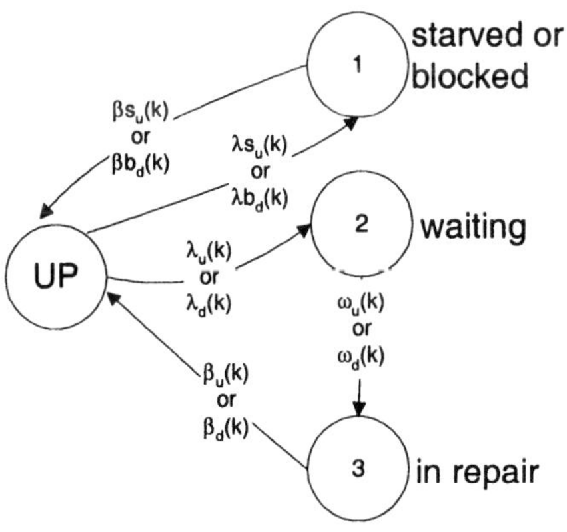

Figure 6. Markov chain for evaluating the isolated efficiencies of the upstream and downstream station of subsystem $L(k)$.

Analyzing this Markov chain leads to the following equations for the isolated efficiencies of the upstream and downstream station of subsystem $L(k)$:

$$e_u(k) = \frac{1}{1 + \frac{\lambda s_u(k)}{\beta s_u(k)} + \frac{\lambda_u(k)}{\omega_u(k)} + \frac{\lambda_u(k)}{\beta_u(k)}} \qquad k = 1, 2, ..., K-1 \tag{7}$$

and

$$e_d(k) = \frac{1}{1 + \frac{\lambda b_d(k)}{\beta b_d(k)} + \frac{\lambda_d(k)}{\omega_d(k)} + \frac{\lambda_d(k)}{\beta_d(k)}} \qquad k = 1, 2, ..., K-1 \tag{8}$$

The Markov chain helps in explaining the modeling philosophy of the approach: The parameters $\lambda s_u(k)$ and $\lambda b_d(k)$ are the transition rates for changing from state "up" to state "starved" or "blocked". They describe the interruption of flow caused by station failures up- or downstream of subsystem $L(k)$. The parameters $\beta s_u(k)$ and $\beta b_d(k)$, however, are the transition rates for going the reverse way, i.e., from state "starved" or "blocked" to state "up". They describe the resumption of flow caused by repairing the failed station. The parameters $\omega_u(k)$ and $\omega_d(k)$, finally, represent the mean waiting rates of the failed stations if a repairman is not available.

Interruption of flow. The transition rate, $\lambda s_u(k)$, for changing the upstream station of subsystem $L(k)$, $M_u(k)$, from state "up" to state "starved" can be evaluated using subsystem $L(k-1)$. Two situations may lead to a starvation of the upstream station of subsystem $L(k)$, $M_u(k)$:

First, if the buffer of subsystem $L(k-1)$ is empty and both stations, the upstream station and the downstream station, are up. Then with the transition rate that the upstream station $M_u(k-1)$ is going to be failed or starved [$\lambda_u(k-1)$ or $\lambda s_u(k-1)$] the upstream station of the considered subsystem $L(k)$, $M_u(k)$, is going to be starved.

Station $M_u(k)$, however, can only starve if it is up and not blocked. Thus the following conditional probability results which describes the probability of realizing the relevant transition rates $\lambda_u(k-1)$ and $\lambda s_u(k-1)$:

$$\frac{\Pr[n_{k-1} = 0, M_u(k-1) = \text{up}, M_d(k-1) = \text{up}]}{\Pr[n_k < N_k \text{ and } M_u(k) = \text{up}]} \tag{9}$$

The term $\Pr[n_{k-1} = 0, M_u(k-1) = \text{up}, M_d(k-1) = \text{up}]$ denotes the probability that the buffer in subsystem $L(k-1)$ is empty and both stations, the upstream station $M_u(k-1)$ and the downstream station $M_d(k-1)$ are in the state "up". The term

$$E_u(k) = \Pr[n_k < N_k \text{ and } M_u(k) = \text{up}], \tag{10}$$

however, describes the probability that the considered station $M_u(k)$ is up and not blocked. Thus the following term A results:

$$A = \frac{[\lambda_u(k-1) + \lambda s_u(k-1)] \cdot \Pr[n_{k-1} = 0, M_u(k-1) = \text{up}, M_d(k-1) = \text{up}]}{E_u(k)} \quad (11)$$

Second, if the upstream station of subsystem $L(k-1)$, $M_u(k-1)$ is not up, one part is left in the buffer of subsystem $L(k-1)$, and the downstream station $M_d(k-1)$ is up – the part in the buffer is the part which is being processed by station $M_d(k-1)$ –, then immediately when the service process is finished the upstream station of the considered subsystem $L(k)$ is starved. Hence, the starving transition rate is equivalent to the service rate $\mu_d(k-1)$.

This is described by the term B:

$$B = \frac{\mu_d(k-1) \cdot \Pr[n_{k-1} = 1, M_u(k-1) = \text{not up}, M_d(k-1) = \text{up}]}{E_u(k)} \quad (12)$$

Term A and term B lead to the unknown transition rate:

$$\lambda s_u(k) = A + B \qquad k = 2, 3, ..., K-1 \quad (13)$$

The unknown probabilities can be approximated using the steady-state probability distribution of the Continuous-Time Markov Chain (CTMC) of the considered subsystem $L(k-1)$ which may be found by solving the probability flow balance equations. The relevant part of the state-transition rate diagram of an arbitrary subsystem is given in Figure 7. Note that the subsystem index k has been omitted for the sake of simplicity.

The subsystem states (i_1, n, i_2) are given by the states of each station and the number of parts finished by the upstream station but not yet finished by the downstream station ($n = 0, 1, ..., N$). The possible states of the upstream and downstream station (i_1 and i_2) are the following:

U	Upstream/downstream station is up
W	Upstream/downstream station is waiting for repair
R	Upstream/downstream station is under repair
S	Upstream station is starved
B	Downstream station is blocked

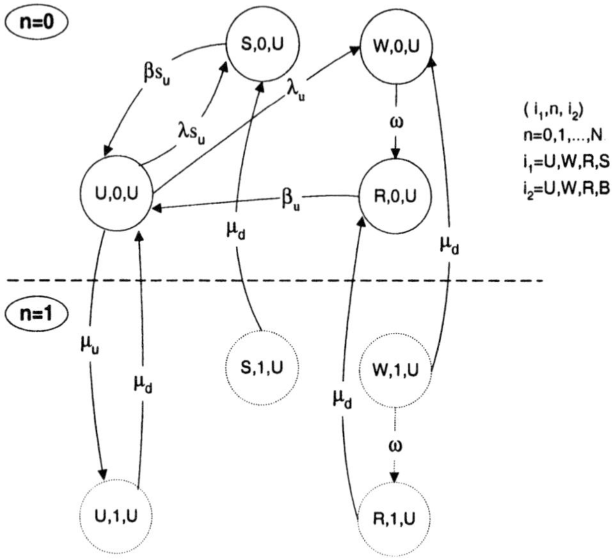

Figure 7. Partial state-transition rate diagram of an arbitrary two-station subsytem, $n = 0$ (starvation of the downstream station).

The balance equations for the steady-state probabilities considering an empty system downstream of the first station ($n = 0$) are given as follows:

$$p(U,0,U) = \frac{1}{\lambda_u + \mu_u + \lambda s_u} \cdot [p(R,0,U) \cdot \beta_u + p(U,1,U) \cdot \mu_d + p(S,0,U) \cdot \beta s_u] \quad (14)$$

$$p(W,0,U) = \frac{1}{\omega} \cdot [p(U,0,U) \cdot \lambda_u + p(W,1,U) \cdot \mu_d] \quad (15)$$

$$p(R,0,U) = \frac{1}{\beta_u} \cdot [p(W,0,U) \cdot \omega + p(R,1,U) \cdot \mu_d] \quad (16)$$

$$p(S,0,U) = \frac{1}{\beta s_u} \cdot [p(U,0,U) \cdot \lambda s_u + p(S,1,U) \cdot \mu_d] \quad (17)$$

From these balance equations the following equation results:

$$p(U,0,U) \cdot [\lambda_u + \lambda s_u] + [p(S,1,U) + p(W,1,U) + p(R,1,U)] \cdot \mu_d = p(R,0,U) \cdot \beta_u + p(S,0,U) \cdot \beta s_u \quad (18)$$

Since the system state $[n_{k-1} = 0, M_u(k-1) = \text{up}, M_d(k-1) = \text{up}]$ of subsystem $L(k-1)$ is equivalent to the state $(U,0,U)$ of the CTMC and the system state $[n_{k-1} = 1, M_u(k-1) = \text{not up}, M_d(k-1) = \text{up}]$ of subsystem $L(k-1)$ are represented by the states $(S,1,U)$, $(W,1,U)$, and $(R,1,U)$ the transition rate $\lambda s_u(k)$ for changing the upstream station of subsystem $L(k)$, $M_u(k)$, from state "up" to state "starved" can be approximated as follows:

$$\lambda s_u(k) = \frac{p(k-1,R,0,U)\cdot\beta_u(k-1) + p(k-1,S,0,U)\cdot\beta s_u(k-1)}{E_u(k)} \qquad k = 2,3,\ldots,K-1 \quad (19)$$

and $\lambda s_u(1) \to 0$

$p(k-1,R,0,U)$	probability that upstream station of subsystem $L(k-1)$, $M_u(k-1)$, is in state "in repair" and the buffer $B(k-1)$ is empty and the downstream station $M_d(k-1)$ is up
$p(k-1,S,0,U)$	probability that upstream station of subsystem $L(k-1)$, $M_u(k-1)$, is in state "starved" and the buffer $B(k-1)$ is empty and the downstream station $M_d(k-1)$ is up
$\beta_u(k-1)$	repair rate of the upstream station of subsystem $L(k-1)$, $M_u(k-1)$
$\beta s_u(k-1)$	rate for the resumption of flow (end of starving) at the upstream station of subsystem $L(k-1)$, $M_u(k-1)$
$E_u(k)$	probability that upstream station of subsystem $L(k)$, $M_u(k)$, is up and buffer $B(k)$ is not full

Another set of equations that describes the transition rates for changing the downstream station of subsystem $L(k)$, $M_d(k)$, from state "up" to state "blocked", $\lambda b_d(k)$ can be evaluated similarly using the steady-state probability distribution of the CTMC of subsystem $L(k+1)$. Figure 8 shows the relevant part ($n = N$) of the state-transition rate diagram of an arbitrary subsystem, again omitting the subsystem index k.

$$\lambda b_d(k) = \frac{p(k+1,U,N,R)\cdot\beta_d(k+1) + p(k+1,U,N,B)\cdot\beta b_d(k+1)}{E_d(k)} \qquad k = 1,2,\ldots,K-2 \quad (20)$$

and $\lambda b_d(K-1) \to 0$

$p(k+1,U,N,R)$	probability that downstream station of subsystem $L(k+1)$, $M_d(k+1)$, is in state "in repair" and the buffer $B(k+1)$ is full and the upstream station $M_u(k+1)$ is up

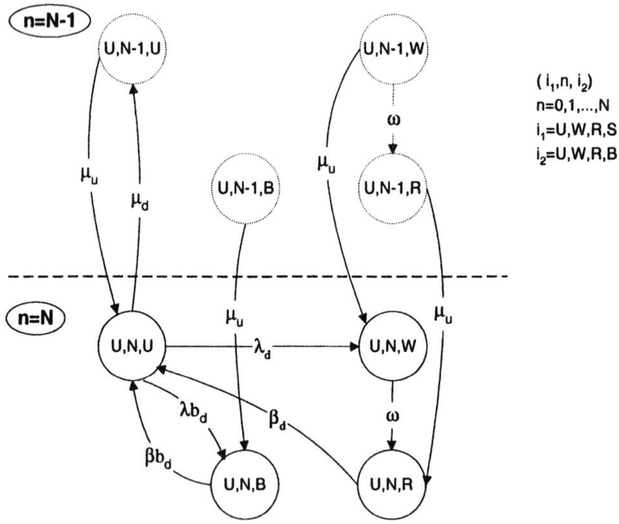

Figure 8. Partial state-transition rate diagram of an arbitrary two-station subsytem, $n = N$ (blocking of the upstream station).

$p(k+1, U, N, B)$	probability that downstream station of subsystem $L(k+1)$, $M_d(k+1)$, is in state "blocked" and the buffer $B(k+1)$ is full and the upstream station $M_u(k+1)$ is up
$\beta_d(k+1)$	repair rate of the downstream station of subsystem $L(k+1)$, $M_d(k+1)$
$\beta b_d(k-1)$	rate for the resumption of flow (end of blocking) at the downstream station of subsystem $L(k+1)$, $M_d(k+1)$
$E_d(k)$	probability that downstream station of subsystem $L(k)$, $M_d(k)$, is up and buffer $B(k)$ is not empty

Resumption of flow. The next two sets of equations define the transition rates for the stations of subsystem $L(k)$ going from state "starved" or "blocked" back to state "up" [$\beta s_u(k)$ or $\beta b_d(k)$]. Thus, the material flow will be resumed.

$$\beta s_u(k) = \frac{\underline{p(k-1,S,0,U)\cdot\beta s_u(k-1)} \\ +p(k-1,R,0,U)\cdot\beta_u(k-1)+p(k-1,W,0,U)\cdot\beta w_u(k-1)}{p(k-1,S,0,U)+p(k-1,R,0,U)+p(k-1,W,0,U)} \quad k=2,3,\ldots,K-1 \quad (21)$$

and $\beta s_u(1) = 1$

$p(k-1, R, 0, U)$	probability that upstream station of subsystem $L(k-1)$, $M_u(k-1)$, is in state "in repair" and the buffer $B(k-1)$ is empty and the downstream station $M_d(k-1)$ is up

$p(k-1, S, 0, U)$	probability that upstream station of subsystem $L(k-1)$, $M_u(k-1)$, is in state "starved" and the buffer $B(k-1)$ is empty and the downstream station $M_d(k-1)$ is up
$p(k-1, W, 0, U)$	probability that upstream station of subsystem $L(k-1)$, $M_u(k-1)$, is in state "waiting" and the buffer $B(k-1)$ is empty and the downstream station $M_d(k-1)$ is up
$\beta_u(k-1)$	repair rate of the upstream station of subsystem $L(k-1)$, $M_u(k-1)$
$\beta s_u(k-1)$	rate for the resumption of flow (end of starving) at the upstream station of subsystem $L(k-1)$, $M_u(k-1)$
$\beta w_u(k-1)$	repair rate of upstream station of subsystem $L(k-1)$, $M_u(k-1)$, given that the station are still waiting for the repairman

$$\beta b_d(k) = \frac{\begin{array}{l} p(k+1, U, N, B) \cdot \beta b_d(k+1) \\ +p(k+1, U, N, R) \cdot \beta_d(k+1) + p(k+1, U, N, W) \cdot \beta w_d(k+1) \end{array}}{p(k+1, U, N, B) + p(k+1, U, N, R) + p(k+1, U, N, W)} \qquad k = 1, 2, ..., K-2 \quad (22)$$

and $\beta b_d(K-1) = 1$

$p(k+1, U, N, R)$	probability that downstream station of subsystem $L(k+1)$, $M_d(k+1)$, is in state "in repair" and the buffer $B(k+1)$ is full and the upstream station $M_u(k+1)$ is up
$p(k+1, U, N, B)$	probability that downstream station of subsystem $L(k+1)$, $M_d(k+1)$, is in state "blocked" and the buffer $B(k+1)$ is full and the upstream station $M_u(k+1)$ is up
$p(k+1, U, N, W)$	probability that downstream station of subsystem $L(k+1)$, $M_d(k+1)$, is in state "waiting" and the buffer $B(k+1)$ is full and the upstream station $M_u(k+1)$ is up
$\beta_d(k+1)$	repair rate of the downstream station of subsystem $L(k+1)$, $M_d(k+1)$
$\beta b_d(k-1)$	rate for the resumption of flow (end of blocking) at the downstream station of subsystem $L(k+1)$, $M_u(k+1)$
$\beta w_d(k+1)$	repair rate of downstream station of subsystem $L(k+1)$, $M_d(k+1)$, given that the station is still waiting for the repairman

The transition rates $\beta w_u(k)$ and $\beta w_d(k)$ represent the repair rates of the upstream and downstream station of subsystem $L(k)$, $M_u(k)$ and $M_d(k)$, respectively given that the stations are still waiting for the repairman.

$$\frac{1}{\beta w_u(k)} = \frac{1}{\beta_u(k)} + \frac{1}{\omega_u(k)} \qquad k = 1, 2, ..., K-1 \quad (23)$$

and

$$\frac{1}{\beta w_d(k)} = \frac{1}{\beta_d(k)} + \frac{1}{\omega_d(k)} \qquad k = 1, 2, ..., K-1 \quad (24)$$

The mean waiting times for the repairman of the upstream and downstream station, $1/\omega_u(k)$ and $1/\omega_d(k)$, respectively, are unknowns. They can be approximated using a machine interference model. This will be described in Chapter 3.4 in more detail.

With the described six sets of equations, the parameters $\mu_u(k)$, $\lambda s_u(k), \beta s_u(k)$, $\mu_d(k)$, $\lambda b_d(k)$, and $\beta b_d(k)$ of subsystem $L(k)$, $k = 1, 2, ..., K - 1$, can be determined using a forward and backward procedure. Note that the boundary conditions of the mean service rates $\mu_u(1)$ and $\mu_d(K-1)$ are given as follows:

$$\mu_u(1) = \mu_1 \tag{25}$$

and

$$\mu_d(K-1) = \mu_K \tag{26}$$

The mean failure and repair rates of the up- and downstream station of subsystem $L(k)$, $k = 1, 2, ..., K-1$, however, are equivalent to the mean failure and repair rates assumed for the stations of the original line:

$$\lambda_u(k) = \lambda_k \qquad k = 1, 2, ..., K-1 \tag{27}$$

$$\beta_u(k) = \beta_k \qquad k = 1, 2, ..., K-1 \tag{28}$$

$$\lambda_d(k) = \lambda_{k+1} \qquad k = 1, 2, ..., K-1 \tag{29}$$

$$\beta_d(k) = \beta_{k+1} \qquad k = 1, 2, ..., K-1 \tag{30}$$

3.4 Determining the mean waiting times for the repairman

Two important parameters of each subsystem $L(k)$ are the mean waiting times of the upstream and downstream station, $1/\omega_u(k)$ and $1/\omega_d(k)$, respectively, considering the fact that all repairmen may be busy repairing other stations of the system when a machine failure of one of the subsystem stations occurs.

The mean waiting times $1/\omega_u(k)$ and $1/\omega_d(k)$ can be determined with a machine interference model as shown in Figure 9.

Note that the mean waiting time of the upstream station of subsystem $L(k)$, $M_u(k)$, are determined by excluding station $(k+1)$ from the machine interference model and vice versa. This is done because the failure and repair processes of station $(k+1)$ as well as of station k are considered explicitly during the evaluation process of subsystem $L(k)$. In addition, the mean waiting times are determined at failure instant of the considered station ([Kuhn, 1998]).

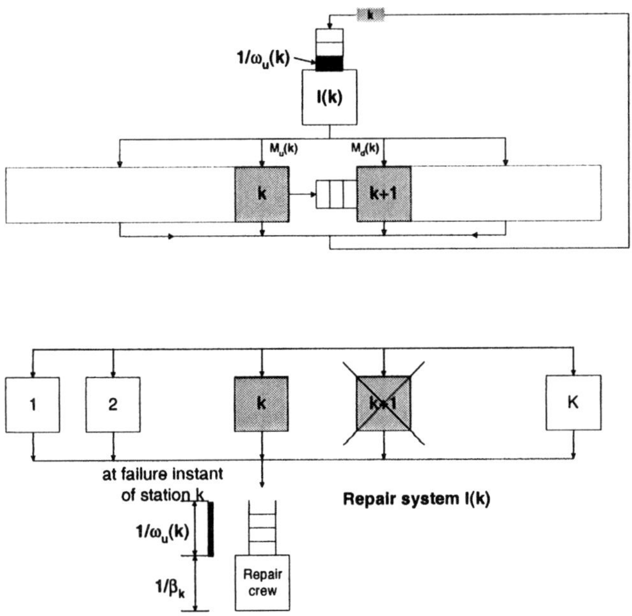

Figure 9. Machine interference model to determine the mean waiting time of a failed station.

4. Results

To demonstrate the approximation quality of the suggested integrative solution approach we consider two examples of automatic flow line systems (AFLS) with limited buffers. The AFLS consist of ten identical stations. The mean processing times, the mean times to failure, and the mean times to repair are assumed as follows: $1/\mu_k = 1.0$, $1/\lambda_k = 10.0$, and $1/\beta_k = 0.1$ (example 1), $1/\beta_k = 1.5$ (example 2), $k = 1, 2, ..., K$. One repairman is available for repairing the failed stations of the AFLS. The number of buffer places in front of each station of the AFLS are varied between 1 and 19 places for each example.

Figure 10 and 11 show the production rates of the AFLS and the utilization of the repair system (example 1), respectively. The values were obtained via simulation and by the help of three different analytical solution procedures. The analytical solution procedures are: the suggested integrative approach, the iterative approach and an approach where the dependency of the production and repair system is fully ignored. This approach is called "independent approach". The production rates approximated by the analytical approaches are almost equal and the deviations to the simulation results are very small. The achieved results, however, are induced by the relative short mean waiting times for the repairman because the repair system has a very low utilization. The utilization of the repair system is shown in Figure 10.

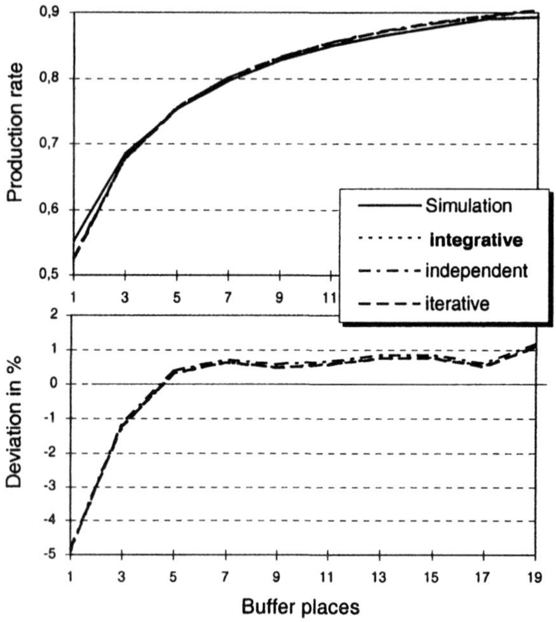

Figure 10. Production rate of the AFLS (example 1).

The chart, however, also shows that the independent approach neglects the utilization of the AFLS, which is strongly influenced by the number of buffer places between the stations. Because operation-dependent failures are assumed, the number of station failures increases with higher station utilization. Thus, the utilization of the repair system increases if the buffer areas of the flow line system are enlarged. The independent approach neglects this phenomenon. The integrative and the iterative approach, however, lead to very similar and exact results.

Different results are achieved if the utilization of the repair system is above 50%. In example 2 the mean time to repair for a failed station is enlarged to 1.5 time units. Figure 12 and 13 show the results obtained with the different solution approaches for the production rate of the line and the utilization of the repairman.

The independent approach overestimates the production rate because the independent approach neglects the waiting times for the repairman. The approximation error increases with increasing buffer areas. Because the production rate of the AFLS increases with larger buffers, the waiting times for the repairman increase, too.

The iterative approach, on the other hand, underestimates the production rates because the failure probabilities of the stations of the AFLS are overesti-

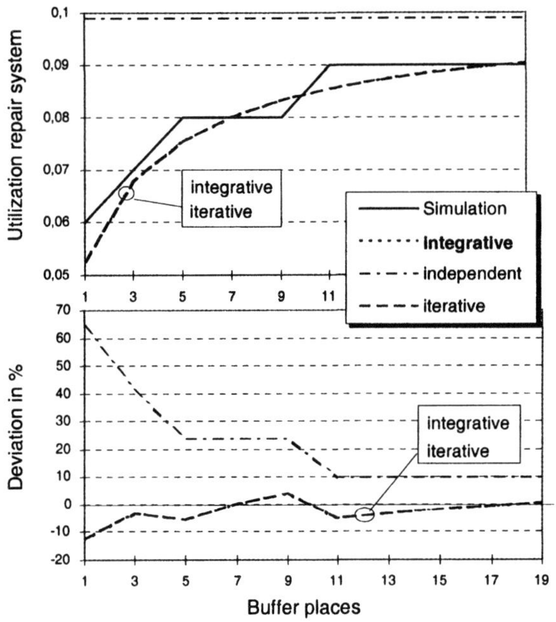

Figure 11. Utilization of the repair system (example 1).

mated. The dependency of neighboring stations leads to a drop of the stations failure rates. The integrative approach, however, leads to nearly exact results.

But, the integrative approach is much more time-consuming than the iterative solution approach. Figure 14 shows the computational time for both procedures versus the number of buffer places in front of each station for example 2 (10-station line).

The computational time of the iterative solution approach increases linearly with the number of buffer places. The computational time of the integrative approach, however, increases exponentially with the number of buffer places. (Note that the time axis, y-axis, is logarithmically scaled.) The exponential increase occurs since the building blocks of the integrative approach are evaluated by solving a finite CTMC with a numerical procedure. The number of states (Y) of the CTMC depends on the number of buffer places ($N - 1$) between the two stations of the line. The number of states are given as follows: $Y = 18 \cdot (N - 1) + 8$.

Note, that N denotes the maximum number of parts processed by the upstream station but not yet processed by the downstream station of the two-station line. Since one unfinished workpiece may be located in the working area of the downstream station – that one, which is just processed by the downstream station – the number of buffer places in the two-station line is ($N - 1$).

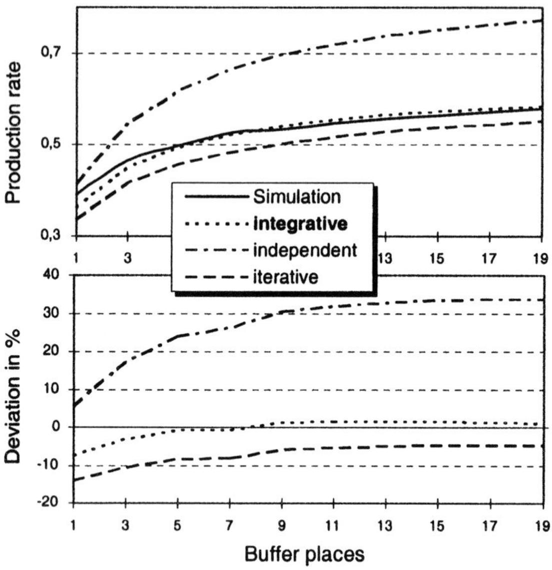

Figure 12. Production rate of the AFLS (example 2).

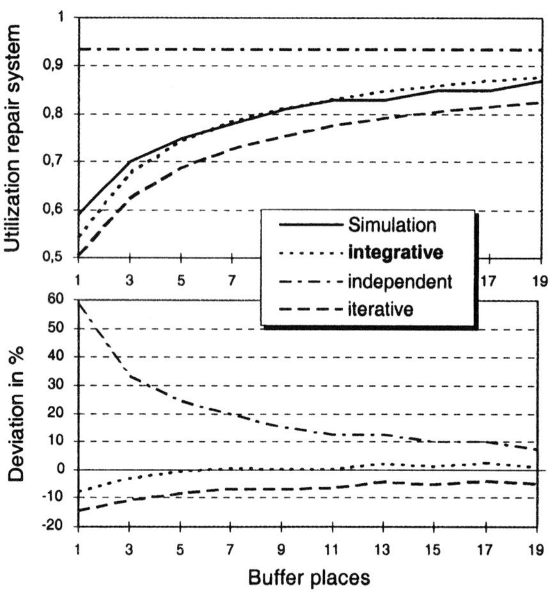

Figure 13. Utilization of the repair system (example 2).

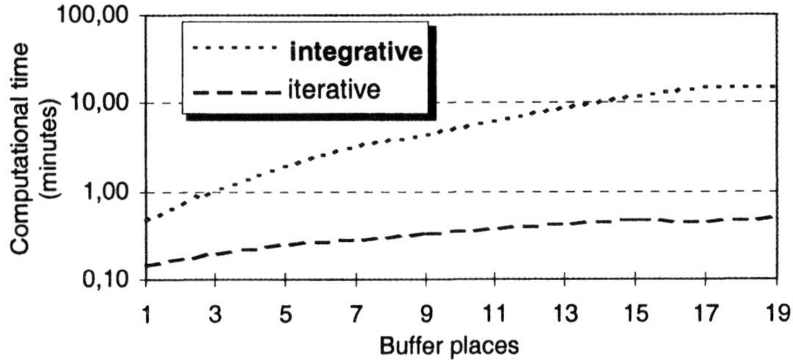

Figure 14. Computational time for the iterative and the integrative procedure versus the number of buffer places in front of each station for a 10-station line (example 2), 486er PC with 66 MHz.

For the 10-station line with 19 buffer places in front of each station the proposed integrative solution approach requires 16 minutes of computational time on an 486er IBM-compatible PC with 66 MHz to achieve results with sufficient consistency among the subsystems (Figure 14). A simulation model, however, requires more than three hours to achieve equivalent results.

References

[Biles and Usher, 1997] Biles, W. E. and Usher, J. S. (1997). Reliability and Maintenance in Automated Production Flow Lines. Technical report, Department of Industrial Engineering, University of Louisville.

[Bonvik et al., 2000] Bonvik, A. M., Dallery, Y., and Gershwin, S. B. (2000). Approximate analysis of production systems operated by a conwip/finite buffer hybrid control policy. *International Journal of Production Research*, 38(3):2845–2869.

[Buzacott and Shanthikumar, 1993] Buzacott, J. A. and Shanthikumar, J. G. (1993). *Stochastic Models of Manufacturing Systems.* Englewood Cliffs: Prentice Hall.

[Dallery et al., 1988] Dallery, Y., David, R., and Xie, X.-L. (1988). An efficient algorithm for analysis of transfer lines with unreliable machines and finite buffers. *IIE Transactions*, 20(3):280–283.

[Dallery and Frein, 1993] Dallery, Y. and Frein, Y. (1993). On decomposition methods for tandem queuing networks with blocking. *Operations Research*, 41(2):386–399.

[Dallery and Gershwin, 1992] Dallery, Y. and Gershwin, S. B. (1992). Manufacturing flow line systems: A review of models and analytical results. *Queueing Systems*, 12(1+2):3–94.

[Di Mascolo et al., 1996] Di Mascolo, M., Frein, Y., and Dallery, Y. (1996). An analytical method for performance evaluation of kanban controlled production systems. *Operations Research*, 44(1):50–64.

[Dudick, 1979] Dudick, D. (1979). *Fixed Cycle Production Systems with In-Line Inventory and Limited Repair Capability*. PhD Thesis, Columbia University, New York.

[Elsayed and Turley, 1980] Elsayed, E. A. and Turley, R. E. (1980). Reliability analysis of production systems with buffer storage. *International Journal of Production Research*, 18(5):637–645.

[Gaver, 1962] Gaver, D. P. (1962). A waiting line with interrupted service, including priorities. *J. Roy. Stat. Soc.*, 24:73–90.

[Gershwin, 1987] Gershwin, S. B. (1987). An efficient decomposition method for the approximate evaluation of tandem queues with finite storage space and blocking. *Operations Research*, 35(2):291–305.

[Gershwin, 1989] Gershwin, S. B. (1989). An efficient decomposition algorithm for unreliable tandem queuing systems with finite buffers. In Perros, H. G. and T., A., editors, *Queueing Networks with Blocking*, pages 127–146. Amsterdam: North Holland.

[Gershwin, 1994] Gershwin, S. B. (1994). *Manufacturing Systems Engineering*. Englewood Cliffs: Prentice Hall.

[Gross and Harris, 1998] Gross, D. and Harris, C. M. (1998). *Fundamentals of Queueing Theory*. New York: Wiley, 3rd edition.

[Helber, 1998] Helber, S. (1998). Decomposition of unreliable assembly/disassembly networks with limited buffer capacity and random processing times. *European Journal of Operational Research*, 109:24–42.

[Helber, 2000] Helber, S. (2000). Approximate analysis of unreliable transfer lines with splits in the flow of material. *Annals of Operations Research*, 93:217–243.

[Kamath and Sanders, 1991] Kamath, M. and Sanders, J. L. (1991). Modelling operator/workstation interference in asyn-chronous automatic assembly systems. *Discrete Event Dynamic Systems: Theory and Applications*, 1(1):93–124.

[Kuhn, 1998] Kuhn, H. (1998). *Fließproduktionssysteme: Leistungsbewertung, Konfigurations- und Instandhaltungsplanung*. Heidelberg: Physica.

[Kuhn, 1999] Kuhn, H. (1999). Analyse des Einflusses des Entstörpersonals auf das Ausbringungsverhalten automatisierter Montageanlagen. *Zeitschrift fuer betriebswirtschaftliche Forschung*, 51(2):107–127.

[Li, 1987] Li, K. F. (1987). Serial production lines with unreliable machines and limited repair. *Naval Research Logistics*, 34:101–108.

[Papadopoulos et al., 1993] Papadopoulos, H. T., Heavey, C., and Browne, J. (1993). *Queueing Theory in Manufacturing Systems Analysis and Design*. London: Chapman&Hall.

[Savsar and Biles, 1984] Savsar, M. and Biles, W. E. (1984). Two-stage production lines with a single repair crew. *International Journal of Production Research*, 22(3):499–514.

[Stecke and Aronson, 1985] Stecke, K. E. and Aronson, J. E. (1985). Review of operator/machine interference models. *International Journal of Production Research*, 23(1):129–151.

[Tempelmeier and Bürger, 2001] Tempelmeier, H. and Bürger, M. (2001). Performance evaluation of unbalanced flow lines with general distributed processing times, failures and imperfect production. *IIE Transactions*, 33:293–302.

[Tempelmeier and Kuhn, 1993] Tempelmeier, H. and Kuhn, H. (1993). *Flexible Manufacturing Systems: Decision Support for Design and Operation*. New York: Wiley.

[Tetzlaff, 1995] Tetzlaff, U. A. W. (1995). Evaluating the effect of tool management on flexible manufacturing system performance. *International Journal of Production Research*, 33(4):877–892.

Chapter 8

PERFORMANCE EVALUATION OF PRODUCTION LINES WITH RANDOM PROCESSING TIMES, MULTIPLE FAILURE MODES AND FINITE BUFFER CAPACITY — PART I: THE BUILDING BLOCK

R. Levantesi
Dipartimento di Meccanica, Politecnico di Milano,
via Bonardi 9, 20133 Milano, Italy
raniero.levantesi@mecc.polimi.it

A. Matta
Dipartimento di Meccanica, Politecnico di Milano,
via Bonardi 9, 20133 Milano, Italy
andrea.matta@mecc.polimi.it

T. Tolio
Dipartimento di Meccanica, Politecnico di Milano,
via Bonardi 9, 20133 Milano, Italy
tullio.tolio@mecc.polimi.it

Abstract The paper presents an exact analytical method for the performance evaluation of two-machine lines with random processing times, multiple failure modes and finite buffer capacity. Unlike previous works on this subject, machines can fail in different modes, each one characterized by exponentially distributed time to failure and time to repair. The steady state probabilities of the Markov chain of the system is obtained in a compact form with a computational effort that depends only on the number of failure modes considered and not on the capacity of the buffer. Finally, the improvements provided by the proposed method over existing techniques are discussed and quantified. The method has been derived to be used as a building block within the development of a new set of decomposition equations for the analysis of larger systems, such long transfer lines and assembly/disassembly systems.

Keywords: exponential lines, two-machine line, multiple failure modes

1. Introduction

The paper presents an exact analytical method for the performance evaluation of two-machine lines with exponentially distributed processing times, multiple failure modes and finite buffer capacity. The expression *multiple failures* means that each machine can fail in more than one mode, each one characterized by a specific time to failure and time to repair.

Indeed, in real factories machines can be affected by different kinds of failures which occur with different frequencies and require different lengths of time to be repaired. For instance, an assembly station may be interrupted by the jamming of a feeder, by electrical or mechanical troubles experienced by the working units, by the breakdown of transfer devices and so forth.

Existing techniques are not able to properly deal with more than one failure mode at each machine/station. Therefore, the different failure modes have to be grouped in an *average failure mode* by means of a first order approximation [8], [9] and, recently, also by a second [3] order approximation. This approximation may not be very accurate when the modes differs significantly. Moreover, the multiple failure feature makes it possible to introduce, at decomposition level (see Part II), some additional failures to represent the interruptions of flow that can take place inside the system as a result of machine disruption propagation.

The first model of unreliable two-machine lines with exponential service times and finite buffer capacity was analyzed by Buzacott in 1972 [1]. He considered the case of machines with the same service rate and subject to failures occurring according to a geometric distribution in the number of operations since the last failure (i.e.*number-ODFs*). In 1980, Gershwin and Berman [8] proposed a model of exponential two-machine lines which differs from the one proposed by Buzacott in that failures are represented by an exponential distribution in time rather than a geometrical distribution in the number of parts produced and in that machines can have different service rates [5].

The model described in this paper is an extension to the case of multiple failure modes of the Gershwin and Berman's model [8]. Some extensions to the case of hyper-exponential distribution of the repair time have been already proposed in literature [3] but they are limited to the case of two stages. The proposed method can properly handle an arbitrary number of stages, that is an arbitrary number of failure modes.

2. Model Description and Assumptions

The proposed method focuses on two-stage production lines with unreliable machines decoupled by a buffer of finite capacity. The system is modelled as a continuous time, discrete state Markov process and is based on the following assumptions (the same as in Gershwin [7] except for the multiple failure modes).

Each part enters the system through the first machine, then goes to the buffer waiting for the availability of the second machine and then leaves the system.

A machine whose upstream buffer is empty is said to be *starved* while a machine whose downstream buffer is full is said to be *blocked.* The upstream machine of the line is never starved while the downstream machine is never blocked.

Processing, failure and repair times for the machines are assumed to be exponential random variables with parameters representing, respectively, processing rate, failure rates and repair rates.

A machine can be in one of the following conditions: *operational* or *failed in a certain mode.* An operational machine which is not processing a workpiece (i.e. *starved* or *blocked*) cannot fail. This means that operation dependent failures (*ODFs*) are assumed.

When a machine is operational, it works on a piece if it is not starved or blocked and it continues working until either it completes the piece or one of the different failures occurs.

A machine cannot fail in more than one mode at the same time. Also, a machine failed in a certain mode cannot experience a different mode before being repaired.

When a machine is under repair, it remains in that state for a period of time which is exponentially distributed with parameters depending on the type of failure that occurred.

Workpieces are not destroyed or rejected at any stage in the line. Partly processed workpieces are not added into the line.

2.1 Notation

The state of the system is given by (n, α_u, α_d) where n is an integer that indicates the number of pieces in the buffer plus the piece on the working area of the downstream machine while α_u and α_d indicate the state of upstream and downstream machine, respectively. The convention for blocking is $n = N = C + 2$, where C is the buffer capacity $(0 \leq n \leq N)$. When the upstream machine is operational $\alpha_u = 1$ while $\alpha_u = u_i$, for $i = 1, ..., s$, means that the upstream machine is down due to failure mode i. Similarly, α_d can assume the values $1, d_1, d_2, ..., d_t$. The steady state probability of the system being in state (n, α_u, α_d) is indicated by $p(n, \alpha_u, \alpha_d)$.

If the upstream machine is working on a piece at time t, during the time interval $(t, t+\delta t)$ it can complete the piece with probability approximately $\mu_u \delta t$ or it can fail in mode u_i with probability approximately $p^{u_i}\delta t$. If the upstream machine is failed in mode u_i at time t, it can be repaired during the time interval $(t, t+\delta t)$ with an approximate probability of $r^{u_i}\delta t$, for small δt. Similarly, μ_d, p^{d_j} and r^{d_j}, for $j = 1, ..., t$, represent the service, failure and repair rates of the downstream machine, respectively.

The total failure rate P^U of the upstream machine, i.e. the approximate probability of a failure during the time interval $(t, t+\delta t)$ regardless of the mode in which the machine fails, is given by $P^U = \sum_{i=1}^{s} p^{u_i}$. Similarly, the total failure rate P^D of the downstream machine is given by $P^D = \sum_{j=1}^{t} p^{d_j}$.

2.2 Performance Measures

The efficiency is defined as the steady state probability that a machine is working on a piece and it can be seen as the fraction of time in which the machine processes parts. Therefore, for the upstream machine it is possible to write:

$$E_u = prob\,[\alpha_u = 1, n < N] = \sum_{n=0}^{N\text{-}1} \left[p(n,1,1) + \sum_{j=1}^{t} p\left(n,1,d_j\right) \right] \tag{1}$$

since the upstream machine, being the first machine of the line, cannot be starved. Similarly, the efficiency of the downstream machine is given by:

$$E_d = prob\,[\alpha_d = 1, n > 0] = \sum_{n=1}^{N} \left[p(n,1,1) + \sum_{i=1}^{s} p\,(n,u_i,1) \right] \tag{2}$$

since the downstream machine, being the last machine of the line, cannot be blocked.

On the basis of the efficiency, it is possible to evaluate the production (throughput) rate of the upstream and downstream machines as:

$$P_u = \mu_u E_u \qquad P_d = \mu_d E_d \tag{3}$$

In addition, no workpieces are created or destroyed. As a result, it can be shown, extending to the case of multiple failures the proof given by Gershwin in [2], that the part flow in the two-machine line is conserved:

$$P = \mu_u E_u = \mu_d E_d \tag{4}$$

Finally, the average buffer level can be written as:

$$\bar{n} = \sum_{n=0}^{N} n \left[p(n,1,1) + \sum_{i=1}^{s} p(n,u_i,1) + \sum_{j=1}^{t} p\left(n,1,d_j\right) + \sum_{i=1}^{s}\sum_{j=1}^{t} p\left(n,u_i,d_j\right) \right] \tag{5}$$

2.3 Balance Equations

Since we have to deal with a continuous time, discrete space Markov process, to obtain the steady state distribution we make use of balance equations, which equate the rate of leaving a state with the rate of entering it.

We distinguish four sets of equations on the basis of the different values of the state variables α_u and α_d. In addition, having defined as *internal* the states (n, α_u, α_d) where $1 \leq n \leq N-1$, we can further classify balance equations that do not involve boundary states as *internal equations*, while all the others are *boundary equations*.

The complete set of balance equations for the Markov chain that models the two-machine line under investigation follows.

$\alpha_u = \mathbf{u}_i$ $(i = 1, \ldots, s)$ **and** $\alpha_d = \mathbf{d}_j$ $(j = 1, \ldots, t)$**:**

$$p\left(n,u_i,d_j\right)\left(r^{u_i} + r^{d_j}\right) = p(n,u_i,1)\,p^{d_j} + p\left(n,1,d_j\right)p^{u_i} \tag{6}$$

for $1 \leq n \leq N-1$.

$$p\left(0,u_i,d_j\right)\left(r^{u_i} + r^{d_j}\right) = p\left(0,1,d_j\right)p^{u_i} \tag{7}$$

$$p\left(N,u_i,d_j\right)\left(r^{u_i} + r^{d_j}\right) = p(N,u_i,1)\,p^{d_j} \tag{8}$$

On the left side of these equations we have the rate at which the system leaves state (n, u_i, d_j). This happens whenever a machine, upstream or downstream, is recovered from a certain failure mode. Conversely, the right side stands for the rate at which the system enter state (n, u_i, d_j). This state can be reached either from state $(n, 1, d_j)$ if the upstream machine fails in mode i or from state $(n, u_i, 1)$ if the downstream machine fails in mode j. If $n = 0$ the downstream machine is starved and, given the assumption of operational dependent failures, cannot fail. Similarly, when $n = N$, the upstream machine is blocked and it cannot be affected by any failure mode.

The remaining three sets of equations can be obtained in a similar way.

$\alpha_u = \mathbf{u}_i(i = 1, \ldots, s)$ **and** $\alpha_d = \mathbf{1}$**:**

$$p(n, u_i, 1)\left(r^{u_i} + \mu_d + \sum_{j=1}^{t} p^{d_j}\right) = \sum_{j=1}^{t} p\left(n, u_i, d_j\right) r^{d_j} + \\ + p(n, 1, 1) p^{u_i} + p(n + 1, u_i, 1)\mu_d \tag{9}$$

for $1 \leq n \leq N - 1$.

$$p(0, u_i, 1)\, r^{u_i} = \sum_{j=1}^{t} p\left(0, u_i, d_j\right) r^{d_j} + p(0, 1, 1) p^{u_i} + p(1, u_i, 1)\mu_d \tag{10}$$

$$p(N, u_i, 1)\left(r^{u_i} + \mu_d + \sum_{j=1}^{t} p^{d_j}\right) = \sum_{j=1}^{t} p\left(N, u_i, d_j\right) r^{d_j} \tag{11}$$

α_u**= 1 and** α_d**=** $\mathbf{d}_j (j = 1, ..., t)$**:**

$$p\left(n, 1, d_j\right)\left(\mu_u + \sum_{i=1}^{s} p^{u_i} + r^{d_j}\right) = \sum_{i=1}^{s} p\left(n, u_i, d_j\right) r^{u_i} + \\ + p(n, 1, 1)\, p^{d_j} + p(n - 1, 1, d_j)\mu_u \tag{12}$$

for $1 \leq n \leq N - 1$.

$$p\left(0, 1, d_j\right)\left(\mu_u + \sum_{i=1}^{s} p^{u_i} + r^{d_j}\right) = \sum_{i=1}^{s} p\left(0, u_i, d_j\right) r^{u_i} \tag{13}$$

$$p\left(N, 1, d_j\right) r^{d_j} = \sum_{i=1}^{s} p\left(N, u_i, d_j\right) r^{u_i} + p(N, 1, 1)\, p^{d_j} + \\ + p(N - 1, 1, d_j)\mu_u \tag{14}$$

α_u**= 1 and** α_d**= 1:**

$$p(n,1,1)\left(\mu_u+\sum_{i=1}^{s}p^{u_i}+\mu_d+\sum_{j=1}^{t}p^{d_j}\right)=p(n+1,1,1)\,\mu_d+ \\ +p(n-1,1,1)\mu_u+ \\ +\sum_{i=1}^{s}p(n,u_i,1)\,r^{u_i}+\sum_{j=1}^{t}p\left(n,1,d_j\right)r^{dj} \tag{15}$$

for $1 \leq n \leq N-1$.

$$p(0,1,1)\left(\mu_u+\sum_{i=1}^{s}p^{u_i}\right)=p(1,1,1)\,\mu_d+\sum_{j=1}^{t}p\left(0,1,d_j\right)r^{d_j}+ \\ +\sum_{i=1}^{s}p(0,u_i,1)\,r^{u_i} \tag{16}$$

$$p(N,1,1)\left(\mu_d+\sum_{j=1}^{t}p^{d_j}\right)=p(N-1,1,1)\mu_u+\sum_{j=1}^{t}p\left(N,1,d_j\right)r^{d_j}+ \\ +\sum_{i=1}^{s}p(N,u_i,1)\,r^{u_i} \tag{17}$$

2.4 Solution Methodology

The solution methodology is an extension to the case of multiple failure modes of the one originally proposed by Gershwin and Berman [8]. The first step consists of the analysis of the Markov chain and the formulation of a guess on the form of the steady state probabilities of the internal states. Then, substituting this guess into the internal balance equations, a set of $s + t + 2$ solutions (i.e. vectors that satisfy the internal equations) can be found. At this point, if the guess is correct it must be possible to find a linear combination of these internal solutions that also satisfies the boundary conditions. The advantage of this approach is that the computational effort depends only on the total number of failure modes taken in account and not on the buffer capacity C.

2.5 Analysis of Internal Equations

Following the analysis carried out in Tolio at al. [13], we guess the following form for the solution of the internal equations:

$$p(n, \alpha_u, \alpha_d) = X^n \Phi_u \Phi_d, 1 \leq n \leq N-1 \tag{18}$$

where X, Φ_u and Φ_d are constants to be identified. Specifically, X refers to the buffer level whereas Φ_u (Φ_d) refers to the upstream (downstream) machine state, taking the value 1 (1) if the machine is operational or the value U_i (D_j) if the machine is down in mode i (j), for $i = 1, ..., s$ ($j = 1, ..., t$).

Substituting equation (18) into the internal equations, after some manipulations we obtain:

$$\frac{p^{u_i}}{U_i} - r^{u_i} = -\left(\frac{p^{d_j}}{D_j} - r^{d_j}\right), i = 1, ..., s, j = 1, ..., t \tag{19}$$

$$\mu_d (X-1) + \frac{p^{u_i}}{U_i} - r^{u_i} + \sum_{j=1}^{t} \left(D_j r^{d_j} - p^{d_j}\right) = 0, i = 1, ..., s \tag{20}$$

$$\mu_u \left(\frac{1}{X} - 1\right) + \frac{p^{d_j}}{D_j} - r^{d_j} + \sum_{i=1}^{s} (U_i r^{u_i} - p^{u_i}) = 0, j = 1, ..., t \tag{21}$$

$$\mu_d (X-1) + \mu_u \left(\frac{1}{X} - 1\right) + \sum_{i=1}^{s} (U_i r^{u_i} - p^{u_i}) + \sum_{j=1}^{t} \left(D_j r^{d_j} - p^{d_j}\right) = 0 \tag{22}$$

Since equation (19) is a linear combination of equations (20), (21) and (22), these three last equations build up a system of $s + t + 1$ equations in $s + t + 1$ unknowns, i.e. the constants X, U_i and D_j we introduced in (18) to formulate the guess.

Considering that in equation (19) all the parameters and unknowns related to the upstream machine are on the left side while all parameters and unknowns referring to the downstream machine are on the right side, it is possible to state:

$$\frac{p^{u_i}}{U_i} - r^{u_i} = K, i = 1, ..., s \tag{23}$$

$$\frac{p^{d_j}}{D_j} - r^{d_j} = -K, j = 1, ..., t \tag{24}$$

where K is not a function of i or j. From equations (23) and (24) we can derive the following expressions for U_i and D_j:

$$U_i = \frac{p^{u_i}}{K + r^{u_i}}, i = 1, ..., s \tag{25}$$

$$D_j = \frac{p^{d_j}}{r^{d_j} - K}, j = 1, ..., t \tag{26}$$

Then, isolating X from equation (20) and from equation (21) and equating the two expressions found out, after the introduction of equations (23) and (24) we obtain:

$$\left[1-\frac{K}{\mu_d}\left(1+\sum_{j=1}^{t}\frac{p^{d_j}}{r^{d_j}-K}\right)\right]\left[1+\frac{K}{\mu_u}\left(1+\sum_{i=1}^{s}\frac{p^{u_i}}{r^{u_i}+K}\right)\right]=1 \qquad (27)$$

which is a polynomial in K of degree $R = s + t + 2$.

If we indicate with K_m ($m = 1, ..., s+t+2$) the mth root of the polynomial, we can define $U_{i,m}, D_{j,m}$ and X_m as the values the internal constants U_i, D_j and X assume with respect to the mth root of the polynomial. These values represent a solution of the internal equations and can be found by means of equations (25), (26) and (20) which can be rewritten as follows:

$$U_{i,m}=\frac{p^{u_i}}{K_m+r^{u_i}},\ i=1,...,s \qquad (28)$$

$$D_{j,m}=\frac{p^{d_j}}{r^{d_j}-K_m},\ j=1,...,t \qquad (29)$$

$$X_m=\left[1-\frac{K_m}{\mu_d}\left(1+\sum_{j=1}^{t}\frac{p^{d_j}}{r^{d_j}-K_m}\right)\right] \qquad (30)$$

Concerning the nature of the roots, an assumption we are claiming is that they all are real. If some roots were complex the steady state probabilities would be either complex, which is impossible, or they would show a periodic pattern not coherent with the behavior of the real system. This statement cannot be formally proved. However, we can provide a sort of justification. Indeed, looking at the structure of the polynomial it is possible to show (as reported in Appendix B) that, regardless of the number of failure modes and their parameters, the polynomial has always a number of intersections with the K axis exactly equal to its degree, thus eliminating the possibility of complex roots. In addition, it is possible to identify the intervals in which each root can be found (see Appendix B). This, in turn, will make it possible to obtain the values of the roots by mean of the bisection algorithm and to avoid all the numerical problems that the root searching in a high degree polynomials usually involves.

One of the roots of the polynomial that arises in the solution of Markovian models of production systems is always zero [7]. It is simple to verify that this is also the case of the polynomial derived in this paper (27). Indeed, if we introduce $K = 0$ in the polynomial we obtain the identity $1 = 1$. Then, defining this root as K_R and substituting it into equations (28), (29) and (30) we obtain:

$$U_{i,R} = \frac{p^{u_i}}{r^{u_i}}, i = 1, ..., s \quad D_{j,R} = \frac{p^{d_j}}{r^{d_j}}, j = 1, ..., t \quad X_R = 1 \tag{31}$$

Since it is possible to find different values for the parameters X, U_i and D_j such that the guess proposed in (18) satisfies the internal equations, also all the possible linear combinations of these solutions satisfy the internal equations. This is because such equations are linear. Therefore, it is possible to write:

$$p\left(n, u_i, d_j\right) = \sum_{r=1}^{R} C_r X_r^n U_{i,r} D_{j,r}, r = 1, ..., R, i = 1, ..., s, j = 1, ..., t \tag{32}$$

The next step consists of finding out if it is possible to set the values of the constant C_r, for $r = 1, ..., R$, so that also the boundary conditions are satisfied.

2.6 Analysis of Boundary Equations

The assumptions of the model imply that certain boundary states are transient, that is they have zero steady state probability and they cannot be reached from any state except possibly other transient states:

$$p\left(0, 1, d_j\right) = p\left(0, u_i, d_j\right) = p\left(N, u_i, d_j\right) = p\left(N, u_i, 1\right) = 0 \tag{33}$$

for $i = 1, ..., s$ and $j = 1, ..., t$.

This result can be obtained extending to the case of multiple failures the proof given by Gershwin in [7] (see Appendix A).

Regarding the non-transient boundary states, since it can be shown that the balance equations related to boundary states $(0, 1, 1)$ (16) and $(N, 1, 1)$ (17) involve only internal states, we guess that the corresponding steady state probabilities can be expressed in internal form, namely:

$$p(0, 1, 1) = \sum_{r=1}^{R} C_r \tag{34}$$

$$p(N, 1, 1) = \sum_{r=1}^{R} C_r X_r^N \tag{35}$$

Then, introducing equations (33) and (34) in (10) and equations (33) and (35) in (14) we obtain the following expressions for the steady state probabilities of boundary states $(0, u_i, 1)$ and $(N, 1, d_j)$:

$$p(0, u_i, 1) = \frac{1}{r^{u_i}} \sum_{r=1}^{R} C_r \left[p^{u_i} + \mu_d X_r U_{i,r}\right], i = 1, ..., s \tag{36}$$

$$p(N,1,d_j) = \frac{1}{r^{d_j}} \sum_{r=1}^{R} C_r X_r^{N-1} \left[X_r p^{d_j} + \mu_u D_{j,r}\right], j = 1, ..., t \tag{37}$$

In this way the steady state probabilities of all the non-transient boundary states can be expressed through the parameters of the two machines, the R different solutions of the internal equations and the constants C_r, for $r = 1, ..., R$, which are the only unknowns that need to be evaluated.

To find out the values of the constant C_r, for $r = 1, ..., R$, we take into account all the boundary equations that are not identically satisfied when the expressions (32), (34), (35), (36) and (37) are introduced. Specifically, equations (16) and (17) become:

$$\sum_{r=1}^{R} C_r \left[\mu_u - \mu_d X_r \left(1 + \sum_{i=1}^{s} U_{i,r}\right)\right] = 0 \tag{38}$$

$$\sum_{r=1}^{R} C_r X_r^{N-1} \left[\mu_d X_r - \mu_u \left(1 + \sum_{j=1}^{t} D_{j,r}\right)\right] = 0 \tag{39}$$

while introducing equation (32) into equations (12) for $n = 1$ and into equations (9) for $n = N - 1$, we obtain:

$$\sum_{r=1}^{R} C_r X_r \left[\left(\mu_u + \sum_{i=1}^{s} p^{u_i} + r^{d_j}\right) D_{j,r} - \sum_{i=1}^{s} U_{i,r} D_{j,r} r^{u_i} - p^{d_j}\right] = 0 \tag{40}$$

for j=1,...,t.

$$\sum_{r=1}^{R} C_r X_r^{N-1} \left[\left(\mu_d + r^{u_i} + \sum_{j=1}^{t} p^{d_j}\right) U_{i,r} - \sum_{j=1}^{t} U_{i,r} D_{j,r} r^{d_j} - p^{u_i}\right] = 0 \tag{41}$$

for i=1,...,s.

Considering also the normalization equation, i.e. the fact that the sum of the steady state probabilities of all the states must be equal to 1, we have:

$$1 = \sum_{n=0}^{N} \left[p(n,1,1) + \sum_{i=1}^{s} p(n,u_i,1) + \sum_{j=1}^{t} p\left(n,1,d_j\right) + \sum_{i=1}^{s} \sum_{j=1}^{t} p\left(n,u_i,d_j\right)\right] \tag{42}$$

Equations (38), (39), (40), (41) and (42) represent a system of $s + t + 3$ equations in the $s + t + 2$ unknowns C_r, for $r = 1, ..., R$, which can be further reduced and simplified.

Indeed, introducing equation (21) in (40) and simplifying every term that can be simplified, we obtain:

$$\sum_{r=1}^{R} C_r D_{j,r} = 0, j = 1, ..., t \tag{43}$$

Similarly, introducing equation (20) in (41) and simplifying every term that can be simplified, we have:

$$\sum_{r=1}^{R} C_r X_r^N U_{i,r} = 0, i = 1, ..., s \tag{44}$$

Then, adding up all equations (43) for $j = 1, .., t$ to equation (38) we obtain:

$$\sum_{r=1}^{R} C_r \left[\mu_u \left(1 + \sum_{j=1}^{t} D_{j,r} \right) - \mu_d X_r \left(1 + \sum_{i=1}^{s} U_{i,r} \right) \right] = 0 \tag{45}$$

Similarly, adding up all equations (44) for $i = 1, ..., s$ to equation (39) we have:

$$\sum_{r=1}^{R} C_r X_r^{N-1} \left[\mu_d X_r \left(1 + \sum_{i=1}^{s} U_{i,r} \right) - \mu_u \left(1 + \sum_{j=1}^{t} D_{j,r} \right) \right] = 0 \tag{46}$$

Since it is possible to show (see Appendix C) that:

$$\mu_u \left(1 + \sum_{j=1}^{t} D_{j,r} \right) - \mu_d X_r \left(1 + \sum_{i=1}^{s} U_{i,r} \right) = 0 \forall r \neq R \tag{47}$$

equations (45) and (46) are both satisfied by setting $C_R = 0$.

At this point, exactly $s + t + 1$ linear conditions, given by the equations (42), (43) and (44), in the $s + t + 1$ unknowns C_r, for $r = 1, ..., R - 1$, have been obtained.

There is a last case to discuss which occurs when the equation (47) is true for $r = R$, that is:

$$\mu_u \left(1 + \sum_{j=1}^{t} D_{j,R} \right) - \mu_d X_R \left(1 + \sum_{i=1}^{s} U_{i,R} \right) = 0 \tag{48}$$

Substituting the values of the internal constants for $r = R$ given by (31), after some manipulations equation (48) assumes the following form:

$$\frac{\mu_d}{(1+\sum_{j=1}^{t}\frac{p^{d_j}}{r^{d_j}})} = \frac{\mu_u}{(1+\sum_{i=1}^{s}\frac{p^{u_i}}{r^{u_i}})} \tag{49}$$

Since the efficiency in isolation of a machine with F failure modes is given by [1]:

$$e = \frac{1}{1+\sum_{i=1}^{F}\frac{p_i}{r_i}} \tag{50}$$

equation (49) can be stated as follows:

$$\mu_u e_u = \mu_d e_d \tag{51}$$

This means that when the two machines have the same production rate in isolation C_R is undefined. However, in this case $K_R = 0$ is a double root of the polynomial (27) and, as a consequence, the solutions of the internal equations linearly independent are $s+t+1$. As a result, also the number of C_R to determine is $s+t+1$ for which the $s+t+1$ linear conditions given by the equations (42), (43) and (44) can be used.

2.7 Algorithm

To obtain the performance measures of the system the following steps must be performed:

(a) Numerical evaluation of the roots K_m ($m = 1, ..., R$) of the polynomial (27). One of these roots is always given by $K_R = 0$.

(b) Evaluation of the constants $U_{i,m}$ ($i = 1, ..., s$), $D_{j,m}$ ($j = 1, ..., t$) and X_m by means of equations (28), (29) and (30) respectively for each root K_m.

(c) Solution of the linear system of $s + t + 1$ equations in the unknowns C_r ($r = 1, ..., R - 1$) formed by equations (42), (43) and (44).

(d) Evaluation of the steady state probability of all the states. For the internal states they are given by equation (32) while for the boundary states they can be obtained using equations (34), (35), (36) and (37).

(e) Evaluation of the performance measures by means of equations (1), (2), (4) and (5).

3. Numerical Results

The possibility to distinguish among different failure modes for each machine leads to a more accurate evaluation of the performance measures of two-machine lines where machines can break down in more than one mode, especially when the various modes differ significantly.

To quantify this improvement, we make a comparison with the method proposed by Gershwin and Berman [8] that considers one failure mode for each machine. Therefore, it requires the introduction of an average failure for each machine whose parameters are chosen in order to preserve the efficiency in isolation of the machine.

The comparison is carried out by considering the behavior of a two-machine line in which the upstream machine can fail in two different ways and the downstream machine has only one failure mode.

The average mode of the first machine can be evaluated preserving the efficiency in isolation and considering that the probability for a machine to be simultaneously blocked and starved is very small [7]:

$$\begin{cases} p^u = p^{u_1} + p^{u_2} \\ \frac{1}{r^u} = \frac{p^{u1}}{p^{u1}+p^{u2}} \cdot \frac{1}{r^{u1}} + \frac{p^{u2}}{p^{u1}+p^{u2}} \cdot \frac{1}{r^{u2}} \end{cases} \tag{52}$$

In the experiments, we consider various cases in which the efficiency in isolation and the failure rates of both machines are kept constant while the repair rates of the two failure modes of the upstream machine are varied. The graph of the average production rate versus r^{u1} is shown in Fig. 1. As r^{u1} increases, r^{u2} decreases in order to maintain e^u constant. It can be observed that the existing model tends to overestimate the throughput rate of the system (Th_GB) in comparison with the exact solution given by the proposed method (Th_LMT). When $r^{u1} = r^{u2} = r^u$ the two methods give the same results. This is because the two failure modes have the same repair rate: in this case there is no need to distinguish among them and the average failure does not introduce any approximation.

Also, as it can be seen in Fig. 2, the existing model tends to underestimate the average buffer level of the system.

Other cases obtained choosing different values for e^u and e^d have been analyzed and they have shown similar results. Therefore, we can conclude that the capability to distinguish among different failure modes can lead to significant improvements in the estimation of both the average production rate (up to 1.4%) and of the average buffer level (up to 16%). It should be noted that the percentages of improvement increase as the complexity of the building block considered, in terms of both number of failure modes and difference among their parameters, increases.

4. Conclusions

An analytical method for the performance evaluation of two-machine lines with random processing times, multiple failure modes and finite buffer capacity has been developed. The method is able to obtain the steady state probability of the Markov chain of the system and, as a consequence, the performance

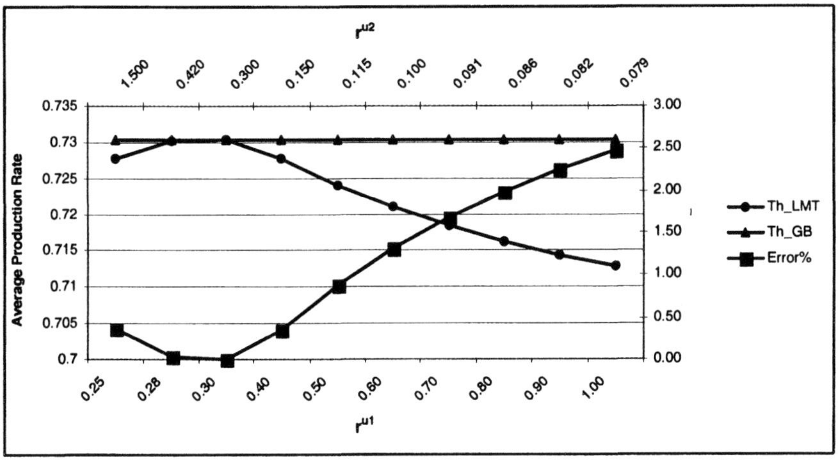

Figure 1. Average Production Rate versus r^{u1} (with $\mu_u = 1.2$, $p^u = 0.1$, $r^u = 0.3$, $\mu_d = 1.0$, $p^d = 0.04$, $r^d = 0.2$, $N = 10$)

measures in a compact form with a computational effort that depends only on the total number of failure modes considered and not on the capacity of the buffer. A comparison with an existing techniques has shown the improvement that can be achieved in performance estimation by distinguishing among different failure modes. The model presented can be used as a building block for the development of a new decomposition method for the analysis of larger systems, such as long transfer lines and assembly/disassembly networks.

Acknowledgments

The authors would like to thank Dr. Stanley B. Gershwin and D. Syrowicz for their useful suggestions. This research has been partially funded by *MIUR* under the projects "Modelli per la Pianificazione della Capacit Produttiva in Sistemi di Produzione Innovativi-Models for capacity planning in advanced manufacturing systems" and "Pianificazione Distribuita del Processo e della Produzione in Ambiente Manifatturiero-Distributed Process and Production Planning in Manufacturing Enterprise Networks".

Appendix: Transient States

The assumptions of the model imply that certain boundary states are transient, that is they have zero steady state probability and they cannot be reached from any state except possibly other transient states:

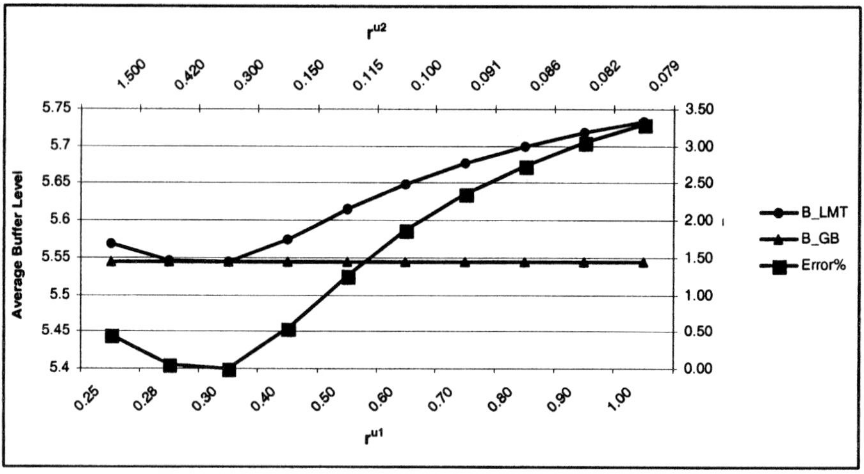

Figure 2. Average Buffer Level versus r^{u1} (with μ_u = 1.2, p^u = 0.1, r^u = 0.3, μ_d = 1.0, p^d = 0.04, r^d = 0.2, N = 10)

$$p(0,1,d_j) = p(0,u_i,d_j) = p(N,u_i,d_j) = p(N,u_i,1) = 0 \tag{1}$$

for $i = 1, ..., s$ and $j = 1, ..., t$.

Adding up equations (7) for $i = 1, ..., s$, combining with equations (13) and simplifying every term that can be simplified we obtain the following equation:

$$p\left(0,1,d_j\right)\left(\mu_u + r^{d_j}\right) + r^{d_j}\sum_{i=1}^{s} p\left(0,u_i,d_j\right) = 0, i = 1,...,s, j = 1,...,t \tag{2}$$

which in order to be satisfied implies:

$$p\left(0,1,d_j\right) = p\left(0,u_i,d_j\right) = 0, i = 1,...,s, j = 1,...,t \tag{3}$$

since the terms $(\mu_u + r^{d_j})$ and r^{d_j}, being parameters of exponential distributions, are positive quantities and the terms $p(0, 1, d_j)$ and $p(0, u_i, d_j)$ are also positive since steady state probabilities cannot be negative.

Similarly, adding up equations (8) for $j = 1, ..., t$, combining with equations (11) and simplifying every term that can be simplified we obtain the following equation:

$$\sum_{j=1}^{t} p\left(N,u_i,d_j\right) r^{u_i} + p\left(N,u_i,1\right)\left(r^{u_i} + \mu_d\right) = 0, i = 1,...,s, j = 1,...,t \tag{4}$$

which implies, for the same reasons above indicated, that:

$$p\left(N,u_i,d_j\right) = p\left(N,u_i,1\right) = 0, i = 1,...,s, j = 1,...,t \tag{5}$$

Appendix: Roots of the polynomial are real

The polynomial derived by the analysis of the internal equations of the two-machine line's Markov chain given by:

$$P(K) = \left[1 - \frac{K}{\mu_d}\left(1 + \sum_{j=1}^{t} \frac{p^{d_j}}{r^{d_j} - K}\right)\right]\left[1 + \frac{K}{\mu_u}\left(1 + \sum_{i=1}^{s} \frac{p^{u_i}}{r^{u_i} + K}\right)\right] - 1 \tag{1}$$

has all real roots.

Looking at the polynomial structure it can be observed that it is built up by two main terms depending exclusively on the parameters of the upstream and downstream machine. Given the nature of the poles of both these terms it is possible to determine also the nature and the position of the zeros, that is of the intersections of the function $P(K)$ with the K axis.

Indeed, the number of the poles characterizing any polynomial is equal to the total number of failure modes in the system, that is $s + t$ and they are given by:

$$\rho^{u_i} = -r^{u_i}, i = 1,...,s \tag{2}$$

for $K < 0$ and by:

$$\rho^{d_j} = r^{d_j}, j = 1,...,t \tag{3}$$

for $K > 0$.

More specifically, the poles related to the upstream machine are in the interval $(-\infty, 0)$ while those related to the downstream machine are in the interval $(0, +\infty)$.

In addition, for any $s \geq 1$ and $t \geq 1$ the following limits hold:

$$\lim_{K\to\rho^{u_i-}} P(K) = +\infty \qquad \lim_{K\to\rho^{u_i+}} P(K) = -\infty \tag{4}$$

$$\lim_{K\to\rho^{d_j-}} P(K) = -\infty \qquad \lim_{K\to\rho^{d_j+}} P(K) = +\infty \tag{5}$$

$$\lim_{K\to-\infty} P(K) = \lim_{K\to+\infty} P(K) = -\infty \tag{6}$$

On the basis of the properties pointed out, it is possible to do the following statements as a general pattern for any polynomial.

There is always a zero before the first pole of the upstream machine (the largest in absolute value on the negative part of K axis) and always a zero after the last pole of the downstream machine (the largest on the positive part of K axis). In both the cases, the function $P(K)$, coming from $-\infty$ and having to reach $+\infty$ as it gets closer to the border poles, is forced to cross the K axis thus determining the zero.

There is always a zero for each pair of consecutive pole except in the case of the pair given by the last pole of the upstream machine and the first pole of the downstream machine. This is because in each of the intervals delimited by two consecutive poles the function $P(K)$ switches from $-\infty$ to $+\infty$ or viceversa. As a result, the function has necessarily to cross the K axis.

Finally, in the interval delimited by the last pole of the upstream machine and the first pole of the downstream machine there are always two roots. One is the known root always equal to zero while the other can vary from zero (this is the case in which zero is a double root that occurs when the two machines presents the same production rate in isolation) to the first pole of the downstream machine.

On the basis of the analysis carried out we can conclude that the number of zeros of $P(K)$ is always equal to $R = s + t + 2$ which is exactly the degree of the polynomial. Therefore all the roots are real.

Appendix: Solution of Internal Equations Implies $C_R = 0$

Writing the equations (19) in the following form:

$$D_j\,(p^{u_i} - r^{u_i} U_i) = -U_i\left(p^{d_j} - r^{d_j} D_j\right), i = 1, ..., s, j = 1, ..., t \tag{1}$$

and addinsg up for $i = 1, ..., s$ and for $j = 1, ..., t$ we obtain:

$$\sum_{j=1}^{t} D_j \sum_{i=1}^{s} (p^{u_i} - r^{u_i} U_i) = -\sum_{i=1}^{s} U_i \sum_{j=1}^{t} \left(p^{d_j} - r^{d_j} D_j\right) \tag{2}$$

which defining:

$$\sum_{i=1}^{s} (p^{u_i} - r^{u_i} U_i) = \gamma \tag{3}$$

and

$$\sum_{j=1}^{t} \left(p^{d_j} - r^{d_j} D_j\right) = \beta \tag{4}$$

assumes the following form:

$$\gamma \sum_{j=1}^{t} D_j = -\beta \sum_{i=1}^{s} U_i \tag{5}$$

Then, substituting equations (19) in equations (20), adding up for $i = 1, ..., s$ and multiplying left and right side for U_i, we obtain:

$$\mu_d (X - 1) \sum_{i=1}^{s} U_i + \sum_{i=1}^{s} (p^{u_i} - r^{u_i} U_i) \left(1 + \sum_{j=1}^{t} D_j\right) = 0 \tag{6}$$

while substituting the same equation in equation (21), adding up for $j = 1, ..., t$ and multiplying left and right side for D_j, we have:

$$\mu_u \left(\frac{1}{X} - 1\right) \sum_{j=1}^{t} D_j + \sum_{j=1}^{t} \left(p^{d_j} - r^{d_j} D_j\right) \left(1 + \sum_{i=1}^{s} U_i\right) = 0 \tag{7}$$

At this point, obtaining X from equation (6) after introducing equation (3):

$$X = 1 - \frac{\gamma\left(1 + \sum_{j=1}^{t} D_j\right)}{\mu_d \sum_{i=1}^{s} U_i} \tag{8}$$

and $\frac{1}{X}$ from equation (7) after introducing equation (4):

$$\frac{1}{X} = 1 - \frac{\beta\left(1 + \sum_{i=1}^{s} U_i\right)}{\mu_u \sum_{j=1}^{t} D_j} \tag{9}$$

and combining the two expression found out for X we obtain:

$$1 = \left[1 - \frac{\gamma\left(1 + \sum_{j=1}^{t} D_j\right)}{\mu_d \sum_{i=1}^{s} U_i}\right]\left[1 - \frac{\beta\left(1 + \sum_{i=1}^{s} U_i\right)}{\mu_u \sum_{j=1}^{t} D_j}\right] \tag{10}$$

Putting in system equations (5) and (10) we obtain the following pair of solutions:

$$\left\{ \begin{array}{l} \gamma_1 = 0 \\ \beta_1 = 0 \end{array} \right\} \tag{11}$$

$$\left\{ \begin{array}{l} \gamma_2 = \dfrac{\mu_d \sum_{i=1}^{s} U_i\left(1+\sum_{i=1}^{s} U_i\right) - \mu_u \sum_{i=1}^{s} U_i\left(1+\sum_{j=1}^{t} D_j\right)}{\left(1+\sum_{j=1}^{t} D_j\right)\left(1+\sum_{i=1}^{s} U_i\right)} \\ \beta_2 = \dfrac{\mu_u \sum_{j=1}^{t} D_j\left(1+\sum_{j=1}^{t} D_j\right) - \mu_d \sum_{j=1}^{t} D_j\left(1+\sum_{i=1}^{s} U_i\right)}{\left(1+\sum_{j=1}^{t} D_j\right)\left(1+\sum_{i=1}^{s} U_i\right)} \end{array} \right\} \tag{12}$$

Substituting the pair of solutions into equation (8) or (9) we have:

$$X_R - 1 \tag{13}$$

$$X_r = \frac{\mu_u\left(1 + \sum_{j=1}^{t} D_j\right)}{\mu_d\left(1 + \sum_{i=1}^{s} U_i\right)}$$

which implies:

$$\mu_u \left(1 + \sum_{j=1}^{t} D_{j,r}\right) - \mu_d X_r \left(1 + \sum_{i=1}^{s} U_{i,r}\right) = 0 \forall r \neq R \tag{14}$$

Therefore, equations (45) and (46) are both satisfied by setting $C_R = 0$.

References

[1] J. Buzacott, The effect of station breakdown and random processing times on the capacity of flow lines. *AIIE Transactions*, 4: 308-312, 1972.

[2] Dallery Y., David R. and Xie X.L. An Efficient Algorithm for Analysis of Transfer Lines With Unreliable Machines and Finite Buffer. *IIE Transactions*, 20(3): 280-283, 1988.

[3] Y. Dallery and H. Le Bihan, An Improved Decomposition method for the Analysis of Production Lines With Unreliable Machines and Finite Buffer. *International Journal of Production Research*, 93:(2000) 265-297.

[4] Y. Dallery, R. David and X.L. Xie, Approximate Analysis of Transfer Lines with Unreliable Machines and Finite Buffers, *IEEE Transactions on Automatic Control*, 34:943-953, 1989.

[5] Y. Dallery and S.B. Gershwin, Manufacturing Flow Line Systems: A Review of Models and Analytical Results,*Queuing System Theory and Applications*, Special Issue on Queuing Model of Manufacturing System, 1992, 12(1-2):3-94.

[6] S.B. Gershwin, An Efficient Decomposition Algorithm for the Approximate Evaluation of Tandem Queueing Systems with Finite Buffers. In H.G. Perros and T. Altiok, editors, Queueing Network with Blocking, pages 127-146, North-Holland, Amsterdam, 1989.

[7] S.B. Gershwin *Manufacturing Systems Engineering* PTR Prentice Hall, Inc., New Jersey, 1994.

[8] S.B. Gershwin and O. Berman, Analysis of Transfer Lines Consisting of Two Unreliable Machines with Random Processing Times and Finite Storage Buffers. *AIIE Transactions*, 13(1), 1981.

[9] Hong, Y. Behavior Analysis of an Unreliable Two-Machine Flow line System with Random Processing Times, POSTECH IE-TR-88-03, 1988.

[10] Hong Y., Glassey C.R. and Seong D. The Analysis of a Production Line with Unreliable Machines and Random Processing Times. *IIE Transaction*, 24(1): 77-83, 1993.

[11] Levantesi R., Matta A. and Tolio T. Exponential Two-Machine Lines with Multiple Failure Modes and Finite Buffer Capacity. Proceeding of the Second International Aegean Conference on the "Analysis and Modeling of Manufacturing Systems", Tinos Island, Greece, May 1999.

[12] Levantesi R., Matta A. and Tolio T. A Decomposition Method for the Perfomance Evaluation of Production Lines with Random Processing Times, Multiple Failure Modes and Finite Buffer Capacity. Proceeding of Second International Aegean Conference on the "Analysis and Modeling of Manufacturing Systems", Tinos Island, Greece, May 1999.

[13] Tolio T., Gershwin S.B. and Matta A., Analysis of two-machine lines with multiple failure modes, IIE Transaction, 34(1): 51-62, 2002.

[14] Tolio T. and Matta A. A Method for Perfomance Evaluation of Automated Flow Lines. *Annals of CIRP* Vol. 47/1, 1998.

Chapter 9

PERFORMANCE EVALUATION OF PRODUCTION LINES WITH RANDOM PROCESSING TIMES, MULTIPLE FAILURE MODES AND FINITE BUFFER CAPACITY — PART II: THE DECOMPOSITION

R. Levantesi
Dipartimento di Meccanica, Politecnico di Milano,
via Bonardi 9, 20133 Milano, Italy
raniero.levantesi@mecc.polimi.it

A. Matta
Dipartimento di Meccanica, Politecnico di Milano,
via Bonardi 9, 20133 Milano, Italy
andrea.matta@mecc.polimi.it

T. Tolio
Dipartimento di Meccanica, Politecnico di Milano,
via Bonardi 9, 20133 Milano, Italy
tullio.tolio@mecc.polimi.it

Abstract The paper presents a decomposition method for the performance evaluation of production lines with exponential processing times, multiple failure modes and finite buffer capacity. The key feature of the proposed method is the capability to model machines subject to different types of failures, each one characterized by exponentially distributed time to failure and time to repair. This feature, being the method based on the approximation of the behavior of a K-machine transfer line by the behavior of K-1 two-machine lines, or building blocks, makes it possible to consider in each building block in which the original line is decomposed real failure and virtual failure modes. While real failures are directly connected with the reliability of the machines of the original line, virtual failures are introduced to model the interruption of flow that take place in the system as a result of machine disruption propagation. Simulation and numerical experiments are carried out to show the performance of the new method in comparison with existing ones. The method can be extended to the analysis of larger open systems, such as assembly/disassembly networks, and systems with loops.

Keywords: exponential lines, decomposition, multiple failure modes

1. Introduction

In this paper an efficient analytical method for the approximate evaluation of production lines with random processing times, multiple failure modes and finite buffer capacity is developed.

In real systems machines can fail in different ways with more or less severe consequences. For instance, a given station can be able to complete a certain operation by means of different devices (feeders, pick & place manipulators, transfer devices etc.) that have different reliability parameters and can cause the failure of the entire station. All these kinds of failures happen with different frequencies and interrupt the station for periods of different lengths. Therefore, in order to obtain a good estimation of the performance measures of a K-machine line, it is important to be able to distinguish among the different breakdown that each station can undergo.

Existing techniques [4] are not able to deal properly with the multiple failure issue and group all the different causes that can interrupt machine operation in an *average failure* by means of a first [2], [9] or second order [5] approximation. This approximation may not be very accurate when the modes differs significantly.

In addition, the multiple failure feature provides, at decomposition level, a very accurate tool to represent the interruptions of flow that can take place inside the system as a result of machine disruption propagation. Indeed, it makes it possible to distinguish among real failures, which are directly connected with the reliability of the physical machines of the original line, and virtual failures which are introduced on the various two-machine lines to take into account the impact of blocking and starvation.

Even limiting the analysis to transfer lines with machines subject to single failure modes, a decomposition scheme based on a multiple failure approach provides deeper insight about the behavior of the system. Indeed, it allows the identification and the distinction of all the different reasons that can lead to the interruption of machine operations in the system. This, in turn, allows the estimation of the impact that each system disruption has on the idle time of a specific machine. Finally, this type of insight has proven to be very useful in the analysis of more general networks, such as assembly/disassembly networks [13] and closed loop systems [8],[14].

The proposed method is based on the decomposition technique originally proposed by Gershwin [6], that is on the approximation of the behavior of a K-machine line by the behavior of $K - 1$ two-machine lines, or building blocks, which are linked together by means of a proper set of equations, called *decomposition equations*. Specifically, it represents an extension to the case of exponential lines of the multiple failures approach developed by Tolio et al. [15] and by Tolio and Matta [16] for the analysis of transfer lines with deterministic processing times.

2. Model Description and Assumptions

This method focuses on transfer lines with unreliable machines separated by buffers of finite capacity. Raw parts come from outside the system to be processed on the first machine, then move to the first buffer, then visit the other machines and buffers in sequence until they reach the last machine and leave the system.

Each machine requires a random amount of time to process a workpiece and, when operational, it can be affected by a set of different and random failures. This means that operation dependent failures, *ODFs*, are assumed. If the machine fails in a certain mode, immediately starts the repair process that also requires a random amount of time. Therefore, each machine is characterized by a set of random variables exponentially distributed, namely the processing time and, for each failure mode, the mean time between failures and the mean time to repair.

When a machine finishes to process a part and the downstream buffer is full the machine is said to be *"blocked"* while if the upstream buffer is empty the machine is said to be *"starved"*. Blocking and starvation mainly occur because of failures in the portion of the line upstream or downstream the machine under examination. However, they may also happen due to the difference in the processing times of the various machines. This is because when a certain machine inside the line takes a long time to process a workpiece, the upstream buffer tends to become full while the downstream buffer tends to become empty. If the phenomenon persists, blocking propagates upstream while starvation propagates downstream.

Finally, we assume there are always raw parts available at the input of the system (i.e. the first machines is never starved) and space available at the output of the system (i.e. the last machine is never blocked).

3. Machine Behavior

A machine M_i can be operational or failed in a certain mode. When a machine is operational, it processes a workpiece if it is not starved or blocked and it goes on working until either it completes the piece or one of the different failures occurs, whichever happen first. Either events can take place during the time interval $(t, t + \delta t)$ with probability approximately $\mu_i \delta t$ or $p_{i,f}\delta t$, for small δt. If a machine is failed in mode f at time t, it can be repaired during the time interval $(t, t + \delta t)$ with probability approximately $r_{i,f}\delta t$.

Since blocked or starved machines are not operating on a piece, given the assumption of operation dependant failures, they are not subject to failures. For the same reason, an already failed machine cannot experience a different down mode before being repaired.

4. Performance Measures

Given the failure and the repair rates for each failure mode of machine M_i it is possible to obtain the *efficiency in isolation*, that is the fraction of time the machine would be productive if it was not subject to blocking and starvation [1], [16]:

$$e_i = \frac{1}{1 + \sum_{i=1}^{F_i} \frac{p_{i,f}}{r_{i,f}}} \tag{1}$$

The *efficiency of the machine inside the line*, that is the fraction of time in which the machine effectively processes workpieces is given by:

$$E_i = e_i(1 - Ps_i - Pb_i) \tag{2}$$

that takes in account the losses caused by blocking (Pb_i) and starvation (Ps_i).

This equation is known in literature as *flow rate-idle time* and it represents an approximate relationship because it is based on the assumption that a machine cannot be starved and blocked at the same time [7].

On the basis of the efficiency it is possible to evaluate the production rate of machine M_i by mean of the relation:

$$P_i = \mu_i E_i \tag{3}$$

5. Outline of the Method

The original transfer line L made of K machines is decomposed, as shown in Fig. 1, into a set of two-machine lines $l(i)^*$, for $i = 1, ..., K - 1$. Each two-machine line is characterized by a buffer $B(i)$ of the same size of the corresponding buffer B_i of the original line, an upstream pseudo machine $M^u(i)$ and a downstream pseudo machine $M^d(i)$. Pseudo machines model the portion of the original line upstream and downstream of buffer B_i. In other words, the

*Integer subscripts refer to machine and buffer of the original line L while symbols followed by an integer in parenthesis refer to the two-machine line indicated by the parenthesis.

upstream machine $M^u(i)$ models how workpieces are transferred into buffer B_i while the downstream machine $M^d(i)$ models how workpieces are taken from buffer B_i.

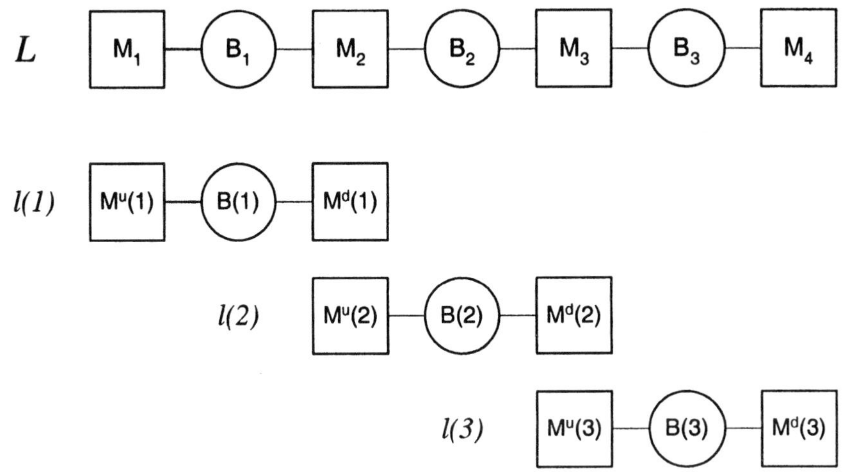

Figure 1. Decomposition of a 4-machine transfer line system

The flow of parts into a buffer of the original line can be interrupted because the upstream machine fails in a certain mode or it takes a long time to process a part or because it becomes starved (this is due, in turn, to a failure or to a long processing time on one of the upstream machines). However, the upstream machine of the corresponding two-machine, being the first machine of the line, cannot be starved. Therefore, to represent the starvation of the upstream machine some *"virtual failures"* are introduced on this machine.

Similarly, the need to represent the blocking of the downstream machine of a two-machine line leads to the introduction of some *"virtual failures"* also on the downstream machine.

Therefore, the key of the decomposition method is to find out the unknown parameters of the virtual modes which are added to pseudo machines so that the flow of parts in the buffer of each two-machine line closely matches the flow of parts in the corresponding buffer of the original line.

5.1 Pseudo Machine Parameters

On the basis of the analysis carried out so far, the upstream pseudo machine $M^u(i)$ of the two-machine line $l(i)$, which models the transfer of material into buffer B_i, results characterized by the following parameters:

- *processing rate* $\mu^u(i)$, representing the rate at which the material is transferred into buffer B_i by $M^u(i)$ when it is processing material.
- *real failure modes* corresponding to the breakdown of the physical machine M_i, with failure and repair rates respectively equal to $p_{i,f}$ and $r_{i,f}$, for $f = 1, ..., F_i$.
- *virtual failure modes* modelling the starvation of the physical machine M_i due to the breakdown of an upstream machine M_j, with failure and repair rates respectively given by $p^u_{j,f}(i)$ and $r^u_{j,f}(i)$, for $j = 1, ..., i-1$ and $f = 1, ..., F_j$.
- *virtual failure modes* representing the starvation of the physical machine M_i because of a long processing time on the upstream machine M_j, with failure and repair rates respectively referred to as $p^u_{j,0}(i)$ and $r^u_{j,0}(i)$, for $j = 1, ..., i-1$.

Similarly, the downstream pseudo machine $M^d(i-1)$ of two-machine line $l(i-1)$, which models the removal of material from buffer B_{i-1}, is characterized by the following parameters:

- *processing rate* $\mu^d(i-1)$, representing the rate at which the material is removed from the buffer B_{i-1} by $M^d(i-1)$ when it is processing material.
- *real failure modes* corresponding to the breakdown of the physical machine M_i, with failure and repair rates respectively equal to $p_{i,f}$ and $r_{i,f}$, for $f = 1, ..., F_i$.
- *virtual failure modes* modelling the blocking of the physical machine M_i due to the breakdown of a downstream machine M_j, with failure and repair rates respectively given by $p^d_{j,f}(i-1)$ and $r^d_{j,f}(i-1)$, for $j = i+1, ..., K$ and $f = 1, ..., F_j$.
- *virtual failure modes* representing the blocking of the physical machine M_i because of a long processing time on the downstream machine M_j, with failure and repair rates respectively referred to as $p^d_{j,0}(i-1)$ and $r^d_{j,0}(i-1)$, for $j = i+1, ..., K$.

Given the parameters of the upstream and downstream pseudo machines, it is possible to obtain, by means of the exact analytical solution provided in

Part I, an evaluation of the two-machine line $l(i)$ performance. Among the various performance measures, of particular interest for the development of decomposition equations are the average production rate of the line, $P(i)$, the efficiency $E^u(i)$ and $E^d(i)$ of the upstream and downstream pseudo machine, the steady state probability $S_{j,f}(i)$ of downstream pseudo machine of being starved due to the failure mode j, f, for $j = 1,...,i$ and $f = 1,...,F_j$ and, finally, the steady state probability $B_{j,f}(i)$ of upstream pseudo machine of being blocked due to failure mode j, f, for $j = i+1,...,K$ and $f = 1,...,F_j$.

6. Decomposition Equations

Since it is assumed that pseudo machines of each two-machine line are characterized by a set of random variables exponentially distributed, that are the processing time and, for each failure mode (real and virtual), the time to failure and the time to repair, the next step consists in finding out the unknown parameters of these distributions by means of a proper set of equation.

The processing rates and the parameters of the real failures of the pseudo machines are known, since they are exactly the same of the corresponding machines in the original line. Conversely, for the parameters of the virtual failures decomposition equations must be derived.

6.1 Pseudo Machine States

Before addressing the derivation of the decomposition equations, it is necessary to indicate the notation we adopt to refer to the probability of finding the pseudo machines in any of the states they can enter.

More specifically, the probability of the upstream pseudo machine $M^u(i)$ of being processing workpieces is $W^u(i)$, of being down in a real mode f is $D^u_f(i)$, for $f = 1,...,F_i$, of being down in a virtual mode j, f is $V^u_{j,f}(i)$, for $j = 1,...,i-1$ and $f = 1,...,F_j$, and, finally, of being down in virtual mode $j,0$ is $V^u_{j,0}(i)$, for $j = 1,...,i-1$. In addition, $M^u(i)$ can be blocked $B_{j,f}(i)$ because of the failure mode j, f of $M^d(i)$, for $j = i+1,...,K$ and $f = 1,...,F_j$, and blocked $B_{j,0}(i)$ because of the mode $j,0$ of $M^d(i)$, for $j = i+1,...,K$.

Similarly, the probability of the downstream pseudo machine $M^d(i-1)$ of being processing a workpiece is given by $W^d(i-1)$, of being down in real mode f by $D^d_f(i-1)$, for $f = 1,...,F_i$, of being down in virtual mode j, f by $V^d_{j,f}(i)$, for $j = i+1,...,K$ and $f = 1,...,F_j$, and of being down in virtual mode $j,0$ by $V^d_{j,0}(i)$, for $j = i+1,...,K$. The machine can then be starved $S_{j,f}(i)$ because of the failure mode j, f of $M^u(i-1)$, for $j = 1,...,i-1$ and $f = 1,...,F_j$, and starved due to failure mode $j,0$ of $M^u(i-1)$ $S_{j,0}(i)$, for $j = 1,...,i-1$.

6.2 Resumption of Flow Equations

The first set of equations discussed are the *Resumption of Flow Equations* that provide the repair rates from the various virtual modes by which the pseudo machines can be affected. In other terms, we have to determine how long does it take for a pseudo machine to be recovered from each interruption of flow it can experience. However, these parameters do not represent effective unknowns.

Indeed, the period in which a machine is under repair does not depend on the states of other machines or buffers. Specifically, the repair rate from the virtual mode j, f of a pseudo-machine corresponds exactly to the repair rate $r_{j,f}$ of the physical machine M_j from the down mode f. This is because the recovery from a certain failure implies the removal of the cause that originated the interruption of flow and the subsequent propagation of blocking and starvation.

Concerning the repair rate from virtual mode $j, 0$ of a pseudo-machines $M^u(i)$ or $M^d(i-1)$, it can be seen that this is just the processing rate μ_j of the corresponding machine M_j of the original line. This is because the probability that machine M_i leaves a condition of starvation (blocking) due to a long processing time on the upstream (downstream) machine M_j corresponds exactly to the probability that machine M_j completes the part it is processing.

6.3 Interruption of Flow Equations

The second set of equations considered are the *Interruption of Flow Equations* that provide the failure rates of the various virtual modes that affect the pseudo machines $M^u(i)$ and $M^d(i-1)$. In other words, we have to establish how often each interruption of flow take place in each building block.

6.3.1 Preliminary Remarks. To obtain this set of equations it is necessary to observe that every time a machine breaks down in a certain mode it gets repaired. On the basis of this observation, it is possible to show, extending to the case of multiple failure modes the proof given by Gershwin in [7], that *failure frequency equals repair frequency for each failure mode, real or virtual.* As a result, we can write for $M^u(i)$ the following equations:

$$p_{i,f} \cdot W^u(i) = r_{i,f} \cdot D^u_f(i), f = 1, ..., F_i \tag{4}$$

$$p^u_{j,f}(i) \cdot W^u(i) = r_{j,f} \cdot V^u_{j,f}(i), j = 1, ..., i-1, f = 1, ..., F_j \tag{5}$$

$$p^u_{j,0}(i) \cdot W^u(i) = \mu_j \cdot V^u_{j,0}(i), j = 1, ..., i-1 \tag{6}$$

Similarly, for $M^d(i-1)$ the following equations hold:

$$p_{i,f} \cdot W^d(i-1) = r_{i,f} \cdot D_f^d(i-1), f = 1, ..., F_i \tag{7}$$

$$p_{j,f}^d(i-1) \cdot W^d(i-1) = r_{j,f} \cdot V_{j,f}^d(i-1), j = i+1, ..., K, f = 1, ..., F_j \tag{8}$$

$$p_{j,0}^d(i-1) \cdot W^d(i-1) = \mu_j \cdot V_{j,0}^d(i-1), j = i+1, ..., K \tag{9}$$

Since the sum on all possible states of $M^u(i)$ must be equal to 1, we can write the following normalization equation:

$$W^u(i) + \sum_{f=1}^{Fi} D_f^u(i) + \sum_{j=1}^{i-1} + \sum_{f=1}^{F_j} V_{j,f}^u(i) + \sum_{j=1}^{i-1} V_{j,0}^u(i) + \\ + \sum_{j=i+1}^{K} \sum_{f=1}^{F_j} B_{j,f}(i) + \sum_{j=i+1}^{K} B_{j,0}(i) = 1 \tag{10}$$

Equation (1), which defines the efficiency in isolation of a machine, can be extended to consider also virtual failures, thus obtaining for the upstream pseudo machine $M^u(i)$ of the two-machine line $l(i)$:

$$e^u(i) = \frac{1}{1 + \sum_{f=1}^{Fi} \frac{p_{i,f}}{r_{i,f}} + \sum_{j=1}^{i-1} \sum_{f=1}^{F_j} \frac{p_{j,f}^u(i)}{r_{j,f}} + \sum_{j=1}^{i-1} \frac{p_{j,0}^u(i)}{\mu_j}} \tag{11}$$

Substituting equations (4), (5), (6) and (11) into equation (10), after some manipulations we obtain:

$$W^u(i) = e^u(i)(1 - Pb(i)) \tag{12}$$

which corresponds to the flow rate-idle time equation (2), since $M^u(i)$, being the first machine of the line $l(i)$, cannot be starved. Therefore, we can conclude:

$$W^u(i) = E^u(i) \tag{13}$$

Similarly, since the sum on all possible states of $M^d(i-1)$ must be equal to 1, we can write:

$$W^d(i-1) + \sum_{f=1}^{Fi} D_f^d(i-1) + \sum_{j=i+1}^{K} \sum_{f=1}^{F_j} V_{j,f}^d(i-1) + \\ + \sum_{j=i+1}^{K} V_{j,0}^d(i-1) + \sum_{j=1}^{i-1} \sum_{f=1}^{F_j} S_{j,f}(i-1) + \sum_{j=1}^{i-1} S_{j,0}(i-1) = 0 \tag{14}$$

Introducing the efficiency in isolation also for the pseudo downstream machine $M^d(i-1)$:

$$e^d(i-1) = \frac{1}{\sum_{f=1}^{Fi} \frac{p_{i,f}}{r_{i,f}} + \sum_{j=i+1}^{K} \sum_{f=1}^{F_j} \frac{p^d_{j,f}(i-1)}{r_{j,f}} + \sum_{j=i+1}^{K} \frac{p^d_{j,0}(i-1)}{\mu_j}} \quad (15)$$

and substituting equations (7), (8), (9) and (15) into equation (14), after some manipulations we obtain:

$$W^d(i-1) = e^d(i-1)(1 - Ps(i-1)) \quad (16)$$

which represents the flow rate-idle time equation (2) since $M^d(i-1)$, being the last machine of the line $l(i)$, cannot be blocked. Therefore, we can state:

$$W^d\,(i-1) = E^d(i-1) \quad (17)$$

6.3.2 Solution Methodology. The virtual failure modes of $M^u(i)$ have been introduced to mimic the different causes of starvation of the physical machine M_i. Therefore, the steady state probability of $M^u(i)$ of being in a specific virtual down state must be equal to the corresponding steady state probability of the starvation of the previous two-machine line $l(i-1)$:

$$V^u_{j,f}(i) = S_{j,f}(i-1),\, j = 1, ..., i-1,\, f = 1, ..., F_j \quad (18)$$

$$V^u_{j,0}\,(i) = S_{j,0}\,(i-1)\,,\, j = 1, ..., i-1 \quad (19)$$

Similarly, the virtual failure modes of $M^d(i-1)$ have been introduced to mimic the different causes of blocking of the physical machine M_i. Therefore, the steady state probability of $M^d(i-1)$ of being in a given virtual down state must be equal to the corresponding steady state probability of blocking of the following two-machine line $l(i+1)$:

$$V^d_{j,f}(i-1) = B_{j,f}(i),\, j = i+1, ..., K,\, f = 1, ..., F_j \quad (20)$$

$$V^d_{j,0}\,(i-1) = B_{j,0}\,(i)\,,\, j = i+1, ..., K \quad (21)$$

At this point, substituting equations (18) and (19), respectively, into equations (5) and (6), and taking in account equation (13), we obtain the following expressions for the virtual failure rates of $M^u(i)$:

$$p^u_{j,f}\,(i) = \frac{S_{j,f}(i-1)}{E^u\,(i)} \cdot r_{j,f},\, j = 1, ..., i-1,\, f = 1, ..., F_j \quad (22)$$

$$p^u_{j,0}(i) = \frac{S_{j,0}(i-1)}{E^u(i)} \cdot \mu_j, j = 1, ..., i-1 \tag{23}$$

while substituting equations (20) and (21) into equations (8) and (9), respectively, and taking in account equation (17), we obtain the following expressions for virtual failure rates of $M^d(i-1)$:

$$p^d_{j,f}(i-1) = \frac{B_{j,f}(i)}{E^d(i-1)} \cdot r_{j,f}, j = i+1, ..., K, f = 1, ..., F_j \tag{24}$$

$$p^d_{j,0}(i-1) = \frac{B_{j,0}(i)(i)}{E^d(i-1)} \cdot \mu_j, j = i+1, ..., K \tag{25}$$

6.4 Algorithm

On the basis of the analysis carried out, an efficient algorithm to solve the decomposition equations has been obtained. The algorithm is based on the DDX scheme [3]) and consists of the following steps:

(a) Initialization: for each two-machine line $l(i)$ in which the original line L is decomposed, the known parameters are introduced and an initial guess for the unknown parameters is made.†

$$\mu^u(i) = \mu_i \tag{26}$$

$$p^u_{i,f}(i) = p_{i,f}, f = 1, ..., F_i \tag{27}$$

$$r^u_{i,f}(i) = r_{i,f}, f = 1, ..., F_i \tag{28}$$

$$p^u_{j,f}(i) = \alpha, j = 1, ..., i-1, f = 1, ..., F_j \tag{29}$$

$$r^u_{j,f}(i) = r_{j,f}, j = 1, ..., i-1, f = 1, ..., F_j \tag{30}$$

$$p^u_{j,0}(i) = \alpha, j = 1, ..., i-1 \tag{31}$$

$$r^u_{j,0}(i) = \mu_j, j = 1, ..., i-1 \tag{32}$$

†In the following we define α this initial guess (for instance α may be set to a value like 0, 01.)

$$\mu^d(i-1) = \mu_i \tag{33}$$

$$p^d_{i,f}(i-1) = p_{i,f}, f = 1, ..., F_i \tag{34}$$

$$r^d_{i,f}(i-1) = r_{i,f}, f = 1, ..., F_i \tag{35}$$

$$p^d_{j,f}(i-1) = \alpha, j = i+1, ..., K, f = 1, ..., F_j \tag{36}$$

$$r^d_{j,f}(i-1) = r_{j,f}, j = i+1, ..., K, f = 1, ..., F_j \tag{37}$$

$$p^d_{j,0}(i-1) = \alpha, j = i+1, ..., K \tag{38}$$

$$r^d_{j,0}(i-1) = \mu_j, j = i+1, ..., K \tag{39}$$

(b) For $i = 2, ..., k-1$: evaluation of the two-machine $l(i)$ using for the parameters of $M^d(i)$ the most recent values from the previous iteration while updating the parameters of $M^u(i)$ with the results of the evaluation of the upstream building block $l(i-1)$:

$$p^u_{j,f}(i) = \frac{S_{j,f}(i-1)}{E^u(i)} \cdot r_{j,f}, j = 1, ..., i-1, f = 1, ..., F_j \tag{40}$$

$$p^u_{j,0}(i) = \frac{S_{j,0}(i-1)}{E^u(i)} \cdot \mu_j, j = 1, ..., i-1 \tag{41}$$

(c) For $i = K-1, ..., 2$: evaluation of the two-machine $l(i-1)$ using for the parameters of $M^u(i-1)$ the most recent values from the previous iteration while updating the parameters of $M^d(i-1)$ with the results of the evaluation of the downstream building block $l(i)$:

$$p^d_{j,f}(i-1) = \frac{B_{j,f}(i)}{E^d(i-1)} \cdot r_{j,f}, j = i+1, ..., K, f = 1, ..., F_j \tag{42}$$

$$p^d_{j,0}(i-1) = \frac{B_{j,0}(i)}{E^d(i-1)} \cdot \mu_j, j = i+1, ..., K \tag{43}$$

(d) Steps 2 and 3 are repeated alternatively until the termination condition is satisfied, that is until the equations converge. The measure of conver-

Case 1							Case 2						
M	*m*	p	r	C	-	Th.	M	*m*	p	r	C	-	Th.
1	1,5	0,03	0,3	8	Sim.	0,9489	1	1,5	0,03	0,3	8	Sim.	0,8026
2	1,2	0,05	0,5	4	Ger.	0,9599	2	1,2	0,05	0,5	4	Ger.	0,8240
3	1,8	0,01	0,2	6	Error	1,16%	3	1,8	0,01	0,2	6	Error	2,67%
4	1,6	0,06	0,4	8	LMT	0,9416	4	1,6	0,06	0,4	8	LMT	0,8073
5	2,0	0,02	0,1	-	Error	-0,77%	5	1,5	0,06	0,1	-	Error	0,59%
Case 3							Case 4						
M	*m*	p	r	C	-	Th.	M	*m*	p	r	C	-	Th.
1	1,5	0,03	0,3	8	Sim.	1,0786	1	1,2	0,03	0,1	8	Sim.	0,6716
2	1,6	0,02	0,4	4	Ger.	1,0886	2	1,0	0,05	0,3	8	Ger.	0,6999
3	1,8	0,01	0,2	6	Error	0,93%	3	1,4	0,01	0,2	8	Error	4,21%
4	1,6	0,06	0,4	8	LMT	1,0678	4	1,2	0,06	0,4	8	LMT	0,6838
5	2,0	0,02	0,1	-	Error	-1,00%	5	1,0	0,02	0,09	-	Error	1,82%
Case 5							Case 6						
M	*m*	p	r	C	-	Th.	M	*m*	p	r	C	-	Th.
1	1,2	0,03	0,01	8	Sim.	0,1817	1	1,2	0,03	0,01	12	Sim.	0,1932
2	1,0	0,05	0,03	8	Ger.	0,1827	2	1,0	0,05	0,03	12	Ger.	0,1953
3	1,4	0,01	0,02	8	Error	0,55%	3	1,4	0,01	0,02	12	Error	1,09%
4	1,2	0,06	0,04	8	LMT	0,182	4	1,2	0,96	0,04	12	LMT	0,1942
5	1,0	0,02	0,09	-	Error	0,17%	5	1,0	0,02	0,09	-	Error	0,52%

Table 1. Five machine lines with one failure mode: parameters and throughput results

gence is given by the maximum relative change in any of the unknown parameters.

7. Numerical Results

In order to show the accuracy of the new decomposition method developed (method *LMT*) a set of numerical tests has been carried out comparing the analytical results with those obtained applying the method developed by Gershwin [7] (method Ger.) and those obtained running simulation experiments (Sim.).

For each simulation experiment, to ensure statistical significance 20 replications have been performed. Each run entails a simulation period of 1.000.000 time units preceded by a warm-up period of 100.000 time units. The first set of experiments considers transfer lines with machines subject to one failure mode to comply with the hypotheses of the existing method. Conversely, the second set of experiments considers machines characterized by two failure modes rather different in terms of repair rates.

In the case of machines characterized by one failure mode the results provided by the two methods are substantially equivalent even if the proposed method seems to be more precise in throughput and buffer level estimation as shown in the examples reported in Table 1 and 2.

In the second set of experiments the capability of the algorithm to deal with machines affected by multiple failure modes has been tested. Specifically, different settings of failure mode parameters have been considered, with repair rates of different orders of magnitude. In this case the existing method, being forced to treat as an average failure the different causes of machine interruption, that is effective machine failure as well as occurring of blocking and starvation, can lead to non-negligible errors in performance estimation with respect to the simulation results (up to 10% as regard to the throughput as shown in Table 3). On the contrary, the proposed method, being able to distinguish among real failure modes and virtual failure modes provides fairly accurate results.

Concerning the speed of execution, the number of iterations is not a good indicator. As a general pattern, we have observed that the proposed technique implies less iterations to converge but takes a little bit longer for each iteration. This is because much of the method complexity is embedded in the two-machine line while the decomposition equations are quite simple. This is not the case of the existing technique which shows an opposite pattern with a rather simple building block but very complex decomposition equations. However, the execution times are comparable since both the methods require few seconds to handle the cases reported in this section.

The building block complexity could appear as a limitation toward the application of the proposed technique to the analysis of larger systems, for instance systems made up by 10 stages or more. Actually, most of the problems of handling large systems are due to the grow of the polynomial degree that makes it difficult to find out the polynomial roots. However, it has been possible to identify the intervals in which each root can be found (see Part I, Appendix B). This, in turn, will simplify significantly the problem of the root searching making the technique able to deal with real size cases.

8. Conclusions

A new decomposition method for the performance evaluation of production lines with exponential processing times, multiple failure modes and finite buffer capacity has been developed.

The main feature of the proposed approach over existing techniques is the capability to take into account machines subject to multiple failure modes. In this way the different types of breakdowns that machines normally undergo in real production systems can be modelled explicitly. In addition, this feature

Case 1					Case 2				
-	B_1	B_2	B_3	B_4	-	B_1	B_2	B_3	B_4
Sim.	6,1604	1,4671	2,3211	1,3811	Sim.	6,6591	1,9417	3,5450	4,4604
Ger.	6,041	1,1610	2,1441	1,4287	Ger.	6,5556	1,8844	3,6959	4,7102
Error	-1,94%	-20,86%	-7,63%	3,45%	Error	-1,55%	-2,95%	4,26%	5,60%
LMT	6,0928	1,2391	2,2326	1,4465	LMT	6,5525	1,8888	3,5489	4,4994
Error	-1,10%	-15,54%	-3,81%	4,74%	Error	-1,60%	-2,72%	0,11%	0,87%
Case 3					Case 4				
-	B_1	B_2	B_3	B_4	-	B_1	B_2	B_3	B_4
Sim.	5,0578	2,0832	3,12	1,8278	Sim.	5,3145	2,700	4,0552	3,8002
Ger.	4,9447	2,0599	3,2011	1,9115	Ger.	5,0711	2,2602	4,1495	4,0705
Error	-2,24%	-1,12%	2,60%	4,58%	Error	-4,58%	-16,29%	2,33%	7,11%
LMT	4,9960	2,0578	3,1112	1,8742	LMT	5,1718	2,4526	4,0565	3,9153
Error	-1,22%	-1,22%	-0,28%	2,54%	Error	-2,69%	-9,16%	0,03%	3,03%
Case 5					Case 6				
-	B_1	B_2	B_3	B_4	-	B_1	B_2	B_3	B_4
Sim.	4,2533	1,795	2,1355	0,6742	Sim.	6,1316	2,3334	3,0929	0,8942
Ger.	4,1752	1,7484	2,1689	0,7431	Ger.	6,0172	2,2028	3,1128	0,9749
Error	-1,84%	-2,60%	1,56%	10,22%	Error	-1,87%	-5,60%	0,64%	9,02%
LMT	4,2243	1,7675	2,1398	0,6997	LMT	6,0914	2,2596	3,0921	0,9262
Error	-0,68%	-1,53%	0,20%	3,78%	Error	-0,66%	-3,16%	-0,03%	3,58%

Table 2. Five machine lines with one failure mode: parameters and buffer results

makes it possible to identify and to keep separate in each building block in which thc original linc is dccomposed all the various causes that lead to an interruption of machine operations.

The performance of the proposed method over existing techniques has been shown through the comparison of simulation experiments and analytical results.

The insight this approach provides has proven to be very useful in the analysis of more general general networks, such as assembly/disassembly networks and systems with loops, which are under development.

Acknowledgments

The authors would like to thank Dr. Stanley B. Gershwin for his useful suggestions. This research has been partially funded by *MIUR* under the projects "Modelli per la Pianificazione della Capacit Produttiva in Sistemi di Produzione Innovativi-Models for capacity planning in advanced manufacturing systems" and "Pianificazione Distribuita del Processo e della Produzione

Case 7								
M	*m*	p_1	p_2	r_1	r_2	*C*	-	*Th.*
1	1,2	0,08	0,02	0,6	0,08	8	Sim.	0,4959
2	1,0	0,09	0,03	0,7	0,09	6	Ger.	0,5454
3	1,2	0,07	0,01	0,9	0,07	8	Error	9,98%
4	1,4	0,06	0,04	0,5	0,06	6	LMT	0,4961
5	1,2	0,05	0,01	0,8	0,10	-	Error	0,04%
Case 9								
M	*m*	p_1	p_2	r_1	r_2	*C*	-	*Th.*
1	0,8	0,09	0,01	0,8	0,073	8	Sim.	0,3148
2	0,6	0,08	0,04	0,8	0,200	8	Ger.	0,3361
3	1,0	0,06	0,03	0,6	0,086	8	Error	6,77%
4	0,8	0,06	0,02	0,9	0,060	8	LMT	0,3181
5	0,6	0,10	0,04	0,5	0,080	-	Error	1,05%
Case 11								
M	*m*	p_1	p_2	r_1	r_2	*C*	-	*Th.*
1	1,0	0,02	0,01	0,2	0,05	6	Sim.	0,6202
2	1,0	0,01	0,01	0,3	0,15	6	Ger.	0,6333
3	1,2	0,02	0,02	0,3	0,60	6	Error	2,11%
4	1,0	0,01	0,03	0,3	0,18	6	LMT	0,6185
5	1,2	0,01	0,02	0,2	0,40	-	Error	0,27%

Table 3. Five machine lines with two failure modes: parameters and throughput results

in Ambiente Manifatturiero-Distributed Process and Production Planning in Manufacturing Enterprise Networks".

References

[1] Buzacott J. The effect of station breakdown and random processing times on the capacity of flow lines. *AIIE Transactions*, 4: 308-312, 1972.

[2] Choong Y.F. and Gershwin S.B. A decomposition method for the approximate evaluation of capacitated transfer lines with unreliable machines and random processing times. *IIE Transactions,* 19: 150-159, 1987.

[3] Dallery Y., David R. and Xie X.L. An Efficient Algorithm for Analysis of Transfer Lines With Unreliable Machines and Finite Buffer. *IIE Transactions*, 20(3): 280-283, 1988.

[4] Y. Dallery and S.B. Gershwin, Manufacturing Flow Line Systems: A Review of Models and Analytical Results,*Queuing System Theory and*

Case 8								
M	m	p_1	p_2	r_1	r_2	C	-	Th.
1	1,8	0,09	0,01	0,8	0,073	8	Sim.	0,7103
2	1,6	0,08	0,04	0,8	0,200	8	Ger.	0,7745
3	2,0	0,06	0,03	0,6	0,086	8	Error	9,04%
4	1,8	0,06	0,02	0,9	0,060	8	LMT	0,7169
5	1,6	0,10	0,04	0,5	0,080	-	Error	0,93%
Case 10								
M	m	p_1	p_2	r_1	r_2	C	-	Th.
1	1,0	0,02	0,01	0,2	0,05	6	Sim.	0,5854
2	1,0	0,01	0,01	0,3	0,15	6	Ger.	0,5935
3	1,2	0,02	0,02	0,3	0,10	6	Error	1,38%
4	1,0	0,01	0,03	0,3	0,18	6	LMT	0,5823
5	1,2	0,01	0,02	0,2	0,40	-	Error	0,53%
Case 11								
M	m	p_1	p_2	r_1	r_2	C	-	Th.
1	1,0	0,02	0,01	0,2	0,05	10	Sim.	0,6680
2	1,0	0,01	0,01	0,3	0,15	10	Ger.	0,6876
3	1,2	0,02	0,02	0,3	0,60	10	Error	2,93%
4	1,0	0,01	0,03	0,3	0,18	10	LMT	0,6749
5	1,2	0,01	0,02	0,2	0,40	-	Error	1,03%

Table 4. Five machine lines with two failure modes: parameters and throughput results

Applications, Special Issue on Queuing Model of Manufacturing System, 1992, 12(1-2):3-94.

[5] Y. Dallery and H. Le Bihan, An Improved Decomposition method for the Analysis of Production Lines With Unreliable Machines and Finite Buffer. *International Journal of Production Research*, Vol 37(5):(1999) 1093-1117.

[6] Gershwin S.B. An Efficient Decomposition Algorithm for the Approximate Evaluation of Tandem Queueing Systems with Finite Buffers. In H.G. Perros and T. Altiok, editors, Queueing Network with Blocking, pages 127-146, North-Holland, Amsterdam, 1989.

[7] Gershwin S.B. *Manufacturing Systems Engineering.* PTR Prentice Hall, Inc., New Jersey, 1994.

[8] S.B. Gershwin, N. Maggio, A. Matta, T. Tolio and L. Werner, Analysis of Loop Networks by Decomposition, *Proceedings of the Third Aegean*

Case 8								
M	*m*	p_1	p_2	r_1	r_2	*C*	-	*Th.*
1	1,8	0,09	0,01	0,8	0,073	8	Sim.	0,7103
2	1,6	0,08	0,04	0,8	0,200	8	Ger.	0,7745
3	2,0	0,06	0,03	0,6	0,086	8	Error	9,04%
4	1,8	0,06	0,02	0,9	0,060	8	LMT	0,7169
5	1,6	0,10	0,04	0,5	0,080	-	Error	0,93%
Case 10								
M	*m*	p_1	p_2	r_1	r_2	*C*	-	*Th.*
1	1,0	0,02	0,01	0,2	0,05	6	Sim.	0,5854
2	1,0	0,01	0,01	0,3	0,15	6	Ger.	0,5935
3	1,2	0,02	0,02	0,3	0,10	6	Error	1,38%
4	1,0	0,01	0,03	0,3	0,18	6	LMT	0,5823
5	1,2	0,01	0,02	0,2	0,40	-	Error	0,53%
Case 11								
M	*m*	p_1	p_2	r_1	r_2	*C*	-	*Th.*
1	1,0	0,02	0,01	0,2	0,05	10	Sim.	0,6680
2	1,0	0,01	0,01	0,3	0,15	10	Ger.	0,6876
3	1,2	0,02	0,02	0,3	0,60	10	Error	2,93%
4	1,0	0,01	0,03	0,3	0,18	10	LMT	0,6749
5	1,2	0,01	0,02	0,2	0,40	-	Error	1,03%

Table 5. Five machine lines with two failure modes: parameters and buffer results

International Conference on "Design and Analysis of Manufacturing Systems", Tinos Island, Greece, May 19-22, 2001.

[9] Hong Y., Glassey C.R. and Seong D. The Analysis of a Production Line with Unreliable Machines and Random Processing Times. *IIE Transaction*, 24(1): 77-83, 1993.

[10] H. Le Bihan and Y. Dallery, A Robust Decomposition Method for the Analysis of Production Lines with Unreliable Machines and Finite Buffer. *Annals of Operations Research special issue on Performance Evaluation of Production Lines*, 93:(2000) 265-297.

[11] Levantesi R., Matta A. and Tolio T. Exponential Two-Machine Lines with Multiple Failure Modes and Finite Buffer Capacity. Second International Aegean Conference on the "Analysis and Modeling of Manufacturing Systems", Tinos Island, Greece, May 1999.

[12] Levantesi R., Matta A. and Tolio T. A Decomposition Method for the Perfomance Evaluation of Production Lines with Random Processing

Times, Multiple Failure Modes and Finite Buffer Capacity. Second International Aegean Conference on the "Analysis and Modeling of Manufacturing Systems", Tinos Island, Greece, May 1999.

[13] R. Levantesi, A. Matta and T. Tolio, Performance Evaluation of Assembly-Disassembly Networks with Multiple Failure Modes and Finite Buffer Capacity, *Proceedings of the 50th International Conference On Production Reserarch*, Bangok, Thailand, August 3-5, 2000.

[14] R. Levantesi, Analysis of multiple loop assembly/disassembly networks, *Phd Thesis, Politecnico di Milano, Milano, Italy, 2002.*

[15] Tolio T., Gershwin S.B. and Matta A., Analysis of two-machine lines with multiple failure modes, IIE Transaction, 34(1): 51-62, 2002.

[16] Tolio T. and Matta A. A Method for Perfomance Evaluation of Automated Flow Lines. *Annals of CIRP* Vol. 47/1, 1998.

Chapter 10

DUE-TIME PERFORMANCE OF PRODUCTION SYSTEMS WITH MARKOVIAN MACHINES*

Jingshan Li
General Motors Research & Development Center
Warren, MI 48090-9055
jingshan.li@gm.com

Semyon M. Meerkov
Department of Electrical Engineering and Computer Science
University of Michigan, Ann Arbor, MI 48109-2122
smm@eecs.umich.edu

Abstract Customer demand satisfaction in serial production lines with Markovian machines and Finished Goods Buffers (FGB) is analyzed. The level of demand satisfaction is measured by the probability to ship to the customer a required number of parts during a fixed interval of time. This performance measure, referred to as Due-Time Performance (DTP), is analyzed using a simplification procedure termed stationarization. As a result, an iteration-based method for DTP calculation is obtained and utilized for analysis of various properties of DTP. In particular, it is shown that when the demand is relatively low (~ 95% of the production capacity), FGB of 2-4 shipments is sufficient to ensure high DTP (~ 0.98). When the demand is high (~ 99%), FGB of 6-9 shipments is required. Further increase of FGB leads to highly diminishing return. Also, it is shown that a ramp, rather than an inverted bowl, of machine efficiency allocation leads to optimization of DTP.

Keywords: Production systems, unreliable machines, due-time performance, finished goods buffer.

*This work was supported by NSF grant No. DMI-9820580.

1. Motivation

The number of parts produced by a manufacturing system with unreliable machines during a fixed interval of time is a random variable. Its distribution defines both production volume and production variability. In practice, the production volume is characterized by the production rate (PR), i.e., the average number of parts produced per unit of time. The production variability is often characterized by the due-time performance (DTP), which is the probability to ship to the customer a required number of parts during a fixed interval of time. This is the case, for instance, in the automotive industry where car and truck assembly plants are expected to be supplied by engine and seat plants on a four-hour delivery schedule. Similarly, engine plants receive ignition and injection units from the second tier suppliers on a fixed shipping schedule, etc.

PR was analyzed in many publications during the last 50 years (see reviews by Dallery and Gershwin 1992, Papadopoulos and Harvey 1996, Govil and Fu 1999, and monographs by Viswanadham and Narahari 1992, Buzacott and Shanthikumar 1993, Papadopoulos et al 1993, Gershwin 1994, Altiok 1997). In contrast, DTP has been addressed just recently. Specifically, Jacobs and Meerkov (1995a) analyzed DTP for a one-machine system without a finished goods buffer (FGB). The case of multiple machines, but also without a FGB, was addressed by Tan (1998, 1999). In practice, however, most production systems do have FGBs, intended to both improve DTP and limit finished goods inventory. Production systems with multiple machines and FGBs were investigated by Li and Meerkov (1998, 2000a, 2000b, 2001). The development was carried out under the assumption that the machines obey the so-called Bernoulli reliability model. According to this model, a machine is up or down during each cycle time (i.e., the time necessary to produce a part) with a probability independent of the status of the machine (up or down) in the previous cycle time. Although this model does take place in some assembly operations, most machining lines are better described by Markovian reliability model whereby the state of the machine in a cycle time is conditioned on the state of the machine in the previous cycle time. DTP in production systems with such machines and FGB has not been analyzed in the literature. This work is intended to contribute to this end.

The outline of the paper is as follows: Section 2 provides the problem formulation. A formal definition of DTP is also introduced. It turns out that to evaluate DTP, the calculation of PR is necessary. Since a method for PR calculation for production system with machines considered in this work is not available in the literature, Section 3 presents an aggregation technique for PR evaluation. In Section 4, a method for DTP calculation in serial lines with Markovian machines and FGB is developed. Section 5 addresses structural properties of DTP. The conclusions are formulated in Section 6. All proofs are given in the Appendix.

2. Problem Formulation

The structure of the production system analyzed in this work is shown in Figure 1, where the circles represent the machines and rectangles are the buffers. Assumptions concerning the machines, buffers, interactions between the machines, and the demand are defined as follows:

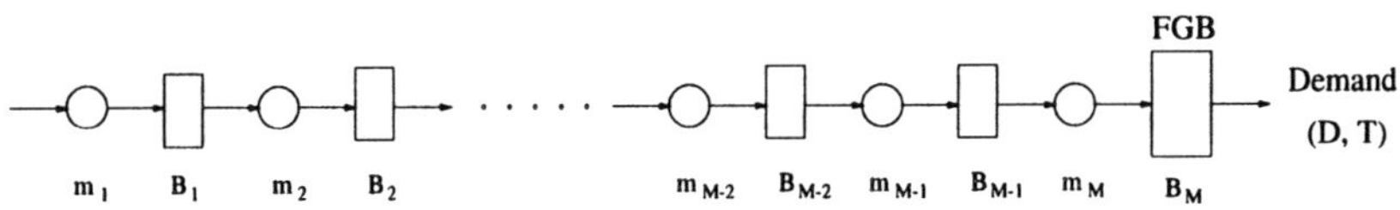

Figure 1. Serial production line with finished goods buffer

Machines:

(i). Each machine, m_i, $i = 1, \cdots, M$, requires a fixed unit of time to process a part. This unit is referred to as the *cycle time*. All machines have identical cycle time. The time axis is slotted with the slot duration equaled to the cycle time.

Remark 1 The assumption that the cycle time is deterministic and identical for all machines in the system is sufficiently realistic for production systems with automated material handling. It may not hold for other systems, e.g., job shop-type environment. Thus, the results of this research contribute to the area of automated production lines, such as typically encountered in the automotive industry. □

(ii). During a cycle time, each machine can be either "up" or "down". When up, the machine could process a part. When down, no processing can take place.

(iii). The state of the machines is defined by a discrete time Markov process: If machine m_i, $i = 1, \cdots, M$, is up, it will be down during the next cycle with probability P_i and up with probability $1 - P_i$; if it is down, it will be up during the next cycle with probability R_i and down with probability $1 - R_i$. In other words, the up- and downtime are distributed geometrically with the parameters P_i and R_i, respectively.

Buffers:

(iv). Each buffer B_i, $i = 1, \cdots, M$, has capacity $1 \leq N_i < \infty$, $i = 1, \cdots, M$. Buffers $B_1, \cdots, B_{M-1}$ are called in-process buffers. Buffer B_M is the finished goods buffer.

Starvation rule:

(v). If B_i, $i = 1, \cdots, M-1$, is empty at the beginning of the time slot, then m_{i+1}, $i = 2, \cdots, M-1$, is starved during this time slot. The first machine is never starved.

Blockage rule:

(vi). If B_i, $i = 1, \cdots, M-1$, is full at the beginning of a time slot and m_{i+1}, $i = 1, \cdots, M-1$, does not take a part from B_i at the beginning of this slot, then m_i, $i = 1, \cdots, M-1$, is blocked during this time slot. The last machine, m_M, can be blocked during a time slot if the FGB is full at the beginning of this time slot.

Remark 2 Assumptions (iii), (v) and (vi) are formulated in terms of the so-called time-dependent failures, i.e., machines can go down even when blocked or starved (Buzacott and Shanthikumar 1993). Another possible model is that of operation-dependent failures, where no breakdowns of starved or blocked machines is possible (Buzacott and Shanthikumar 1993, Gershwin 1994). Both models are practical, depending on the production system at hand: For automated palletized material handling systems, the time-dependent model is more applicable. In case of manual material handling, operation-dependent failures often take place. Both failure modes, however, result in similar behavior. Studies show that throughputs of a line with time-dependent or operation-dependent failures differ at most by 3% - 4% (Buzacott and Shanthikumar 1993), which is well within the accuracy of the data describing production lines. □

Demand:

(vii). From the point of view of the demand, the time axis is divided into "epochs", each containing T time slots.

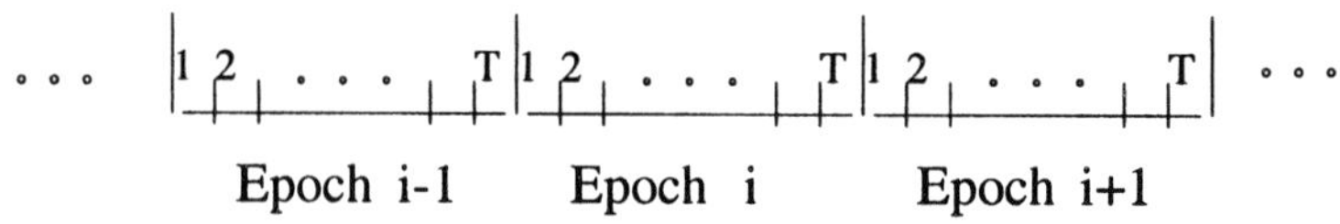

Figure 2. Epochs

(viii). At the end of each epoch, a shipment of D parts has to be available for the customer. If p_a is the production rate of the system (without FGB), then

$$D \leq T p_a. \tag{1}$$

Demand satisfaction policy:

(ix). At the beginning of epoch i, parts are removed from the FGB in the amount of $\min(H(i-1), D)$, where $H(i-1)$ is the number of parts in the FGB at the end of $(i-1)$-th epoch. If $H(i-1) \geq D$, the shipment is complete; if $H(i-1) < D$, the balance of the shipment, i.e., $D - H(i-1)$ parts, is to be produced by m_M during the shipping period T. Parts produced are immediately removed from the FGB and prepared for shipment, until the shipment is complete, i.e., D parts are available. If the shipment is complete before the end of the epoch, the system continues operating, but with the parts being accumulated in the FGB, either until the end of the epoch or until the last machine, m_M, is blocked, whatever occurs first. If the shipment is not complete by the end of the epoch, an incomplete shipment is sent to the customer. No backlog is allowed.

Remark 3 Assumption of no backlog is introduced to simplify the analysis. However, since backlog is typically satisfied by overtime, this assumption does not reduce the generality. □

Assumptions (i)-(ix) define the serial production line under consideration. In the time scale of the epoch and in an appropriately defined state space, the system (i)-(ix) is a stationary ergodic Markov chain. We refer to its steady state as the "normal system operation".

Let $\bar{t}_i$ be the number of parts produced by machine m_M in epoch i during the normal system operation. Then DTP can be expressed as

$$DTP = Pr\big(\bar{t}_i + H(i-1) \geq D\big). \tag{2}$$

In the framework of model (i)-(ix), DTP is a function of all system parameters. In other words,

$$DTP = DTP(P, R, N, N_M, D, T), \tag{3}$$

where N_M, T, and D are defined in (iv), (vii), and (viii), respectively, P, R and N are vectors of the machines and in-process buffers parameters,

$$P = [P_1, \cdots, P_M], \qquad R = [R_1, \cdots, R_M], \qquad N = [N_1, \cdots, N_{M-1}].$$

The goal of this paper is to develop a method evaluating function (3) and to investigate its structural (e.g., monotonicity) properties.

3. Production Rate Evaluation

This Section describes a method for PR calculation in serial lines defined by assumptions (i)-(vi), i.e., having no FGB. In Section 4, this method is used for DTP calculation in corresponding systems with FGBs.

3.1 Two-machine Line

3.1.1 Calculations.

Theorem 1 *The production rate in a serial production line (i)-(vi) with $M = 2$ is given by*

$$PR = e_2\left[1 - Q(P_1, R_1, P_2, R_2, N)\right], \tag{4}$$

where

$$\begin{aligned} e_i &= \frac{R_i}{P_i + R_i}, \quad i = 1, 2, \\ Q(P_1, R_1, P_2, R_2, N) &= \begin{cases} \frac{P_1\beta_2}{(R_1+R_2-R_1R_2)(R_1+P_1)}, & \text{if } N = 1, \\ \frac{P_1\alpha_1\alpha_2\beta_2^2(R_2+P_2)}{A+B+C+D}, & \text{if } N > 1, \end{cases} \end{aligned} \tag{5}$$

and

$$\begin{aligned} \alpha_1 &= P_1 + P_2 - P_1P_2 - R_1P_2, \\ \alpha_2 &= P_1 + P_2 - P_1P_2 - R_2P_1, \\ \beta_1 &= R_1 + R_2 - R_1R_2 - P_1R_2, \\ \beta_2 &= R_1 + R_2 - R_1R_2 - P_2R_1, \\ \sigma &= \frac{\alpha_2\beta_1}{\alpha_1\beta_2}, \\ A &= P_1R_2\alpha_1\alpha_2\beta_2(P_2 + \beta_2), \\ B &= P_1R_1R_2\alpha_2\left[\beta_2^2 + P_2(\alpha_1 + \beta_1)(\alpha_2 + 2\beta_2)\right], \\ C &= \sum_{k=2}^{N-1} P_1P_2R_1R_2(\alpha_2 + \beta_2)^3\sigma^{k-1}, \\ D &= P_2R_1\alpha_1\beta_2\left[R_2(\alpha_1 + \beta_1) + \alpha_2(P_1 + R_1)\right]\sigma^{N-1}. \end{aligned}$$

Moreover, the average probability of buffer occupancy, $E(h)$, and the probabilities of manufacturing blockage of m_1, mb_1, and starvation of m_2, ms_2, are *given by*

$$E(h) = \begin{cases} \frac{P_1[(R_1+R_2-R_1R_2)(P_2+R_2)+P_1P_2]}{(R_1+R_2-R_1R_2)(R_1+P_1)(R_2+P_2)} & N = 1, \\ \frac{B+\sum_{k=2}^{N-2} P_1P_2R_1R_2k(\alpha_1+\beta_1)^3\sigma^{k-1}+ND}{A+B+C+D} & N > 1, \end{cases}$$

$$ms_2 = \begin{cases} \frac{P_1R_2\beta_2}{(R_1+R_2-R_1R_2)(R_1+P_1)(R_2+P_2)} & N = 1, \\ \frac{P_1R_2\alpha_2\alpha_2\beta_2^2}{A+B+C+D} & N > 1, \end{cases} \quad (6)$$

$$mb_1 = \begin{cases} \frac{P_2R_1\beta_1}{(R_1+R_2-R_1R_2)(R_1+P_1)(R_2+P_2)} & N = 1, \\ \frac{P_2R_1\alpha_1\alpha_2\beta_1\beta_2\sigma^{N-1}}{A+B+C+D} & N > 1. \end{cases} \quad (7)$$

Proof: See Appendix. ■

Remark 4 The PR calculation for two-machine line with time- dependent failure was studied by Buzacott and Shanthikumar (1993), Sheskin (1976), however the assumptions defined were different from (i)-(vi). More specifically, it is either assumed that machine m_2 could produce a part if both machines are up and buffer is empty (Buzacott and Shanthikumar 1993), or an additional constraint that $P_i + R_i = 1$ is needed (Sheskin 1976). □

3.1.2 Monotonicity.

Proposition 1 *For $N = 1$, function $Q(P_1, R_1, P_2, R_2, N)$ is monotonically increasing with respect to P_1, R_2, and decreasing with respect to P_2, R_1.*

Proof: See Appendix. ■

Unfortunately, the proof of this property for $N > 1$ seems to be all but impossible. Therefore, based on numerical simulation, we formulate

Hypothesis 1 *Function $Q(P_1, R_1, P_2, R_2, N)$, $N > 1$, is monotonically increasing with respect to P_1, R_2, and decreasing with respect to P_2, R_1.*

An illustration is shown in Figure 3.

3.2 M-machine Line

3.2.1 Aggregation. No closed form expression for *PR* in M-machine line is available. Therefore, we develop an aggregation procedure, based on

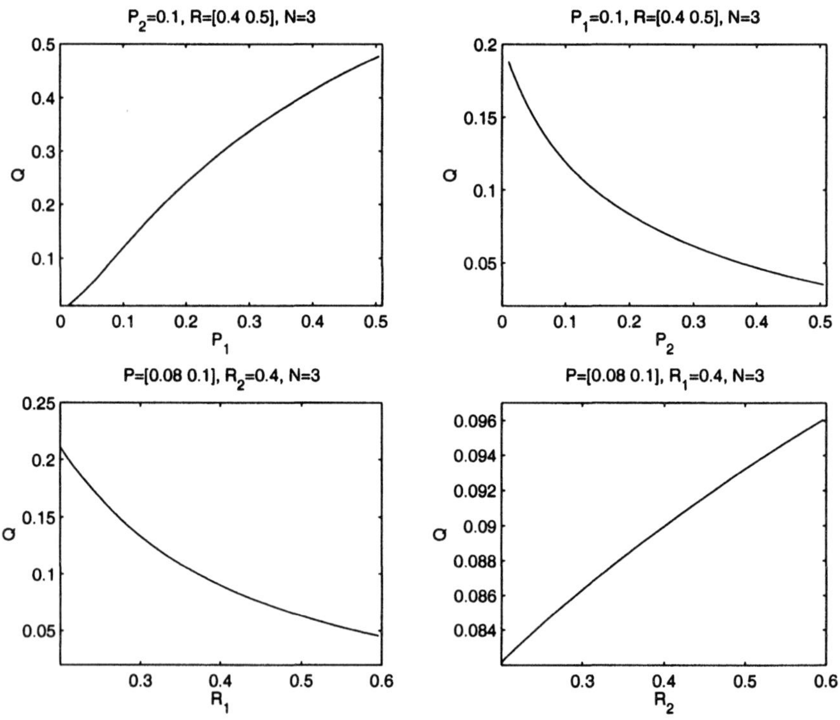

Figure 3. Monotonicity of $Q(\cdot)$

the result of the previous Subsections. Specifically, we aggregate the first two machines into a single machine, m_2^f, with parameter R_2^f defined as

$$R_2^f = R_2[1 - Q(P_1, R_1, P_2, R_2, N_1)],$$

and P_2^f selected so that

$$\frac{R_2^f}{P_2^f + R_2^f} = \frac{R_2}{P_2 + R_2}[1 - Q(P_1, R_1, P_2, R_2, N_1)],$$

i.e.,

$$P_2^f = P_2 + R_2 Q(P_1, R_1, P_2, R_2, N_1),$$

where $Q(\cdot)$ is defined in (5). Next m_2^f is a aggregated with m_3 to result in m_3^f, with the parameters defined as shown above, and so on until all M machines are aggregated in a single one, m_M^f. This constitutes the forward aggregation (superscript f is used to denote this fact). Then, in the backward aggregation, the last machine, m_M, is aggregated with m_{M-1}^f to result in m_{M-1}^b and so on until all machines are again aggregated in a single machine, m_1^b. Then the procedure is repeated again. Formally, this process can be represented as follows:

Procedure 1

$$\begin{aligned}
R_i^b(s+1) &= R_i - R_i Q\big(P_{i+1}^b(s+1), R_{i+1}^b(s+1), P_i^f(s), R_i^f(s), N_i\big), \\
&\qquad\qquad i = 1, \cdots, M-1, \\
P_i^b(s+1) &= P_i + R_i Q\big(P_{i+1}^b(s+1), R_{i+1}^b(s+1), P_i^f(s), R_i^f(s), N_i\big), \\
&\qquad\qquad i = 1, \cdots, M-1, \\
R_i^f(s+1) &= R_i - R_i Q\big(P_{i-1}^f(s+1), R_{i-1}^f(s+1), P_i^b(s+1), R_i^f(s+1), N_{i-1}\big), \\
&\qquad\qquad i = 2, \cdots, M, \qquad (8) \\
P_i^f(s+1) &= P_i + R_i Q\big(P_{i-1}^f(s+1), R_{i-1}^f(s+1), P_i^b(s+1), R_i^f(s+1), N_{i-1}\big), \\
&\qquad\qquad i = 2, \cdots, M,
\end{aligned}$$

with boundary conditions

$$\begin{aligned}
P_1^f(s) = P_1, &\qquad R_1^f(s) = R_1, \\
P_M^b(s) = P_M, &\qquad R_M^b(s) = R_M, \\
s &= 0, 1, 2, \cdots,
\end{aligned}$$

and initial conditions

$$P_i^f(0) = P_i, \quad R_i^f(0) = R_i, \qquad i = 2, \cdots, M-1$$

and function $Q(\cdot)$ defined in (5). □

The question of convergence of the resulting sequences $P_i^b(s)$, $R_i^b(s)$, $P_i^f(s)$, $R_i^f(s)$, $i = 1, \cdots, M$, $s = 0, 1, \cdots$, is answered in the following:

Theorem 2 *Under Hypothesis 1, recursive procedure 1 is convergent and, therefore, the following limits exist:*

$$\begin{aligned} \lim_{s\to\infty} P_i^f(s) &:= P_i^f, \qquad \lim_{s\to\infty} P_i^b(s) := P_i^b, \\ \lim_{s\to\infty} R_i^f(s) &:= R_i^f, \qquad \lim_{s\to\infty} R_i^b(s) := R_i^b, \\ i &= 1, 2, \cdots, M. \end{aligned} \tag{9}$$

Moreover, the following relationship holds:

$$\frac{R_M^f}{P_M^f} = \frac{R_1^b}{P_1^b}. \tag{10}$$

Proof: See Appendix. ■

The limits in (9) can be used to define estimates of performance measures for line (i)-(vi) without FGB. Indeed, since the last machine is not blocked and the first is not starved, production rate can be estimated as

$$\widehat{PR}(P_1, R_1, \cdots, P_M, R_M, N_1, \cdots, N_{M-1}) = \frac{R_M^f}{P_M^f + R_M^f} = \frac{R_1^b}{P_1^b + R_1^b}. \tag{11}$$

The accuracy of the estimate (11) is discussed next.

3.2.2 Accuracy. The accuracy of the estimate (11) has been evaluated numerically. We simulate dozens of systems defined by assumptions (i)-(vi) without FGB with various machine and buffer parameters assumed. Twenty of them with 3 - 8 machines, are shown in Table 1. In each simulation run, zero initial occupancy of all buffers has been assumed, and 50,000 time slots of warm up period has been carried out. The next 500,000 slots of stationary regime have been used to statistically evaluate the PR. Confidence intervals have been evaluated with 20 runs. The 95% confidence intervals were consistently around ±0.0015. Same procedures are followed for all numerical simulations throughout this paper. In Table 1, *PR* denotes the actual production rate obtained by simulation, whereas $\widehat{PR}$ denotes the estimate of production rate calculated according to (11). As it can be seen from Table 1, the estimate results in relatively high precision, comparable with that of Gershwin (1994), Dallery et al (1978), Jacobs (1993), Jacobs and Meerkov (1995b), Chiang (1999).

Table 1. Numerical justification of production rate estimation for serial lines (err=$\frac{|PR-\widehat{PR}|}{PR}\cdot$ 100%)

P_i	R_i	N_i	PR	$\widehat{PR}$	err
0.06 0.07 0.08	0.27 0.28 0.29	1 2	0.561	0.562	0.21
0.12 0.15 0.10	0.43 0.46 0.50	2 2	0.606	0.605	0.18
0.05 0.10 0.15	0.50 0.45 0.40	2 2	0.627	0.629	0.32
0.10 0.10 0.10	0.42 0.42 0.42	3 3	0.668	0.668	0.04
0.10 0.05 0.20	0.90 0.85 0.75	3 2	0.776	0.786	1.24
0.10 0.02 0.06	0.60 0.04 0.09	1 1	0.360	0.349	3.11
0.11 0.08 0.08 0.11	0.40 0.41 0.41 0.40	2 3 2	0.608	0.606	0.39
0.11 0.12 0.13 0.10	0.39 0.38 0.36 0.43	4 5 3	0.606	0.605	0.20
0.08 0.09 0.07 0.06	0.37 0.43 0.41 0.39	3 2 2	0.644	0.645	0.22
0.15 0.04 0.30 0.02	0.50 0.80 0.40 0.70	3 2 3	0.550	0.557	1.26
0.06 0.08 0.05 0.10	0.36 0.39 0.42 0.37	3 3 4	0.682	0.689	0.94
0.04 0.07 0.10 0.13	0.40 0.43 0.37 0.46	2 3 3	0.646	0.645	0.13
0.10 0.07 0.09 0.12 0.11	0.40 0.35 0.33 0.42 0.39	4 4 4 3	0.607	0.607	0.15
0.10 0.12 0.13 0.11 0.12	0.45 0.42 0.43 0.46 0.44	4 3 4 3	0.613	0.612	0.03
0.12 0.09 0.12 0.09	0.41 0.36 0.41 0.36	3 4 4 3	0.627	0.631	0.54
0.05 0.09 0.13 0.17 0.21	0.42 0.45 0.48 0.51 0.54	2 2 2 2	0.542	0.540	0.31
0.80 0.80 0.80 0.80 0.80 0.80	0.42 0.42 0.42 0.42 0.42 0.42	3 3 3 3 3	0.638	0.644	0.97
0.06 0.08 0.07 0.10 0.12 0.09	0.43 0.46 0.45 0.48 0.47 0.44	2 2 3 2 3	0.616	0.617	0.13
0.09 0.08 0.07 0.10 0.12 0.10 0.07	0.35 0.37 0.32 0.38 0.39 0.41 0.36	3 2 3 4 3 2	0.544	0.547	0.58
0.06 0.07 0.09 0.10 0.12 0.08 0.11 0.09	0.43 0.42 0.41 0.41 0.43 0.45 0.44 0.40	3 3 2 3 4 3 2	0.575	0.582	1.34

4. DTP Calculation

The Due Time Performance measure, introduced in Section 2, is defined as

$$DTP = Pr\left(\bar{t}_i + H(i-1) \geq D\right),$$

where $\bar{t}_i$ be the number of parts produced during i-th epoch, $H(i-1)$ is the number of parts that remained in the FGB at the end of $(i-1)$-th epoch. Unlike the case of production rate, the calculation of DTP even for a one-machine system is a nontrivial problem. Therefore, before addressing the general case, we study a single machine system.

4.1 One-machine Case

Consider a one-machine production system with the FGB of capacity N (Figure 4). Introduce notations:

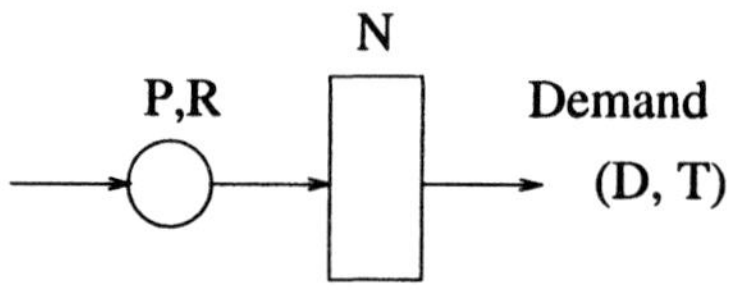

Figure 4. One-machine system with FGB

$t_i =$ number of parts produced during epoch i if no blockage occurs.

$\pi(j, s, n) =$ Probability that in the time slot j of an epoch ($j = 1, \cdots, T$) the machine is in state s ($s = 0$ when the machine is down and $s = 1$ when machine is up and n parts ($n = 0, 1, \cdots, j$) have been produced during $[0, j]$.

$z_k = Pr\big(H(i-1) = k\big),\ k = 0, 1, \cdots, N.$

$r_{k,j} = Pr(t_i = D + k - j),\ k = 1, \cdots, N-1, j = 0, 1, \cdots, N.$

$\bar{r}_{N,j} = Pr(t_i \geq D + N - j),\ \ j = 0, 1, \cdots, N.$

Probabilities $\pi(j, s, n)$ can be calculated recursively as follows (see Gershwin 1993):

$$\begin{aligned}
\pi(j, 0, n) &= (1-R)\pi(j-1, 0, n) + P\pi(j-1, 1, n), \\
&\qquad 0 < n < T, \quad 1 \leq j \leq T, \\
\pi(j, 1, n) &= R\pi(j-1, 0, n-1) + (1-P)\pi(j-1, 1, n-1), \\
&\qquad 0 < n < T, \quad 1 \leq j \leq T,
\end{aligned} \tag{12}$$

with initial conditions

$$\begin{aligned}
&\pi(j, 1, 0) = 0, && \pi(j, 0, j) = 0, \\
&\pi(j, 0, 0) = \tfrac{P}{P+R}(1-R)^{j-1}, && \pi(j, 1, j) = \tfrac{R}{P+R}(1-P)^{j-1}, \\
& && j = 1, \cdots, T.
\end{aligned} \tag{13}$$

Then, $r_{k,j}$ and $\bar{r}_{N,j}$ can be calculated as follows:

$$r_{k,j} = \sum_{s=0}^{1} \pi(T, s, D + k - j),$$
$$\bar{r}_{k,j} = \sum_{n=D+k-j}^{T} \sum_{s=0}^{1} \pi(T, s, n), \qquad (14)$$

Using the above notations, the DTP of a one-machine system with the FGB can be calculated as follows:

Theorem 3 *Let $Z = [z_1, \cdots, z_N]^T$ be a vector defined by*

$$Z = -\mathbf{R}^{-1} Z_0, \qquad (15)$$

with matrix **R** *and vector Z_0 as follows:*

$$\mathbf{R} = \begin{pmatrix} r_{1,1} - r_{1,0} - 1 & r_{1,2} - r_{1,0} & \cdots & r_{1,N} - r_{1,0} \\ r_{2,1} - r_{2,0} & r_{2,2} - r_{2,0} - 1 & \cdots & r_{2,N} - r_{2,0} \\ \cdots & \cdots & \cdots & \cdots \\ \bar{r}_{N,1} - \bar{r}_{N,0} & \bar{r}_{N,2} - \bar{r}_{N,0} & \cdots & \bar{r}_{N,N} - \bar{r}_{N,0} - 1 \end{pmatrix}, \qquad (16)$$

$$Z_0 = \begin{pmatrix} r_{1,0} \\ r_{2,0} \\ \cdots \\ r_{N-1,0} \\ \bar{r}_{N,0} \end{pmatrix}. \qquad (17)$$

Then, under assumptions (i)-(ix) with $M = 1$, the DTP is given by:

$$DTP = \sum_{k=0}^{N} z_k \sum_{n=D-k}^{T} \sum_{s=0}^{1} \pi(T, s, n), \qquad (18)$$

where $\pi(T, s, n)$ is calculated by (12).

Proof: Similar to the proof for the Bernoulli case of Li and Meerkov (2000a). ■

4.2 Approach to *M*-machine Line

4.2.1 Idea of the approach. The main difficulty in developing a method for calculating DTP is that the performance of a line within an epoch (i.e., in the time scale of the machine cycle time) is not described by a stationary random process. For instance, the probability that the last machine, m_M is blocked during a time slot j, $j = 1, \cdots, T$, clearly depends on j and is a monotonically increasing function of j. This makes exact calculations of DTP all but impossible. Therefore, a simplification is necessary. The simplification used in this work is based on two ideas: stationarization and subsequent recursive iterations. These are described below.

4.2.2 Stationarization. The stationarization procedure is intended to approximate the nonstationary behavior of the production line within an epoch (i.e., in the time scale of the cycle time) by a stationary process, so that the behavior of the system in the time scale of the epoch is not changed too dramatically. To accomplish this, consider a one-machine line, shown in Figure 4. Assume that the probability that the machine, m, is up and FGB is full, during a time slot j, $j = 1, \cdots, T$, is known. Denoted this probability as $P_b(j)$. Then the probability, P_{k_b}, that m is up and FGB is full during an epoch exactly k_b times, $k_b = 1, \cdots, T$, can be expressed as

$$P_{k_b} = \begin{cases} P_b(T - k_b + 1) - P_b(T - k_b), & 1 \le k_b \le T - 1, \\ 0, & k_b = T. \end{cases} \tag{19}$$

Thus, the expected number in an epoch, $\bar{k}_b$, is

$$\bar{k}_b = \sum_{k_b=0}^{T} k_b P_{k_b} = \sum_{k_b=1}^{T-1} k_b P_{k_b}. \tag{20}$$

Using this expression, the approximate probability that m is up and FGB is full during *any* slot j can be introduced as follows:

$$\widehat{mb} = \frac{\bar{k}_b}{T}. \tag{21}$$

Then the approximate probability that FGB is full during any time slot j is:

$$\widehat{P}_f = \frac{\widehat{mb}}{\frac{R}{P+R}} = \frac{\bar{k}_b}{T\frac{R}{P+R}}. \tag{22}$$

We use these expressions as the stationarized probabilities of the appropriate events.

To evaluate $\widehat{mb}$ and $\widehat{P}_f$, the distribution $P_b(j)$ is necessary. It can be calculated as follows:

Introduce the probabilities:

$$
\begin{aligned}
h(i,j,s) &= \text{Prob}\{i \text{ parts in FGB at the beginning of slot } j, \text{ and} \\
&\qquad \text{the machine is in state } s\}, \\
&\qquad i = 1,\cdots,N, j = 1,\cdots,T, s = 0,1, \\
h_0(0,j,s,k) &= \text{Prob}\{\text{FGB is empty at the beginning of slot } j, \text{ the} \\
&\qquad \text{machine is in state } s, \text{ and } k \text{ parts have been} \\
&\qquad \text{sent to customer}\}, \\
&\qquad i = 1,\cdots,N, j = 1,\cdots,T, k = 0,\cdots,D, s = 0,1.
\end{aligned}
$$

Under assumptions (i)-(ix) and $M = 1$, function $h(i,j,s)$ is defined recursively by

$$
h(i,j,0) = \begin{cases} h(1,j-1,0)(1-R) + h_0(0,j-1,1,D)P, & \text{for } i = 1, \\ h(i,j-1,0)(1-R) + h(i-1,j-1,1)P, & \text{for } 1 < i < N, \\ h(N,j-1,0)(1-R) + h(N,j-1,1)P & \\ \quad +h(N-1,j-1,1)P, & \text{for } i = N, \end{cases} \tag{23}
$$

$$
h(i,j,1) = \begin{cases} h(1,j-1,0)R + h_0(0,j-1,1,D)(1-P), & \text{for } i = 1, \\ h(i,j-1,0)R + h(i-1,j-1,1)(1-P), & \text{for } 1 < i < N, \\ h(N,j-1,0)R + h(N,j-1,1)(1-P) & \\ \quad +h(N-1,j-1,1)(1-P), & \text{for } i = N, \end{cases} \tag{24}
$$

where

$$
h_0(0,j,\lambda,k) = \begin{cases} h_0(0,j-1,0,k)(1-R) + h_0(0,j-1,1,k-1)P, & \text{for } \lambda = 0, k > 0, \\ h_0(0,j-1,1,k-1)(1-P) + h_0(0,j-1,0,k)R, & \text{for } \lambda = 1, k > 0, \end{cases} \tag{25}
$$

with initial conditions

$$
h_0(0,j,\lambda,0) = \begin{cases} z_0 \frac{P}{P+R}(1-R)^{j-1}, & \text{for } \lambda = 0, \\ z_0 \frac{R}{P+R}(1-R)^{j-1}, & \text{for } \lambda = 1, \end{cases} \tag{26}
$$

$$
h_0(0,1,\lambda,k) = \begin{cases} z_k \frac{P}{P+R}, & \text{for } \lambda = 0, 0 \le k \le D, N > D, \\ z_k \frac{R}{P+R}, & \text{for } \lambda = 1, 0 \le k \le D, N > D, \\ 0, & \text{for } N+1 \le k \le D, N < D, \end{cases} \tag{27}
$$

$$
h(i,1,0) = \begin{cases} \frac{P}{P+R}, & \text{for } i = 0, D \ge N, \\ \sum_{i=1}^{D} z_i \frac{P}{P+R}, & \text{for } i = 0, D < N, \\ z_{D+i} \frac{P}{P+R}, & \text{for } 0 < i \le N-D, \\ 0, & \text{for } i > N-D > 0, \text{ or } i > 0, D \ge N, \end{cases} \tag{28}
$$

$$
h(i,1,1) = \begin{cases} \frac{R}{P+R}, & \text{for } i = 0, D \ge N, \\ \sum_{i=1}^{D} z_i \frac{R}{P+R}, & \text{for } i = 0, D < N, \\ z_{D+i} \frac{R}{P+R}, & \text{for } 0 < i \le N-D, \\ 0, & \text{for } i > N-D > 0, \text{ or } i > 0, D \ge N, \end{cases} \tag{29}
$$

and z_i, $i = 0, 1, \cdots, N$, can be calculated from (15). Therefore, the probability that m is up and FGB is full during slot j is:

$$P_b(j) = \sum_{s=0}^{1} h(N, j, s)\frac{R}{P+R}. \tag{30}$$

Thus, $\widehat{mb}$ can be calculated as follows:

$$\begin{aligned}\widehat{mb} &= \frac{1}{T}\sum_{k_b=1}^{T-1} k\big[P_b(T-k_b+1) - P_b(T-k_b)\big] \\ &= \frac{R}{T(P+R)}\sum_{k_b=1}^{T-1}\Big[\sum_{s=0}^{1} h(N, T-k_b+1, s) - \sum_{\lambda=0}^{1} h(N, T-k_b, s)\Big]k_b \\ &= \frac{R}{T(P+R)}\sum_{j=2}^{T}\Big[\sum_{s=0}^{1} h(N, j, s) - \sum_{s=0}^{1} h(N, j-1, \lambda)\Big](T-j+1).\end{aligned} \tag{31}$$

Analogously,

$$\widehat{P}_f = \frac{1}{T}\sum_{j=2}^{T}\Big[\sum_{s=0}^{1} h(N, j, s) - \sum_{s=0}^{1} h(N, j-1, s)\Big](T-j+1). \tag{32}$$

To summarize, it is convenient to introduce an operator of stationarization, Φ_1, defined as follows:

$$\begin{aligned}\widehat{P}_f &= \Phi_1(P, R, N, T, D) \\ &:= \frac{1}{T}\sum_{j=2}^{T}\Big[\sum_{s=0}^{1} h(N, j, s) - \sum_{s=0}^{1} h(N, j-1, s)\Big](T-j+1).\end{aligned} \tag{33}$$

where $h(N, j, s)$ is calculated according to (23)-(29). Expression (33) is used in subsequent subsections for DTP calculation.

4.2.3 Iterations. Consider the serial or assembly line shown in Figures 1. To adopt the above stationarization procedure, assume that the probability that the last machine m_M is starved in the communication sense is known. Denote the estimate of this probability as $\widehat{p_s}$. Modify R_M by multiplying it by $1 - \widehat{p_s}$ and P_M by adding it by $R_M\widehat{p_s}$, denote the modified machine as m'_M. Using this machine and procedure (23)-(29) and (33), calculate the stationarized probability, $\widehat{P}_f$. Introduce now another fictitious machine, m''_M, defined by $R_M(1 - \widehat{P}_f)$ and $P_M + R_M\widehat{P}_f$. Using this machine and deleting FGB, calculate a new $\widehat{p}_s$ by employing the recursive procedure for performance analysis of serial lines developed in Section 3. Having this probability, repeat the process described above anew.

In other words, in the case of serial lines, the iterations have the form:

$$\begin{aligned}
R'_M(n+1) &= R_M[1-\widehat{p}_s(n)], \\
P'_M(n+1) &= P_M + R_M\widehat{p}_s(n), \\
\widehat{P}_f(n+1) &= \Phi_1\big(P'_M(n+1), R'_M(n+1), N_M, T, D\big), \\
R''_M(n+1) &= R_M[1-\widehat{P}_f(n+1)], \\
P''_M(n+1) &= P_M + R_M\widehat{P}_f(n+1), \\
\widehat{p}_s(n+1) &= \Phi_2\big(P_1, R_1, \cdots, P''_M(n+1), R''_M(n+1), N_1, \cdots, N_{M-1}\big),
\end{aligned} \tag{34}$$

where Φ_1 is the stationarization operator defined in (33) and Φ_2 is the operator that represents the mapping from P''_M, R''_M to $\widehat{p}_s$. This operator is described Section 4.3.

If procedure (34) is convergent and the following limits exist,

$$\lim_{n\to\infty} \widehat{p}_s(n) := \widehat{p}_s,$$

the DTP of a M-machine serial line can be evaluated, using (18), as the DTP of a one-machine system defined by $R_M(1-\widehat{p}_s)$, $P_M + R_M\widehat{p}_s$ and FGB of capacity N_M, i.e.,

$$\widehat{DTP}_M = DTP_1(P_M + R_M\widehat{p}_s, R_M(1-\widehat{p}_s), N_M, D, T). \tag{35}$$

In Subsections 4.3-4.5, operators Φ_2 is described, convergence of the iterations is considered, and the accuracy of DTP calculations is investigated.

4.3 Operator Φ_2

Operator Φ_2 can be defined through the aggregation procedure for performance analysis of serial lines developed in Section 3. Indeed, assume that $P''_M(n+1)$ and $R''_M(n+1)$, calculated according to (34), are known. Consider the serial line consisting of M machines with parameters $P_1, R_1, \cdots, P_{M-1}, R_{M-1}$, $P''_M(n+1)$, $R''_M(n+1)$ and $M-1$ in-process buffers $N_1, \cdots, N_{M-1}$. According to Section 3, its performance can be analyzed using the convergent recursive procedure 1. In terms of the steady state of this procedure, the probability that buffer B_{M-1} is empty can be estimated as

$$\widehat{p}_s(n+1) = 1 - \frac{\frac{R^f_M(n+1)}{P^f_M(n+1)+R^f_M(n+1)}}{\frac{R_M}{P_M+R_M}}. \tag{36}$$

4.4 Convergence

To prove the convergence of recursive procedure (34), it is important to know whether function $\widehat{P}_f$, defined by (23)–(27) and (33), is monotonic with respect to P and R. Intuitively, this property clearly takes place. However, a rigorous proof of this fact seems to be all but impossible. Therefore, although in every example numerically analyzed this function was found to be monotonically increasing, we introduce this property as a hypothesis:

Hypothesis 2 *For system (i)-(ix) with $M = 1$, the stationarized probability that FGB is full, $\widehat{P}_f$, defined by (23)-(27) and (33), is a monotonically increasing function of R and monotonically decreasing function of P.*

An illustration of this property is given in Figure 5.

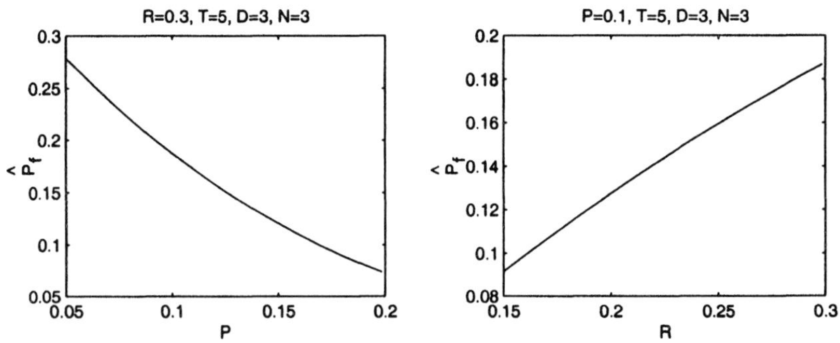

Figure 5. $\widehat{P}_f$: Monotonicity with respect to P and R

Theorem 4 *Under Hypothesis 2, iteration procedure (34), (36) is convergent, i.e., the following limits exist:*

$$\begin{aligned} \lim_{n\to\infty} \widehat{P}_f(n) &:= \widehat{P}_f, \\ \lim_{n\to\infty} \widehat{p}_s(n) &:= \widehat{p}_s. \end{aligned} \tag{37}$$

Proof: See Appendix. ∎

Thus, the DTP_M of a serial production defined by (i)-(ix) can be calculated using (35) with $\widehat{p_s}$ defined by (37).

4.5 Accuracy

The accuracy of $\widehat{DTP}_M$ has been evaluated using discrete event simulation. We simulated dozens of systems defined by (i)-(ix) with various machine, buffer, demand, and epoch parameters. Table 2 presents 20 of them with 2 - 6 machines. In each run of the corresponding discrete event model, zero initial conditions for all buffers have been assumed and $10,000T$ time slots of warm up period has been carried out (T, as before, is the length of the epoch). The next $100,000T$ slots of stationary regime have been used to statistically evaluate the DTP. In Table 2, DTP denotes the actual DTP obtained by simulation, whereas $\widehat{DTP}$ denotes the estimate of DTP calculated according to (8), (34)-(36). As it can be seen from Table 2, the estimates result in a relatively high precision, with errors ranging from 0.26% to 2.82%.

Remark 5 Calculation of DTP using expressions (8), (34)-(36) is orders of magnitude faster than using discrete event simulations. For example, discrete event simulation of every system included in Table 2 takes, on the average, 40 minutes using HP C360 workstation, whereas calculation according to (8), (34)-(36) takes about 1 second. □

5. Structural Properties

5.1 Load Factor

This Section illustrates the utility of DTP calculations by analyzing structural properties of production lines with FGBs. Two questions are addressed: Which capacity of FGB is necessary to ensure sufficiently high DTP? What are monotonicity properties of DTP, in particular with respect to the shipping period T? It turns out that answers to both questions depend on how large is the demand, D, vis-a-vis the production capacity of the system, Tp_a, where, as before, p_a is the production rate of the corresponding system without the FGB. We formalize this relationship by the load factor, L, defined as follows:

$$L = \frac{D}{Tp_a}. \tag{38}$$

Due to (1), $0 \leq L \leq 1$. Obviously, large L implies that the demand is heavy, relative to the average production capacity of the system; small L means that the production capacity is under-utilized.

The load factor (38) was introduced and analyzed by Jacobs and Meerkov (1995a) for a single machine production system without FGB. Here we use this notion in the general case.

Table 2. Numerical justification of DTP estimation for serial line (err=$\frac{|DTP-\widehat{DTP}|}{DTP} \cdot 100\%$)

P_i	R_i	N_i	T	D	DTP	$\widehat{DTP}$	err
0.12 0.14	0.43 0.38	1 1	5	3	0.713	0.705	1.05
0.05 0.10	0.37 0.42	2 1	4	3	0.796	0.798	0.33
0.12 0.12	0.76 0.76	2 2	5	4	0.907	0.904	0.26
0.10 0.10	0.40 0.40	1 1	3	2	0.774	0.771	0.34
0.05 0.05	0.30 0.30	1 2	4	3	0.805	0.817	1.45
0.12 0.08	0.35 0.39	2 3	3	2	0.875	0.892	2.00
0.10 0.12	0.33 0.40	2 2	5	3	0.848	0.850	0.26
0.10 0.05 0.20	0.90 0.85 0.75	3 2 3	4	3	0.962	0.968	0.60
0.10 0.20 0.06	0.41 0.58 0.32	2 2 3	5	3	0.879	0.855	2.82
0.10 0.10 0.10	0.42 0.42 0.42	3 3 3	3	2	0.907	0.915	0.80
0.12 0.15 0.10	0.43 0.46 0.50	2 2 2	5	3	0.857	0.849	0.93
0.08 0.12 0.10	0.45 0.44 0.42	1 2 2	5	3	0.835	0.820	1.77
0.10 0.12 0.10 0.09	0.46 0.49 0.47 0.48	3 3 3 2	3	2	0.912	0.911	0.30
0.11 0.08 0.08 0.11	0.40 0.41 0.41 0.40	2 3 2 2	5	3	0.850	0.831	2.22
0.06 0.08 0.05 0.10	0.36 0.39 0.42 0.37	3 3 4 3	3	2	0.918	0.935	1.88
0.15 0.04 0.30 0.02	0.50 0.80 0.40 0.70	3 2 3 2	6	3	0.948	0.965	1.83
0.06 0.07 0.09 0.08	0.42 0.41 0.44 0.43	2 3 2 3	3	2	0.908	0.915	0.77
0.10 0.07 0.09 0.12 0.11	0.40 0.35 0.33 0.42 0.39	4 4 4 3 2	5	3	0.867	0.843	2.43
0.10 0.10 0.10 0.10 0.10	0.41 0.41 0.41 0.41 0.41	3 3 3 3 3	5	3	0.901	0.892	1.07
0.08 0.10 0.12 0.13 0.09 0.11	0.45 0.46 0.44 0.47 0.43 0.42	3 2 4 3 4 3	5	3	0.924	0.918	0.62

5.2 FGB Capacity for High DTP

Consider a serial line with five machines and four in-process buffers, each with capacity 5. Assume that parameters P_i of the machines are identical ($P_i = P = 0.05, i = 1, \cdots, 5$), while R_i are constrained by $\sum_{i=1}^{5} R_i = R^*$ and consider four types of R_i allocations:

uniform: $R_i = R, \quad \forall i, i = 1, \cdots, 5;$

inverted bowl: $R_3 > R_2 = R_4 > R_1 = R_5;$

ramp: $R_1 < R_2 < R_3 < R_4 < R_5;$

inverted ramp: $R_1 > R_2 > R_3 > R_4 > R_5.$

Specific values of R_i for each of these allocations are shown in Figures 6 and 7 with $R^* = 2$ and 1.75, respectively, along with the corresponding production rate, *PR* (i.e., the average number of parts produced by the last machine per cycle time when no FGB is present), and the load factor of each line. Note that in Figure 6 the load is relatively low whereas in Figure 7 it is high.

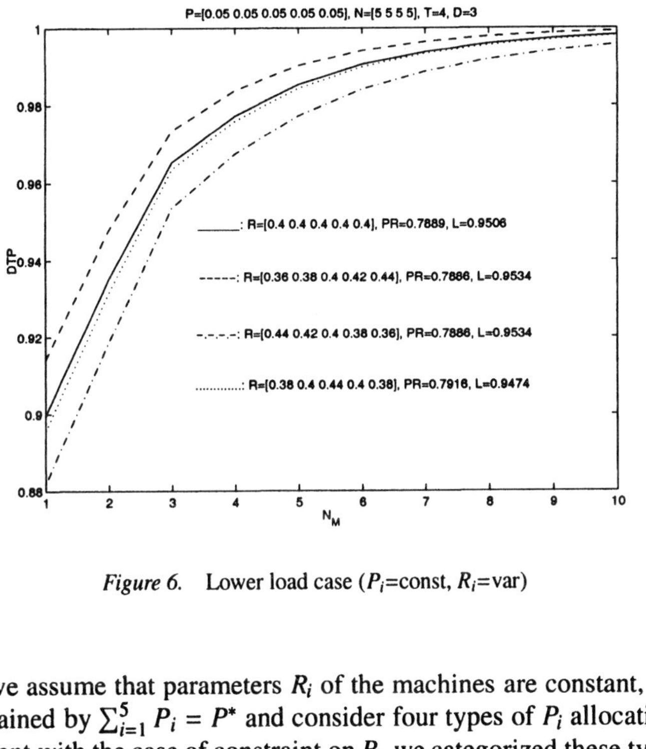

Figure 6. Lower load case (P_i=const, R_i=var)

Next, we assume that parameters R_i of the machines are constant, while P_i are constrained by $\sum_{i=1}^{5} P_i = P^*$ and consider four types of P_i allocations. (To be consistent with the case of constraint on R_i, we categorized these types from the point of view of machine efficiency: $e_i = \frac{R_i}{P_i+R_i}$.)

$$\begin{aligned}
&\text{uniform: } P_i = P, \quad \forall i, i = 1, \cdots, 5;\\
&\text{inverted bowl: } P_3 < P_2 = P_4 < P_1 = P_5;\\
&\text{ramp: } P_1 > P_2 > P_3 > P_4 > P_5;\\
&\text{inverted ramp: } P_1 < P_2 < P_3 < P_4 < P_5.
\end{aligned}$$

Specific values of P_i for each of these allocations are shown in Figures 8 and 9 with $P^* = 0.25$, but $R_i = 0.4$, $i = 1, \cdots, 5$, and $R_i = 0.35$, $i = 1, \cdots, 5$, respectively, along with the corresponding production rate, *PR*, of each line. Figures 8 and 9 correspond to relatively low and relatively high loads, respectively.

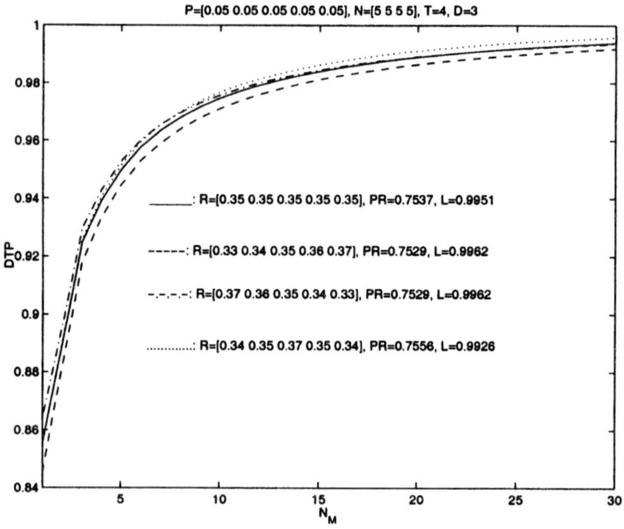

Figure 7. Higher load case (P_i=const, R_i=var)

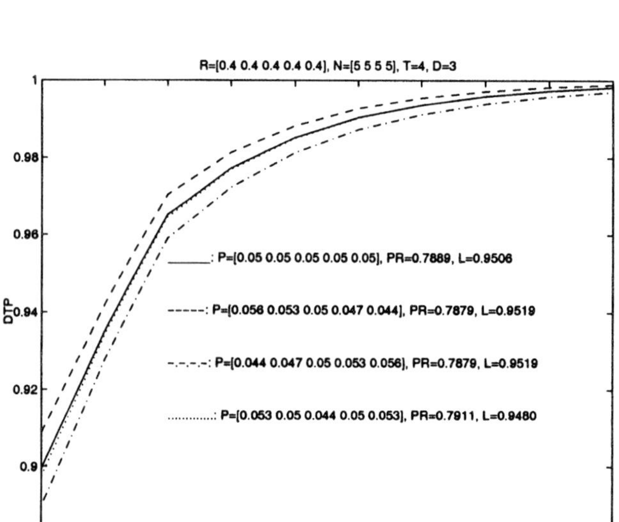

Figure 8. Lower load case (R_i=const, P_i=var)

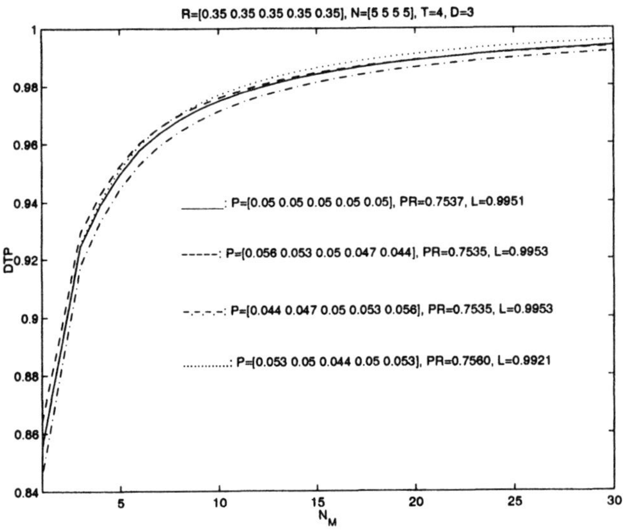

Figure 9. Higher load case (R_i=const, P_i=var)

Within this scenario, using (8), (34)-(36), we calculate the $\widehat{DTP}$ for all lines involved. The results are shown in Figures 6 - 9. Examining these data, we arrive at the following observations:

1. For lighter load and for all R_i and P_i allocations, FGB capacity of about 2 - 4 shipments is sufficient to provide high DTP.

2. For the heavy load, FGB of capacity 6 - 9 shipments is necessary to provide high DTP for all R_i and P_i allocations.

3. Although, as it is well known (Hiller and Boling 1966), *PR* is maximized by the inverted bowl e_i allocation (where $e_i = \frac{R_i}{P_i+R_i}$), the DTP is maximized by the ramp allocation, with the exception of relatively large N_M in the case of heavy load. Thus, a ramp, rather than an inverted bowl, is the appropriate e_i allocation if DTP is to be optimized. The advantage of ramp allocation is particularly clear for relatively small N_M, i.e., in the case of the so-called lean operation.

Remark 6 Although the above observations are based on the data obtained via approximate calculations (8), (34)-(36), we believe that relative properties of various allocations remain valid for the real systems as well. Discrete event simulation support these conclusions. □

5.3 Monotonicity

Intuitively, it is clear that DTP is monotonically decreasing as a function of the load factor, L, and P_i, and increasing as a function of R_i and N_i, i =

$1, \cdots, M$. Although analytical proof of these facts seems to be impossible, all systems analyzed in this research, using (8), (34)-(36), exhibited these properties. Examples are shown in Figures 10 and 11. (In Figure 10, values of $L > 1$ are included for the sake of mathematical completeness.)

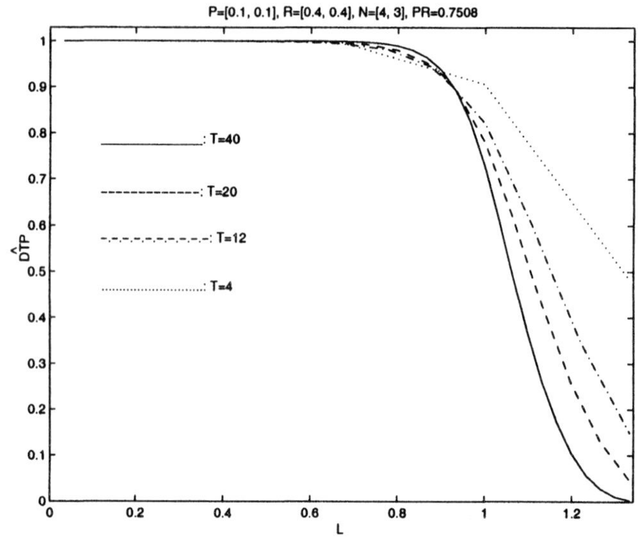

Figure 10. Monotonicity with respect to load factor

On the other hand, the intuitively expected property of DTP monotonicity with respect to the shipping period, T, does not take place. Several examples are shown in Figures 12 and 13. (In Figure 12, the value of D/T is kept constant, while T is changing; in Figure 13, the value of PR is fixed.) The lack of monotonicity is especially pronounced for high load factors and decreased for relatively small L. Discrete event simulations also support this conclusion.

The counterintuitive behavior of Figures 12 and 13 may be explained as follows: For smaller T (which implies smaller D), the ratio N_M/D is larger, and the probability to satisfy the demand by the contents of FGB is larger than that for relatively larger T. This explains the decrease of DTP. Only when T becomes so large that the averaging effect of the shipping period becomes dominant, the system ships reliably all demands, D, at or below its capacity.

Thus, smaller but more frequent shipments are easier to meet than larger but less frequent ones. This conclusion is important for lean operations, since the regime when DTP is increasing as a function of T occurs for relatively large T, typically associated with mass production operations.

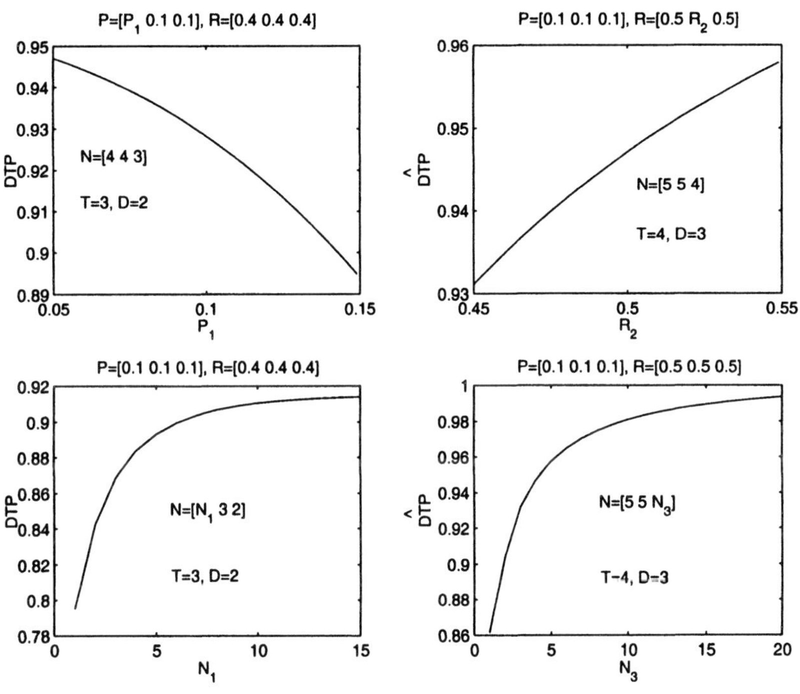

Figure 11. Monotonicity with respect to P_i, R_i and N_i

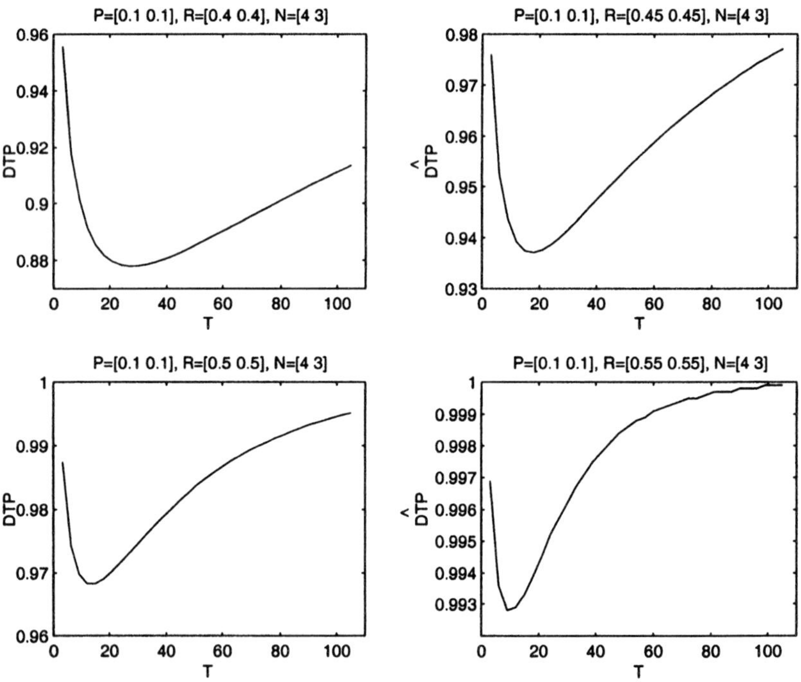

Figure 12. Lack of monotonicity with respect to shipping period (the case of D/T=constant)

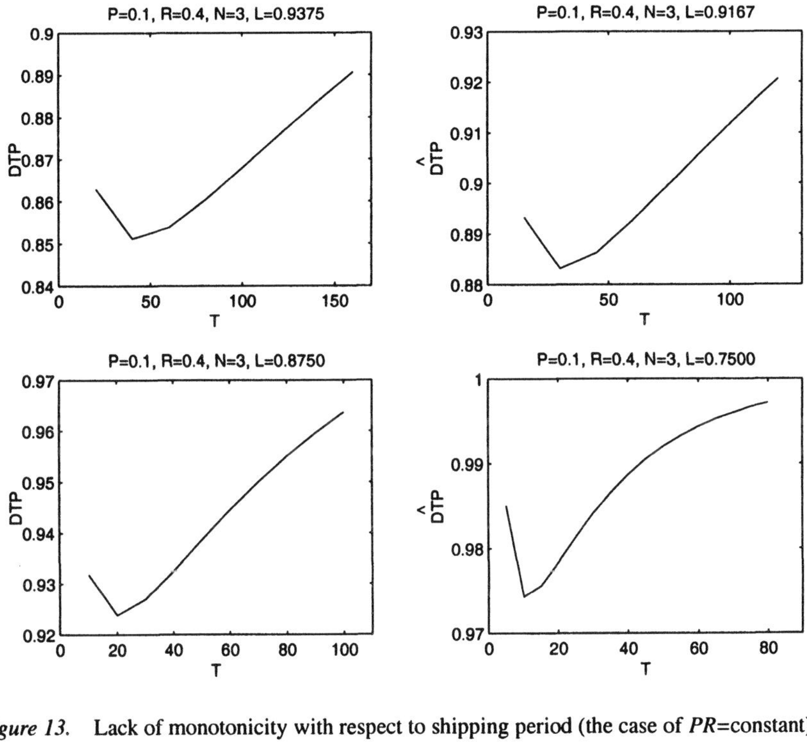

Figure 13. Lack of monotonicity with respect to shipping period (the case of *PR*=constant)

6. Conclusions

This paper provides a method for calculating Due-Time Performance measure in serial production lines with Markovian machines, finite in-process buffering, and finite Finished Goods Buffer. The method is computationally efficient and results in sufficiently high accuracy. Based on this method, system properties of DTP are analyzed, and FGB capacity necessary for high DTP is investigated.

The results obtained may be useful for analysis and design of production systems from the point of view of customer demand satisfaction and for design of efficient supply chains.

Appendix: Proofs

Due to page limitation, we provide here only sketch of the proofs. Complete proofs can be found in Li and Meerkov (2000c).

Proof of Theorem 1

The proof of Theorem 1 consists of the following three steps:

Step 1: Derivation of the steady state balance equations.

First, introduce the following steady state probabilities

$$Y_{k,s_1s_2} = \text{Prob}\{k \text{ parts in the buffer}, m_1 \text{ and } m_2 \text{ are in states } s_1 \text{ and } s_2 \text{ respectively at beginning of the slot}\}, \quad k = 0, 1, \cdots, N,$$

where

$$s_i = \begin{cases} 1, & m_i \text{ is up}, \\ 0, & m_i \text{ is down}, \quad i = 1, 2. \end{cases}$$

Next, write the balance equations for empty buffer, buffer occupancy equaled to 1 ($N = 1$, $N > 1$, respectively), buffer occupancy equaled to k, $1 < k < N$, and the full buffer, respectively.

Step 2: Analysis of case $N = 1$.

- Write $Y_{0,00}$, $Y_{0,10}$, $Y_{0,01}$, $Y_{1,11}$, $Y_{1,10}$, $Y_{1,01}$, $Y_{1,00}$ in terms of $Y_{0,11}$.
- From the fact that the total probability is equaled to 1, calculate $Y_{0,11}$ and $Q(P_1, R_1, P_2, R_2, N)$.

$$Y_{0,11} = \frac{R_1R_2P_1\beta_2}{(R_1 + R_2 - R_1R_2)(R_1 + P_1)(R_2 + P_2)}.$$

It follows then that

$$Q(P_1, R_1, P_2, R_2, N) = \frac{Y_{0,11}}{e_2R_1} = \frac{P_1\beta_2}{(R_1 + R_2 - R_1R_2)(R_1 + P_1)}.$$

Step 3: Analysis of the case $N > 1$.

- Write $Y_{k,11}$, $Y_{k,10}$, $Y_{k,01}$, $Y_{k,00}$, $k = 1, \cdots, N$, in terms of $Y_{0,11}$.

• From the fact that the total probability is equaled to 1, calculate $Y_{0,11}$ and $Q(P_1, R_1, P_2, R_2, N)$.

$$\begin{aligned} Y_{0,11} &= P_1R_1R_2\alpha_1\alpha_2\beta_2^2\Big[P_1R_2\alpha_1\alpha_2\beta_2(P_2+\beta_2) \\ &\quad +P_1R_1R_2\alpha_2\big[\beta_2^2+P_2(\alpha_1+\beta_1)(\alpha_2+2\beta_2)\big] \\ &\quad +\sum_{k=2}^{N-1}P_1P_2R_1R_2(\alpha_2+\beta_2)^3\sigma^{k-1} \\ &\quad +P_2R_1\alpha_1\beta_2\big(R_2(\alpha_1+\beta_1)+\alpha_2(P_1+R_1)\big)\sigma^{N-1}\Big]^{-1} \\ &= \frac{P_1R_1R_2\alpha_1\alpha_2\beta_2^2}{A+B+C+D}, \end{aligned}$$

where,

$$\begin{aligned} A &= P_1R_2\alpha_1\alpha_2\beta_2(P_2+\beta_2), \\ B &= P_1R_1R_2\alpha_2\big[\beta_2^2+P_2(\alpha_1+\beta_1)(\alpha_2+2\beta_2)\big], \\ C &= \sum_{k=2}^{N-1}P_1P_2R_1R_2(\alpha_2+\beta_2)^3\sigma^{k-1}, \\ D &= P_2R_1\alpha_1\beta_2\big[R_2(\alpha_1+\beta_1)+\alpha_2(P_1+R_1)\big]\sigma^{N-1}. \end{aligned}$$

It follows then that,

$$Q(P_1,R_1,P_2,R_2,N)=\frac{P_1\alpha_1\alpha_2\beta_2^2(R_2+P_2)}{A+B+C+D}.$$

• Calculate $E(h)$, ms_2, mb_1, where

$$\begin{aligned} E(h) &= \sum_{k=1}^{N}k(Y_{k,11}+Y_{k,10}+Y_{k,01}+Y_{k,00}), \\ ms_2 &= Y_{0,11}+Y_{0,01}, \\ mb_1 &= Y_{N,10}. \end{aligned}$$

Theorem 1 is proved. ■

Proof of Proposition 1

Rewrite (5) as

$$Q(P_1,R_1,P_2,R_2,N)=\frac{R_1+R_2-R_1R_2-P_2R_1}{(R_1+R_2-R_1R_2)(1+\frac{R_1}{P_1})}.$$

It follows then that $Q(P_1,R_1,P_2,R_2,N)$ is monotonically increasing with respect to P_1 and monotonically decreasing with respect to P_2.

Rewrite (5) as

$$Q(P_1,R_1,P_2,R_2,N)=\frac{P_1}{R_1+P_1}(1-\frac{P_2}{1+\frac{R_2}{R_1}-R_2}).$$

It follows then that $Q(P_1,R_1,P_2,R_2,N)$ is monotonically decreasing with respect to R_1, and monotonically increasing with respect to R_2. Proposition 1 is proved. ■

The proof of Theorem 2 is based on the following 5 Lemmas, the proofs of which can be found in Li and Meerkov (2000c).

Lemma 1 *Function $Q(P_1, R_1, P_2, R_2, N)$, defined in (5) is monotonically decreasing with respect to N.*

Lemma 2 *Function $Q(P_1, R_1, P_2, R_2, N)$, defined in (5) takes values on* $(0, 1)$.

Lemmas 3 and 4 are used to analyze the monotonicity properties of sequences $P_j^b(s)$, $R_j^b(s)$ and $P_j^f(s)$, $R_j^f(s)$, as formulated next.

Lemma 3 *Consider $P_i^f(s)$, $R_i^f(s)$, $P_i^b(s)$ and $R_i^b(s)$, $i = 1, \cdots, M$, defined by recursive procedure 1. If for all $j = 2, \cdots, M$, $R_j^f(s) < R_j^f(s-1)$ and $P_j^f(s) > P_j^f(s-1)$, then for all $j = 1, \cdots, M-1$, $R_j^b(s) > R_j^b(s-1)$ and $P_j^b(s) < P_j^b(s-1)$.*

Lemma 4 *If for all $j = 1, \cdots, M-1$, $R_j^b(s+1) > R_j^b(s)$ and $P_j^b(s+1) < P_j^b(s)$, then for all $j = 2, \cdots, M$, $R_j^f(s+1) < R_j^f(s)$ and $P_j^f(s+1) > P_j^f(s)$.*

Lemma 5 *Sequences $P_j^f(s)$ and $P_j^b(s)$ are monotonically increasing and sequences $R_j^f(s)$ and $R_j^b(s)$ are monotonically decreasing.*

Proof of Theorem 2: Under the assumptions of the Theorem, since the sequences $P_i^f(s)$, $R_i^f(s)$, $P_i^b(s)$ and $R_i^b(s)$, $i \leq j \leq M$, are monotonic (Lemma 5) and bounded from above and below (Lemma 2), they are convergent. This proves (9).

To prove (10), consider the steady state equations of the recursive procedure (1) and define

$$
\begin{aligned}
e_i^f &= \frac{R_i^f}{R_i^f + P_i^f}, \qquad i = 1, \cdots, M, \\
e_i^b &= \frac{R_i^b}{R_i^b + P_i^b}, \qquad i = 1, \cdots, M.
\end{aligned}
$$

The following property holds (see Li and Meerkov 2000c):

$$
\frac{e_i^f e_i^b}{e_i} = \frac{e_j^f e_j^b}{e_j}, \qquad i, j = 1, \cdots, M, \forall i \neq j.
$$

Therefore, $\frac{R_M^f}{P_M^f} = \frac{R_1^b}{P_1^b}$. Theorem 2 is proved. ■

Proof of Theorem 4

By induction: For $n = 0$, $\widehat{p}_s(0) = 0$, then

$$
\begin{aligned}
&P_M'(1) = P_M, \qquad R_M'(1) = R_M, \\
&\widehat{P}_f(1) = \Phi_1(P_M, R_M, N_M, T, D), \\
&R_M''(1) = R_M[1 - \widehat{P}_f(1)], \\
&P_M''(1) = P_M + R_M \widehat{P}_f(1).
\end{aligned}
$$

Thus,

$$
\begin{aligned}
\widehat{p}_s(1) &= \Phi_2(P_1, R_1, \cdots, P_M + R_M \widehat{P}_f(1), R_M[1 - \widehat{P}_f(1)], N_1, \cdots, N_{M-1}) > 0, \\
R_M'(2) &= R_M[1 - \widehat{p}_s(1)] < R_M'(1), \\
P_M'(2) &= P_M + R_M \widehat{p}_s(1) > P_M'(1).
\end{aligned}
$$

From Hypothesis 2, we have

$$\widehat{P}_f(2) < \widehat{P}_f(1).$$

Assume that for $n > 0$,

$$\widehat{p}_s(n) > \widehat{p}_s(n-1).$$

Then

$$\begin{aligned}\widehat{P}_f(n+1) &< \widehat{P}_f(n),\\ R''_M(n+1) &> R''_M(n),\\ P''_M(n+1) &< P''_M(n).\end{aligned}$$

Due to the monotonicity of operator Φ_2, we have

$$\widehat{p}_s(n+1) > \widehat{p}_s(n),$$

which implies that

$$\widehat{P}_f(n+2) < \widehat{P}_f(n+1).$$

Therefore, the sequences $\widehat{p}_s(n)$ and $\widehat{P}_f(n)$ are monotonically increasing and decreasing, respectively. From (5) and (31), $\widehat{p}_s(n)$ and $\widehat{P}_f(n)$ are bounded from above and below. Therefore, they are convergent. Theorem 4 is proved. ∎

References

[1] Altiok, T. (1997). *Performance Analysis of Manufacturing Systems*, Springer.

[2] Buzacott, J. A., and J. G. Shanthikumar. (1993). *Stochastic Models of Manufacturing Systems*, Prentice Hall.

[3] Chiang, S. -Y. (1999). "Bottlenecks in Production Systems with Markovian Machines: Theory and Applications", *Ph.D Thesis*, Dept. of EECS, Univ. of Michigan, Ann Arbor, MI.

[4] Dallery, Y., R. David and X. L. Xie. (1988). "An Efficient Algorithm for Analysis of Transfer Lines with Unreliable Machines and Finite Buffers", *IIE Transactions*, Vol. 20, pp. 280-283.

[5] Dallery, Y., and S. B. Gershwin. (1992). "Manufacturing Flow Line Systems: A Review of Models and Analytical Results", *Queuing Systems*, Vol. 12, pp. 3-94.

[6] Gershwin, S. B. (1993). "Variance of Output of a Tandem Production System", in *Queuing Networks with Finite Capacity*, pp. 291-304, ed. R. O. Onvaral and I. F. Akyildi, Elsevier Science Publishers.

[7] Gershwin, S. B. (1994). *Manufacturing Systems Engineering*, Prentice Hall.

[8] Govil, M. C., and M. C. Fu. (1999). "Queueing Theory in Manufacturing: A Survey", *Journal of Manufacturing Systems*, Vol. 18, pp. 214-240.

[9] Hillier, F. S., and R. W. Boling. (1966). "The Effect of Some Design Factors on the Efficiency of Production Lines with Variable Operation Times", *Journal of Industrial Engineering*, Vol. 17, pp. 651-658.

[10] Jacobs, D. A. (1993). "Improvability in Production Systems: Theory & Case Studies", *Ph.D. Thesis*, Dept. of EECS, Univ. of Michigan, Ann Arbor, MI.

[11] Jacobs, D. A., and S. M. Meerkov. (1995a). "System-Theoretic Analysis of Due-Time Performance in Production Systems", *Mathematical Problems in Engineering*, Vol. 1, pp. 225-243.

[12] Jacobs, D. A., and S. M. Meerkov. (1995b). "A System-theoretic Property of Serial Production Lines: Improvability", *International Journal of System Sciences*, Vol. 26, pp. 755-785.

[13] Li, J., and S. M. Meerkov. (1998). "Production Variability in Manufacturing Systems: Problem Formulation and Performance Bounds", *Proc. of 37th IEEE CDC*, pp. 2730-2735, Tampa, FL.

[14] Li, J., and S. M. Meerkov. (2000a). "Production Variability in Manufacturing Systems: Bernoulli Reliability Model", *Annals of Operations Research*, Vol. 93, pp. 299-324.

[15] Li, J., and S. M. Meerkov. (2000b). "Bottleneck with respect to Due-Time Performance in Pull Serial Production Lines", *Mathematical Problems in Engineering*, Vol. 6, pp. 479-498.

[16] Li, J., and S. M. Meerkov. (2000c). "Due-Time Performance in Markovian Production Systems with Finished Goods Buffers", *Control Group Report No. CGR 00-05*, EECS Dept., Univ. of Michigan.

[17] Li, J., and S. M. Meerkov. (2001). "Customer Demand Satisfaction in Production Systems: A Due-Time Performance Approach", *IEEE Transactions on Robotics and Automation*, Vol. 17, pp. 472-482.

[18] Papadopoulos, H. T., and C. Harvey. (1996). "Queueing Theory in Manufacturing Systems Analysis and Design: A Classification of Models for Production and Transfer Lines", *European Journal of Operational Research*, Vol. 92, pp. 1-27.

[19] Papadopoulos, H. T., C. Harvey and J. Browne. (1993). *Queueing Theory in Manufacturing Systems Analysis and Design*, Chapman & Hall.

[20] Sheskin, T. J. (1976). "Allocation of Interstage Along an Automatic Production Line", *AIIE Transactions*, Vol. 8, pp. 146-152.

[21] Tan, B. (1998). "Effects of Variability on the Due-Time Performance of Production Lines", *International Journal of Production Economics*, Vol. 54, pp. 87-100.

[22] Tan, B. (1999). "Variance of the Output as a Function of Time: Production Line Dynamics", *European Journal of Operational Research*, Vol. 117, pp. 470-484.

[23] Viswanadham, N., and Y. Narahari. (1992). *Performance Modeling of Automated Manufacturing System*, Practice Hall.

Chapter 11

ANALYSIS OF TWO-VALVE FLUID-FLOW SYSTEMS WITH GENERAL REPAIR TIMES

Ünsal Özdoğru
Transport Dynamics, Inc.
Princeton, NJ 08640
ozdogru@transdynamics.com

Tayfur Altiok
Rutgers University
Piscataway, NJ 08854
altiok@rci.rutgers.edu

Abstract A two-valve fluid-flow system with finite storage in between is considered. Valves are subject to failures with exponential up times and phase type repair times. We have developed a continuous time Markov Chain approach to study the steady-state behavior of the valves and the material in the storage. The Markov chain has both continuous and discrete states. We have constructed a system of linear equations for the boundary states and a set of differential equations for the interior states, that are solved using the eigenvalue method. We have also investigated the impact of down time variability on the system performance metrics.

1. Introduction

Models of continuous-material flow systems naturally exist in chemical industries as well as in mineral extraction and transport systems such as bulk ports where ships are unloaded or loaded by coal, cement, salt, bauxite or other materials. Motivation of the present study stems from the latter where bauxite is brought in by trains to a port, stored in a yard, dried and eventually loaded on ships to be transported to destinations with aluminum production facilities. Material in the yards which are viewed as buffers in the system is transferred from one location to another by some type of equipment. Machinery involved in the entire operation includes train dumpers, stacker/reclaimers, conveyors, dryers and ship loaders, all prone to failures causing stoppages in the material flow. Since a failure on a machine instantly stops the material flow, in our analysis each machine in the system is considered as a simple valve operating at a constant rate measured in tons per unit time and experiencing on/off periods.

Clearly, it is difficult to analyze systems with more than two valves. A common approach is decomposition in which the system is divided into smaller subsystems with two valves that can be analyzed with less difficulty and can be related to the original system. Briefly, the first and second valves in each two-valve subsystem respectively replicate the impact of the upstream (starvation effect) and the downstream (blocking effect) valves on the system behavior. Thus, the simplest abstraction out of a material flow system to be analyzed is a two-valve continuous-flow system with a finite capacity buffer in between as shown in Figure 1. Valves 1 and 2 are represented by V_i, $i = 1, 2$, respectively.

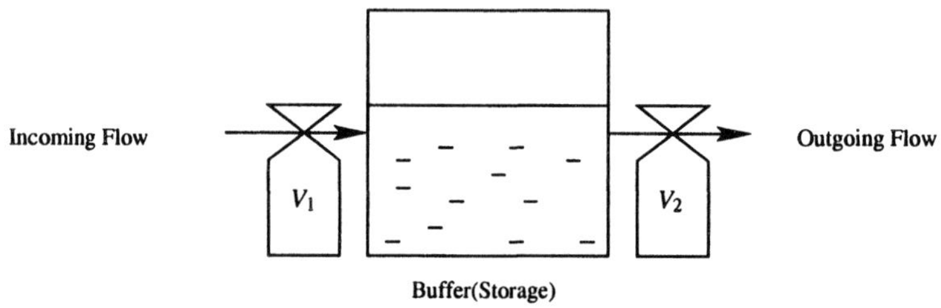

Figure 1. A two-valve fluid-flow system

The two-valve fluid-flow system with exponentially distributed failure and repair times has already been extensively studied. One of the earliest studies was reported by Koenigsberg [5]. In the study, failures are assumed to be operation-dependent and both machines may fail at the same time. Koenigsberg used a Markov Chain approach to find the throughput of the system. Sevast'Yanov [7] analyzed a continuous-flow system with unreliable machines under operation-dependent failures and assumed that only one machine may fail at a time.

The ones which extensively studied two-valve fluid-flow systems include Wijngaard [8], Gershwin and Schick [4], Yeralan, Franck and Quasem [9] and Mitra [6]. Wijngaard [8] investigated the effect of the buffer capacity on the throughput of a two-valve continuous-material flow line for the case where the valves have different flow rates, failure rates and repair rates. Wijngaard used renewal theory arguments to obtain a closed form solution for the throughput of the system. Gershwin and Schick [4] analyzed the behavior of the system in two classes: the boundary behavior and the behavior when the system is in intermediate states. They studied continuous time Markov process with a mixture of continuous and discrete states to obtain the steady-state probabilities of the system behavior. Yeralan, Franck and Quasem [9] employed a combination of the two methods introduced by Wijngaard[8] and Gershwin and Schick [4]. In particular, Yeralan et al. [9] used a Markovian approach coupled with a renewal argument to analyze two-valve continuous-material flow systems with valves operating at the same rate. The system analyzed by Mitra [6] is more involved than the ones previous studies. Mitra [6] considered the first and the second valve consisting of m and n machines coupled with a buffer in between. This study is different than the previous ones in the sense that each machine can be considered as a valve. Consequently, there are multiple valves at each end of the buffer. Initially, Mitra's model isolates the valves as sets of input and output valves and analyzes their behavior as a continuous time Markov chain, one for each set of valves. Recently, Bihan and Dallery [2] attempted to analyze systems with continuous flow and identical processing rates of the machines under operation-dependent failures assumption. Their approach to solving systems of continuous material flow differs from the earlier work in a sense that the repair time distribution of the machines are assumed to be of two-phase hyperexponential.

In the previous models of two-valve continuous-flow systems, the common assumption was that time to failures and repair times of the valves were exponentially distributed. In this research, phase-type repair times are assumed due to their flexibility and analytical tractability. In particular, the Mixture of Generalized Erlang Distribution with two phases (MGE-2) is assumed for the repair times. The choice of MGE-2 is due to the squared coefficient of variation of the MGE distributions which can assume any value greater than equal to 0.5. However, the approach we employ to obtain the density functions defining the system behavior can be extended for the case where the repair time distribution of the valves is MGE with any number of phases. The approach in this analysis is based on continuous-time Markov process with a mixture of continuous and discrete states. Note that the fluid-flow model we propose in this paper can be used as an approximation to model asynchronous flow lines with deterministic processing times. Let us next discuss some important aspects the system behavior before going into the details of the model. During those periods where both valves are operating, the material level will exhibit the behavior shown in Figure 2. Let c_i be the flow rate capacity of V_i, such as ton/hour. Let us assume that $c_1 < c_2$ and both valves are operational and there is some material in the buffer. If both valves remain operational for a long time, material level in the buffer decreases gradually at the rate of $(c_2 - c_1)$, per unit time; and at one point in time it hits zero level. Since $c_1 < c_2$, the process continues with the valves operating at the rate of c_1 until one of the valves fails. If V_1 fails first, V_2 starts starving due to lack of material in the buffer. Since we assume that failures are operation-dependent, V_2 does not fail once it enters starving state. The system remains there until V_1 recovers from the failure. If V_2 fails before V_1 does, the buffer level starts increasing by c_1 per unit time. Then, V_1 may fail before V_2 gets repaired; or V_2 becomes functional and consequently the level in the buffer starts decreasing by $(c_2 - c_1)$ per unit time. Therefore, once the system enters intermediate states, depending on the status of the valves, the buffer level increases or decreases. Similar behavior can be observed when $c_1 > c_2$.

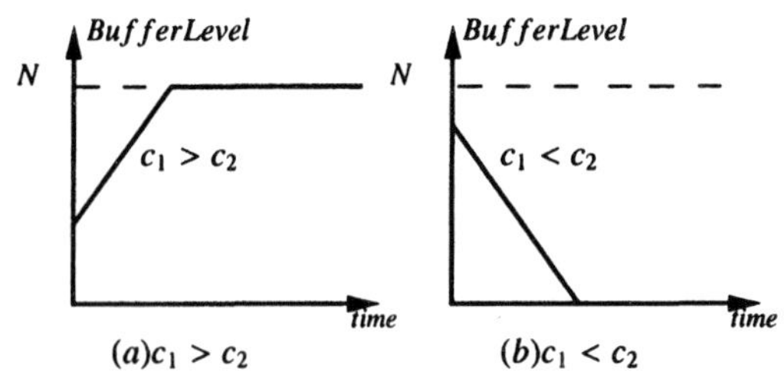

Figure 2. Behavior of the material level in the buffer

Clearly, the level in the buffer does not change at all if both valves are down; or both valves are operational given that $c_1 = c_2$; or both valves are operating with zero material in the buffer given that $c_1 < c_2$; or both valves are operating with full buffer given that $c_1 > c_2$. The last two events indicate that the capacity of valves reduced to the minimum of the two capacities, c_1 and c_2. More precisely, in the former case, V_2 starts operating at the rate of c_1 since there is no material in the buffer and the amount of material provided by V_1 is no more than c_1. Similarly, in the latter case, V_1 starts operating at the rate of c_2 since the buffer reaches up to its capacity and the amount of flow through V_2 can not be more than c_2.

2. The Model

Let us study the stochastic process $S_t = (\alpha_1(t), \alpha_2(t), Z_t)$ where $\alpha_i(t)$ indicates the status of V_i and Z_t designates the random variable representing the material level in the buffer with $0 \leq Z_t \leq N$ at time t.

Below are the set of assumptions we consider in this study.

- There is a continuous supply of incoming flow at V_1 and infinite buffer space is provided after V_2 such that V_1 never starves and the V_2 never gets blocked.

- The buffer in between the valves (container, yard or simply a storage place) has finite capacity denoted by N in units of the material (tons, e.g.).
- Valves can operate at different rates that are c_1 and c_2 for V_1 and V_2, respectively. Material flow rate through the system is reduced to the minimum of the valve rates when the buffer is either empty and $c_1 < c_2$ or full and $c_1 > c_2$. Let c be

$$c = \min(c_1, c_2). \tag{1}$$

- Valves are subject to operation-dependent failures. Up time of valve V_i, $i = 1, 2$ is assumed to be exponentially distributed with rate δ_i, $i = 1, 2$. Down time of valve V_i, $i = 1, 2$ is assumed to be a Mixture of Generalized Erlang random variable with two phases (MGE-2) with parameters $(\gamma_i^1, \gamma_i^2, a_i)$. A pictorial representation of an MGE-2 distribution is given in Figure 3. The time V_i, $i = 1, 2$ spends in the first and the second phases is exponentially distributed random variable with rate γ_i^j, $j = 1, 2$. The probability density function of the MGE-2 distribution is as follows.

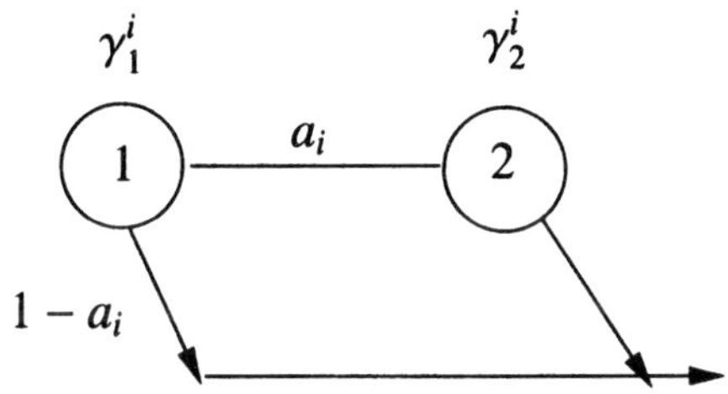

Figure 3. A graphical representation of an MGE-2 distribution

$$f(x) = k_1 \gamma_i^1 e^{-\gamma_i^1 x} + k_2 \gamma_i^2 e^{-\gamma_i^2 x}, \quad x \geq 0$$

$$\text{where } k_1 = \frac{\gamma_i^1 (1 - a) - \gamma_i^2}{-\gamma_i^1 \gamma_i^2}, \; k_2 = 1 - k_1 \quad \text{with } \gamma_i^1 \neq \gamma_i^2 \tag{2}$$

Briefly, when V_i, $i = 1, 2$ fails, it may become operational with probability $(1 - a_i)$ right after the first phase of the repair time, or with probability a_i it also goes through the second phase and then becomes operational.

Let α_i, $i = 0, 1, 2$ indicate the status of V_1 and V_2, respectively. More specifically, $\alpha_i = 0, 1, 2$, $i = 1, 2$, corresponds to the valve state that the ith valve is operating, in the first down phase or in the second down phase, respectively. At any given time, the buffer can be in three possible class of states: full, empty or partially full. While the buffer remains full, V_1 may be blocked. Similarly, while the buffer remains empty V_2 may be starved. While the buffer is partially full, there are nine valve states, as described below.

Valve States (α_1, α_2)

(a) $(0,0)$: Both V_1 and V_2 are operating.

(b) $(0,1)$: V_1 is operating and V_2 is in the repair phase 1.

(c) $(0,2)$: V_1 is operating and V_2 is in the repair phase 2.

(d) $(1,0)$: V_1 is in the repair phase 1 and V_2 is operating.

(e) $(2,0)$: V_1 is in the repair phase 2 and V_2 is operating.

(f) $(1,1)$: Both V_1 and V_2 are in the first repair phase.

(g) $(1,2)$: Both V_1 and V_2 are in the first and the second repair phases, respectively.

(h) $(2,1)$: Both V_1 and V_2 are in the second and the first repair phase, respectively.

(i) $(2,2)$: Both V_1 and V_2 are in the second repair phase.

Furthermore, there is four more valve states the system enters and stays with nonzero probability when the buffer is not partially full. These are,

(j) $(B,1)$ and $(B,2)$ are the two boundary states denoting that V_1 is blocked when V_2 is in its first and second repair phase, respectively.

(k) $(1,S)$ and $(2,S)$ are the two additional boundary states denoting that V_2 is starved when V_1 is in its first and second repair phase, respectively.

Since the material level is a continuous random variable, the internal behavior of the system is described by using probability density functions. Let the time dependent cumulative distribution function of the buffer level be (when the buffer is partially full),

$$F_{\alpha_1(t),\alpha_2(t)}(z,t) = P[\text{ Material Level } \leq z, (\alpha_1(t), \alpha_2(t)) \text{ at } t,]\ 0 \leq z \leq N. \quad (3)$$

The time dependent density function of Z_t at different valve states is given by

$$f_{\alpha_1(t),\alpha_2(t)}(z,t) = \frac{\partial F_{\alpha_1(t),\alpha_2(t)}(z,t)}{\partial z},$$
$$\alpha_1(t),\ \alpha_2(t) = 0,1,2,\ 0 \leq z \leq N,\ t \geq 0. \quad (4)$$

Then, the steady-state density function of the material level Z for the valve states α_1 and α_2 will be

$$f_{\alpha_1,\alpha_2}(z) = \lim_{t\to\infty} f_{\alpha_1,\alpha_2}(z,t),\ \alpha_1,\ \alpha_2 = 0,1,2,\ 0 \le z \le N,\ t \ge 0 \tag{5}$$

that can be used to obtain many of the performance measures such as the average material level, the throughput and the probabilities that the valves being up and down.

Note that, the steady-state density function of the material level in the buffer when it is partially full can be stated as follows.

$$f_Z(z) = f_{0,0}(z) + f_{1,0}(z) + f_{2,0}(z) + f_{0,1}(z) + f_{0,2}(z) + f_{1,1}(z)$$
$$+ f_{1,2}(z) + f_{2,1}(z) + f_{2,2}(z),\ \ 0 \le z \le N. \tag{6}$$

Furthermore, let $P(\alpha_1(t), \alpha_2(t), z_t)$ denote the probability mass function defining the boundary behavior of the system at time t for valve states $\alpha_1 = B, 0, 1, 2$, $\alpha_2 = S, 0, 1, 2$; and for the material level $z = 0, N$ with $P(\alpha_1, \alpha_2, z)$ being its steady-state probability.

In this study, we are interested in the steady-state behavior of the two valve system. Particularly, we are interested in finding the system performance metrics such as the distribution and the average buffer level ($\bar{N}$), the throughput ($\bar{o}$), and the probabilities of the valves being up and down. Some of these measures are given below.

$$\bar{o} = c_1[\int_0^N (f_{0,0}(z) + f_{0,1}(z) + f_{0,2}(z))dz + P(0,0,0)] + cP(0,0,N)$$

$$\bar{o} = c_2[\int_0^N (f_{0,0}(z) + f_{1,0}(z) + f_{2,0}(z))dz + P(0,0,N)] + cP(0,0,0) \tag{7}$$

where c is given by (1), and

$$\bar{N} = \int_0^N z \sum_{\alpha_1=0}^{2} \sum_{\alpha_2=0}^{2} f_{\alpha_1,\alpha_2}(z)dz + N[P(B,1,N) + P(B,2,N)]. \tag{8}$$

The analysis of the system is divided in two parts: the boundary behavior and the behavior when the system is in intermediate states. As shown in Figure 2, while $c_1 < c_2$ and both valves remain operational for a long time, the level of the buffer decreases gradually and thereof the buffer gets empty at some point in time. Nevertheless, processing continues at a reduced rate of q in state $(0,0,0)$ until one of the valves fails. Note that since both valves operate at the same rate, the buffer level does not change. This behavior is completely different than that of an internal state $(0,0,z)$, $0 \leq z \leq N$, since the system can not stay at a particular level due to the change in the level in the buffer per unit time. To further clarify the system behavior in the boundary states, let us demonstrate the circumstances the system may go through after it enters state $(0,0,0)$. Once the system enters state $(0,0,0)$, it stays there until one of the two events occurs: either V_1 or V_2 fails. Note that the time until the system state changes is the minimum of the exponential failure times of the two valves. In case V_1 fails before V_2 does, the system enters another boundary state, $(1,S,0)$. The time the system spends in that state is an exponential random variable with rate γ_1^1. Note that, due to operation-dependent failures assumption, V_2 does not fail while it is starving. However, after the system spends a random amount of time in $(1,S,0)$, it may either go back to state $(0,0,0)$ if V_1 recovers from the first phase of the repair time distribution with probability $(1-a_1)$, or enters another boundary state which is $(2,S,0)$ if V_1 goes through the second phase of the repair time distribution with probability a_1. Therefore, probability mass functions adequately define the behavior of the system at boundary states. On the other hand, if V_2 fails before V_1 does given that the system has been in $(0,0,0)$, the system visits intermediate states again since the level in the buffer starts increasing at the rate of c_1 per unit time.

Hence, we studied the continuous-time Markov process with a mixture of continuous and discrete states to obtain the steady-state probabilities of the system states. The analysis when the system is in intermediate states yields a set of linear and homogeneous partial differential equations. On the other hand, the boundary analysis gives rise to a set of equations linear in the probability mass functions which are expressed in terms of the density functions defining the system's interior behavior. In the following section, we provide a continuous-time Markovian approach to analyze the steady-state behavior of the system.

2.1 Flow-balance equations

Flow-balance equations are constructed to derive the steady-state density functions explicitly by restricting attention to the states of the process at time t and $(t + h_t)$ where h_t designate the time increment. Let h_z denote the material increment in the buffer level. Furthermore, a valve being under repair is assumed to be effective in the same time interval it is repaired. In other words, material is assumed to flow through a valve within the time interval of length h_t it recovers from the failure.

Let us next show how a flow-balance equation is derived. For example, both valves will be operating with the material level z at time $(t + h_t)$ (the system will be in state $(0,0,z,t + h_t)$),

- if both valves were operating at material level $(z - h_z(c_1 - c_2))$ with probability $(1 - \delta_1 h_t)(1 - \delta_2 h_t)$, that is the system is in state $(0, 0, z - h_z(c_1 - c_2))$ at t or
- if V_1 was operating and V_2 was in down phase 1 at material level $(z - h_z(c_1 - c_2))$ with probability $(1 - \delta_1 h_t)(1 - a_2)\gamma_2^1 h_t$, that is the system is in state $(0, 1, z - h_z(c_1 - c_2))$ at t or
- if V_1 was operating and V_2 was in down phase 2 at material level $(z - h_z(c_1 - c_2))$ with probability $(1 - \delta_1 h_t)\gamma_2^2 h_t$, that is the system is in state $(0, 2, z - h_z(c_1 - c_2))$ at t or
- if V_1 was in down phase 1 and V_2 was operating at material level $(z - h_z(c_1 - c_2))$ with probability $(1 - a_1)\gamma_1^1 h_t(1 - \delta_2 h_t)$, that is the system is in state $(1, 0, z - h_z(c_1 - c_2))$ at t or
- if V_1 was in down phase 2 and V_2 was operating at material level $(z - h_z(c_1 - c_2))$ with probability $\gamma_1^2 h_t(1 - \delta_2 h_t)$, that is the system is in state $(2, 0, z - h_z(c_1 - c_2))$ at t.

Consequently, we have

$$\begin{aligned} f_{0,0}(z, t + h_t) = (1 - \delta_1 h_t)(1 - \delta_2 h_t) f_{0,0}(z - h_z(c_1 - c_2), t) + \\ (1 - \delta_1 h_t)(1 - a_2)\gamma_2^1 h_t f_{0,1}(z - h_z(c_1 - c_2), t) + \\ (1 - \delta_1 h_t)\gamma_2^2 h_t f_{0,2}(z - h_z(c_1 - c_2), t) + \\ (1 - a_1)\gamma_1^1 h_t(1 - \delta_2 h_t) f_{1,0}(z + h_z(c_1 - c_2), t) + \\ \gamma_1^2 h_t(1 - \delta_2 h_t) f_{2,0}(z + h_z(c_1 - c_2), t) \end{aligned} \tag{9}$$

Let us next collect all the second order terms under $o(h_t^2)$, and move $f_{0,0}(z - h_z(c_1 - c_2), t)$ to the left hand side of the equation (9). Furthermore, to obtain the rate of changes in time and material, we also add and then subtract $f_{0,0}(z, t)$ to and from the left hand side of the equation.

$$\begin{aligned} f_{0,0}(z, t + h_t) + f_{0,0}(z, t) - f_{0,0}(z, t) - f_{0,0}(z - h_z(c_1 - c_2), t) = \\ -(\delta_1 + \delta_2) h_t f_{0,0}(z - h_z(c_1 - c_2), t) + (1 - a_2)\gamma_2^1 h_t f_{0,1}(z - h_z(c_1 - c_2), t) + \\ \gamma_2^2 h_t f_{0,2}(z - h_z(c_1 - c_2), t) + (1 - a_1)\gamma_1^1 h_t f_{1,0}(z - h_z(c_1 - c_2), t) + \\ \gamma_1^2 h_t f_{2,0}(z - h_z(c_1 - c_2), t) + o(h_t^2) \end{aligned} \tag{10}$$

which can be written as

$$\frac{f_{0,0}(z, t + h_t) - f_{0,0}(z, t)}{h_t} + \frac{f_{0,0}(z, t) - f_{0,0}(z - h_z(c_1 - c_2), t)}{h_z} =$$

$$-(\delta_1+\delta_2)f_{0,0}(z-h_z(c_1-c_2),t)+(1-a_2)\gamma_2^1 f_{0,1}(z-h_z(c_1-c_2),t)+$$
$$\gamma_2^2 f_{0,2}(z-h_z(c_1-c_2),t)+(1-a_1)\gamma_1^1 f_{1,0}(z-h_z(c_1-c_2),t)$$
$$+\gamma_1^2 f_{2,0}(z-h_z(c_1-c_2),t)+\frac{o(h_t^2)}{h_t} \tag{11}$$

Then, by letting $h_t \to 0$ and $h_z \to 0$ in both sides of (11), the following differential equation is obtained.

$$\frac{\partial f_{0,0}(z,t)}{\partial t}+(c_1-c_2)\frac{\partial f_{0,0}(z,t)}{\partial z} = -(\delta_1+\delta_2)f_{0,0}(z,t)+(1-a_2)\gamma_2^1 f_{0,1}(z,t)+$$
$$\gamma_2^2 f_{0,2}(z,t)+(1-a_1)\gamma_1^1 f_{1,0}(z,t)+\gamma_1^2 f_{2,0}(z,t)$$

Finally, let $t \to \infty$ and obtain

$$(c_1-c_2)\frac{df_{0,0}(z)}{dz} = -(\delta_1+\delta_2)f_{0,0}(z)+(1-a_2)\gamma_2^1 f_{0,1}(z,t)+\gamma_2^2 f_{0,2}(z,t)+$$
$$(1-a_1)\gamma_1^1 f_{1,0}(z)+\gamma_1^2 f_{2,0}(z). \tag{12}$$

The flow-balance equations below are constructed in the same manner by considering possible transitions among the system states when the buffer is partially full.

$$f_{1,0}(z,t+h_t)=\delta_1 h_t(1-\delta_2 h_t)f_{0,0}(z+h_z c_2,t)+$$
$$(1-\gamma_1^1 h_t)(1-\delta_2 h_t)f_{1,0}(z+h_z c_2,t)+(1-\gamma_1^1 h_t)(1-a_2)\gamma_2^1 h_t f_{1,1}(z+h_z c_2,t)+$$
$$(1-\gamma_1^1 h_t)\gamma_2^2 h_t f_{1,2}(z+h_z c_2,t)+o(h_t^2) \tag{13}$$
$$f_{2,0}(z,t+h_t)=\gamma_1^1 a_1^1 h_t(1-\delta_2 h_t)f_{1,0}(z+h_z c_2,t)+$$
$$(1-\gamma_1^2 h_t)(1-\delta_2 h_t)f_{2,0}(z+h_z c_2,t)+(1-\gamma_1^2 h_t)(1-a_2)\gamma_2^1 h_t f_{2,1}(z+h_z c_2,t)+$$
$$(1-\gamma_1^2 h_t)\gamma_2^2 h_t f_{2,2}(z+h_z c_2,t)+o(h_t^2) \tag{14}$$
$$f_{0,1}(z,h_t)=(1-\delta_1 h_t)\delta_1^2 h_t f_{0,0}(z-h_z c_1,t)+$$
$$(1-\gamma_1^2 h_t)\gamma_1^2 h_t f_{2,1}(z-h_z c_1,t)+(1-a_1)\gamma_1^1 h_t(1-\gamma_2^1 h_t)f_{1,1}(z-h_z c_1,t)+$$
$$(1-\delta_1 h_t)(1-\gamma_2^1 h_t)f_{0,1}(z-h_z c_1,t)+o(h_t^2) \tag{15}$$
$$f_{0,2}(z,t+h_t)=(1-\delta_1 h_t)(1-\gamma_2^2 h_t)f_{0,2}(z-h_z c_1,t)$$
$$+(1-\delta_1 h_t)a_2\gamma_2^1 h_t f_{0,1}(z-h_z c_1,t)+(1-a_1)\gamma_1^1 h_t(1-\gamma_2^2 h_t)f_{1,2}(z-h_z c_1,t)+$$
$$\gamma_1^2 h_t(1-\gamma_2^2 h_t)f_{2,2}(z-h_z c_1,t)+o(h_t^2) \tag{16}$$

For each flow-balance equation given above, we first reorganize the coefficients of the density functions and collect all the second order terms under $o(h_t^2)$. Then, we divide both sides of the resultant equations obtained from (13)-(16) by h_t and sequentially let $h_t \to 0$, $h_z \to 0$ and $t \to \infty$. Thus, the following are obtained in addition to the one we obtained in (12).

$$-c_2 \frac{df_{1,0}(z)}{dz} = -(\gamma_1^1 + \delta_2) f_{1,0}(z) + \delta_1 f_{0,0}(z) + (1 - a_2)\gamma_2^1 f_{1,1}(z) + \gamma_2^2 f_{1,2}(z) \quad (17)$$

$$-c_2 \frac{df_{2,0}(z)}{dz} = -(\gamma_1^2 + \delta_2) f_{2,0}(z) + \gamma_1^1 a_1 f_{1,0}(z) + (1 - a_2)\gamma_2^1 f_{2,1}(z) + \gamma_2^2 f_{2,2}(z) \quad (18)$$

$$c_1 \frac{df_{0,1}(z)}{dz} = -(\delta_1 + \gamma_2^1) f_{0,1}(z) = \delta_2 f_{0,0}(z) + (1 - a_1)\gamma_1^1 f_{1,1}(z) + \gamma_1^2 f_{2,1}(z) \quad (19)$$

$$c_1 \frac{df_{0,2}(z)}{dz} = -(\delta_1 + \gamma_2^2) f_{0,2}(z) + a_2 \gamma_2^1 f_{0,1}(z) + (1 - a_1)\gamma_1^1 f_{1,2}(z) + \gamma_1^2 f_{2,2}(z) \quad (20)$$

Also, the possible transitions between the system states when both V_1 and V_2 are down give us the following flow-balance equations.

$$f_{1,1}(z, t + h_t) = (1 - \gamma_1^1 h_t)(1 - \gamma_2^1 h_t) f_{1,1}(z, t) + \delta_1 h_t (1 - \gamma_2^1 h_t) f_{0,1}(z, t) +$$
$$(1 - \gamma_1^1 h_t)\delta_2 h_t f_{1,0}(z, t)$$

$$f_{1,1}(z, t + h_t) - f_{1,1}(z, t) = -(\gamma_1^1 + \gamma_2^1) h_t f_{1,1}(z, t) + \delta_1 h_t f_{0,1}(z, t) +$$
$$\delta_2 h_t f_{1,0}(z, t) + o(h_t^2)$$

$$\frac{f_{1,1}(z, t + h_t) - f_{1,1}(z, t)}{h_t} = -(\gamma_1^1 + \gamma_2^1) f_{1,1}(z, t) + \delta_1 f_{0,1}(z, t) +$$
$$\delta_2 f_{1,0}(z, t) + \frac{o(h_t^2)}{h_t}$$

By letting $h_t \to 0$, in both sides of the above equation, we obtain

$$\frac{\partial f_{1,1}(z, t)}{\partial t} = -(\gamma_1^1 + \gamma_2^1) f_{1,1}(z, t) + \delta_1 f_{0,1}(z, t) + \delta_2 f_{1,0}(z, t) \quad (21)$$

and letting $t \to \infty$ provides us with

$$0 = -(\gamma_1^1 + \gamma_2^1) f_{1,1}(z) + \delta_1 f_{0,1}(z) + \delta_2 f_{1,0}(z) \quad (22)$$

which is linear in density functions $f_{1,1}(z)$, $f_{0,1}(z)$, $f_{1,0}(z)$.

The following flow-balance equations are constructed in the same manner when V_1 and V_2 are both in their respective down phases;

$$f_{2,2}(z,t+h_t) = (1-\gamma_1^2 h_t)(1-\gamma_2^2 h_t)f_{2,2}(z,t) + a_2\gamma_2^1 h_t f_{2,1}(z,t) + a_1\gamma_1^1 h_t(1-\gamma_2^2 h_t)f_{1,2}(z,t) \quad (23)$$

$$f_{2,1}(z,t+h_t) = (1-\gamma_1^2 h_t)(1-\gamma_2^1 h_t)f_{2,1}(z,t) + a_1\gamma_1^1 h_t f_{1,1}(z,t) + (1-\gamma_1^2 h_t)\delta_2 h_t f_{2,0}(z,t) \quad (24)$$

$$f_{1,2}(z,t+h_t) = (1-\gamma_1^1 h_t)(1-\gamma_2^2 h_t)f_{1,2}(z,t) + a_2\gamma_2^1 h_t f_{1,1}(z,t) + \delta_1 h_t(1-\gamma_2^2 h_t)f_{0,2}(z,t) \quad (25)$$

with each equation (23)-(25) having $o(h_t^2)$ corresponding to the events occurring with a negligible probability. By using appropriate transformations and sequentially letting $h_t \to 0$ and $t \to \infty$ in both sides of (23)-(25), the following set of linear equations are obtained.

$$0 = -(\gamma_1^2 + \gamma_2^2)f_{2,2}(z) + a_2\gamma_2^1 f_{2,1}(z) + a_1\gamma_1^1 f_{1,2}(z) \quad (26)$$

$$0 = -(\gamma_1^2 + \gamma_2^1)f_{2,1}(z) + a_1\gamma_1^1 f_{1,1}(z) + \delta_2 f_{2,0}(z) \quad (27)$$

$$0 = -(\gamma_1^1 + \gamma_2^2)f_{1,2}(z) + a_2\gamma_2^1 f_{1,1}(z) + \delta_1 f_{0,2}(z) \quad (28)$$

Equations (26)-(28) and (22) can be put into the following form.

$$f_{2,2}(z) = \frac{a_2\gamma_2^1}{(\gamma_1^2+\gamma_2^2)}f_{2,1}(z) + \frac{a_1\gamma_1^1}{(\gamma_1^2+\gamma_2^2)}f_{1,2}(z) \quad (29)$$

$$f_{2,1}(z) = \frac{a_1\gamma_1^1}{(\gamma_1^2+\gamma_2^1)}f_{1,1}(z) + \frac{\delta_2}{(\gamma_1^2+\gamma_2^1)}f_{2,0}(z) \quad (30)$$

$$f_{1,2}(z) = \frac{a_2\gamma_2^1}{(\gamma_1^1+\gamma_2^2)}f_{1,1}(z) + \frac{\delta_1}{(\gamma_1^1+\gamma_2^2)}f_{0,2}(z) \quad (31)$$

$$f_{1,1}(z) = \frac{\delta_1}{(\gamma_1^1+\gamma_2^1)}f_{0,1}(z) + \frac{\delta_2}{(\gamma_1^1+\gamma_2^1)}f_{1,0}(z). \quad (32)$$

Notice that function $f_{2,2}(z)$ is a mixture of $f_{2,1}(z)$ and $f_{1,2}(z)$. On the other hand, functions $f_{2,1}(z)$ and $f_{1,2}(z)$ include $f_{1,1}(z)$ which itself is the mixture of $f_{0,1}(z)$ and $f_{1,0}(z)$. Therefore, functions $f_{2,2}(z)$, $f_{2,1}(z)$ and $f_{1,2}(z)$ can be expressed in terms of $f_{1,0}(z)$, $f_{2,0}(z)$, $f_{0,1}(z)$ and $f_{0,2}(z)$. Furthermore, equations (29)-(32) can be substituted into the differential equations given in (17)-(20) along with the one given in (12). Hence, the result is a system of 5×5 homogeneous differential equations expressed as functions $f_{0,0}(z)$, $f_{0,1}(z)$, $f_{0,2}(z)$, $f_{1,0}(z)$ and $f_{2,0}(z)$.

$$\frac{df_{0,0}(z)}{dz} = \frac{-(\delta_1+\delta_2)}{c_1-c_2}f_{0,0}(z) + \frac{(1-a_2)\gamma_2^1}{c_1-c_2}f_{0,1}(z) + \frac{\gamma_2^2}{c_1-c_2}f_{0,2}(z) + \frac{(1-a_1)\gamma_1^1}{c_1-c_2}f_{1,0}(z) + \frac{\gamma_1^2}{c_1-c_2}f_{2,0}(z)$$

$$\frac{df_{0,1}(z)}{dz} = [\frac{-(\delta_1+\gamma_2^1)}{c_1} + \frac{\gamma_1^1\,(1-a_1)\,\delta_1}{c_1(\gamma_1^1+\gamma_2^1)} + \frac{\gamma_1^2\,a_1\,\gamma_1^1\,\delta_1}{c_1(\gamma_1^2+\gamma_2^1)\,(\gamma_1^1+\gamma_2^1)}]f_{0,1}(z) + [\frac{\gamma_1^1\,(1-a_1)\,\delta_2}{c_1(\gamma_1^1+\gamma_2^1)} + \frac{\gamma_1^2\,a_1\,\gamma_1^1\,\delta_2}{c_1(\gamma_1^2+\gamma_2^1)\,(\gamma_1^1+\gamma_2^1)}]f_{1,0}(z) + \frac{\delta_2}{c_1}f_{0,0}(z) + \frac{\gamma_1^2\,\delta_2}{c_1(\gamma_1^2+\gamma_2^1)}f_{2,0}(z)$$

$$\frac{df_{0,2}(z)}{dz} = [\frac{a_2\,\gamma_2^1}{c_1} + \frac{(1-a_1)\,\gamma_1^1\,a_2\,\gamma_2^1\,\delta_1}{c_1(\gamma_1^1+\gamma_2^2)\,(\gamma_1^1+\gamma_2^1)} + \frac{\gamma_1^2\,A\,\delta_1}{c_1(\gamma_1^2+\gamma_2^2)}]f_{0,1}(z) + [\frac{(1-a_1)\,\gamma_1^1\,a_2\,\gamma_2^1\,\delta_2}{c_1(\gamma_1^1+\gamma_2^2)\,(\gamma_1^1+\gamma_2^1)} + \frac{\gamma_1^2\,A\,\delta_2}{c_1(\gamma_1^2+\gamma_2^2)}]f_{1,0}(z) + \frac{\gamma_1^2\,a_2\,\gamma_2^1\,\delta_2}{c_1(\gamma_1^2+\gamma_2^2)\,(\gamma_1^2+\gamma_2^1)}f_{2,0}(z) + [\frac{-(\delta_1+\gamma_2^2)}{c_1} + \frac{(1-a_1)\,\gamma_1^1\,\delta_1}{c_1(\gamma_1^1+\gamma_2^2)} + \frac{\gamma_1^2\,a_1\,\gamma_1^1\,\delta_1}{c_1(\gamma_1^2+\gamma_2^2)\,(\gamma_1^1+\gamma_2^2)}]f_{0,2}(z)$$

$$\frac{df_{1,0}(z)}{dz} = \frac{-\delta_1}{c_2}f_{0,0}(z) + \frac{\gamma_2^2\,\delta_1}{c_2(\gamma_1^1+\gamma_2^2)}f_{0,2}(z) - [\frac{(1-a_2)\,\gamma_2^1\,\delta_1}{c_2(\gamma_1^1+\gamma_2^1)} + \frac{\gamma_2^2\,a_2\,\gamma_2^1\,\delta_1}{c_2(\gamma_1^1+\gamma_2^2)\,(\gamma_1^1+\gamma_2^1)}]f_{0,1}(z) - [-\frac{(\gamma_1^1+\delta_2)}{c_2} + \frac{(1-a_2)\,\gamma_2^1\,\delta_2}{c_2(\gamma_1^1+\gamma_2^1)} + \frac{\gamma_2^2\,a_2\,\gamma_2^1\,\delta_2}{c_2(\gamma_1^1+\gamma_2^2)\,(\gamma_1^1+\gamma_2^1)}]f_{1,0}(z)$$

$$\frac{df_{2,0}(z)}{dz} = [\frac{(1-a_2)\,\gamma_2^1\,a_1\,\gamma_1^1\,\delta_1}{c_2(\gamma_1^2+\gamma_2^1)\,(\gamma_1^1+\gamma_2^1)} + \frac{\gamma_2^2\,A\,\delta_1}{c_2(\gamma_1^2+\gamma_2^2)}]f_{1,0}(z) -$$

$$\frac{\gamma_2^2\, a_1\, \gamma_1^1\, \delta_1}{c_2(\gamma_1^2+\gamma_2^2)\,(\gamma_1^1+\gamma_2^2)} f_{0,2}(z)$$

$$+[\frac{\gamma_1^1\, a_1}{c_2} + \frac{(1-a_2)\,\gamma_2^1\, a_1\, \gamma_1^1\, \delta_2}{c_2(\gamma_1^2+\gamma_2^1)\,(\gamma_1^1+\gamma_2^1)} + \frac{\gamma_2^2\, A\, \delta_2}{c_2(\gamma_1^2+\gamma_2^2)}] f_{1,0}(z)$$

$$-[\frac{-(\gamma_1^2+\delta_2)}{c_2} + \frac{(1-a_2)\,\gamma_2^1\,\delta_2}{c_2(\gamma_1^2+\gamma_2^1)} + \frac{\gamma_2^2\, a_2\, \gamma_2^1\, \delta_2}{c_2(\gamma_1^2+\gamma_2^2)\,(\gamma_1^2+\gamma_2^1)}] f_{2,0}(z) \qquad (33)$$

where $A = \frac{a_1\gamma_1^1 a_2\gamma_2^1}{\gamma_1^1+\gamma_2^1}[\frac{1}{(\gamma_1^1+\gamma_2^2)} + \frac{1}{(\gamma_1^2+\gamma_2^1)}]$.
The sum of the differential equations in (33) gives the following relationship among $f_{0,0}(z)$, $f_{0,1}(z)$, $f_{0,2}(z)$, $f_{1,0}(z)$, and $f_{2,0}(z)$.

$$c_1[f_{0,0}(z) + f_{1,0}(z) + f_{2,0}(z)] = c_2[f_{0,0}(z) + f_{0,1}(z) + f_{0,2}(z)],\ 0 \le z \le N. \quad (34)$$

Notice that expressing one of the functions in (34) in terms of the others reduces the system of equations (33) to a system of 4×4 set of homogeneous differential equations. To facilitate the solution, we choose function $f_{0,0}(z)$ to be expressed in terms of the others.

$$f_{0,0}(z) = -\frac{c_1}{c_1-c_2} f_{0,1}(z) - \frac{c_1}{c_1-c_2} f_{0,2}(z) + \frac{c_2}{c_1-c_2} f_{1,0}(z) +$$

$$\frac{c_2}{c_1-c_2} f_{2,0}(z),\ 0 \le z \le N. \qquad (35)$$

We now eliminate the first differential equation and substitute (35) into the remaining ones in (33). The resultant system is as follows.

$$\begin{pmatrix} \frac{df_{0,1}(z)}{dz} \\ \frac{df_{0,2}(z)}{dz} \\ \frac{df_{1,0}(z)}{dz} \\ \frac{df_{2,0}(z)}{dz} \end{pmatrix} = \begin{pmatrix} c_{11} & c_{12} & c_{13} & c_{14} \\ c_{21} & c_{22} & c_{23} & c_{24} \\ c_{31} & c_{32} & c_{33} & c_{34} \\ c_{41} & c_{42} & c_{43} & c_{44} \end{pmatrix} \begin{pmatrix} f_{0,1}(z) \\ f_{0,2}(z) \\ f_{1,0}(z) \\ f_{2,0}(z) \end{pmatrix}$$

$$= CF(Z),\ 0 \le z \le N \qquad (36)$$

where the expressions for c_{ij}, $i, j = 1, .., 4$ are given in Appendix 2.1. When $c_1 = c_2$, the linear relationship expressed among the density functions in (34) turns out to be of the following form. Notice that, since $c_1 = c_2$ density function $f_{0,0}(z)$ disappears from both sides of (34). We have

$$f_{0,1}(z) + f_{0,2}(z) = f_{1,0}(z) + f_{2,0}(z),\ \ 0 \le z \le N \qquad (37)$$

Now, equation (37) can be rewritten in such a way that one of the functions becomes a function of the other density functions. We choose density function $f_{0,1}(z)$ to be expressed in term of the other density functions.

$$f_{0,1}(z) = -f_{0,2}(z) + f_{1,0}(z) + f_{2,0}(z), \quad 0 \le z \le N \tag{38}$$

Incorporating the dependency given in (38) among the density functions reduces the dimension of the system of differential equations system given in (36) by one when $c_1 = c_2$. Also, this special case forces (12) to be a linear equation in the density functions. Notice that the left hand side of the equation becomes zero since there is no rate of change of $f_{0,0}(z)$ with respect to z when $c_1 = c_2$. Then, one of the density functions on the right hand side can be expressed in terms of the other densities. We choose $f_{0,0}(z)$; and below is the equation acquired.

$$f_{0,0}(z) = \frac{1}{(\delta_1 + \delta_2)}[(1 - a_2)\gamma_2^1 f_{0,1}(z,t) + \gamma_2^2 f_{0,2}(z,t) + (1 - a_1)\gamma_1^1 f_{1,0}(z) + \gamma_1^2 f_{2,0}(z)], \quad 0 \le z \le N. \tag{39}$$

Notice that $f_{0,0}(z)$ can also be expressed in terms of $f_{0,2}(z)$, $f_{1,0}(z)$ and $f_{2,0}(z)$ by substituting (38) into (39), when $c_1 = c_2$, and obtain

$$\begin{pmatrix} \frac{df_{0,2}(z)}{dz} \\ \frac{df_{1,0}(z)}{dz} \\ \frac{df_{2,0}(z)}{dz} \end{pmatrix} = \begin{pmatrix} c_{11} & c_{12} & c_{13} \\ c_{21} & c_{22} & c_{23} \\ c_{31} & c_{32} & c_{33} \end{pmatrix} \begin{pmatrix} f_{0,2}(z) \\ f_{1,0}(z) \\ f_{2,0}(z) \end{pmatrix} = CF(Z), \quad 0 \le z \le N \tag{40}$$

where the expressions for c_{ij}, $i, j = 1, 2, 3$ are given in Appendix 2.2.

2.2 Boundary Conditions

The rate at which the material flows through the system may be reduced to the minimum of the valve capacities, c as given by (1), while the buffer is either empty $c_1 < c_2$ or full with $c_1 > c_2$. The following boundary flow-balance equations are constructed by considering possible transitions among the valve states from boundary to boundary, interior to boundary and boundary to interior material levels when the buffer is empty and when it is full.

- Buffer is Empty :

  (a) *Boundary to Boundary Equations*

  – Consider the state where the buffer level is zero and both V_1 and V_2 are in their down phase 1 at time $t + h_t$ such that we have*

$$P(1,1,0,t+h_t) = \delta_1 h_t(1-\gamma_2^1 h_t)P(0,1,0,t)+$$
$$(1-\gamma_1^1 h_t)\delta_2 h_t P(1,0,0,t)$$
$$+(1-\gamma_1^1 h_t)(1-\gamma_2^1 h_t)P(1,1,0,t)+o(h_t^2). \quad (41)$$

- Consider the state where the buffer level is zero and both V_1 and V_2 are in their down phases 2 and 1, respectively, at time $t+h_t$ such that we have

$$P(2,1,0,t+h_t) = a_1\gamma_1^1 h_t(1-\gamma_2^1 h_t)P(1,1,0,t)+$$
$$(1-\gamma_1^2 h_t)\delta_2 h_t P(2,0,0,t)$$
$$+(1-\gamma_1^2 h_t)(1-\gamma_2^1 h_t)P(2,1,0,t)+o(h_t^2). \quad (42)$$

- Consider the state where the buffer level is zero and both V_1 and V_2 are in their down phases 1 and 2 respectively, at time $t+h_t$ such that we have

$$P(1,2,0,t+h_t) = (1-\gamma_1^1 h_t)a_1\gamma_1^1 h_t P(1,1,0,t)+$$
$$(1-\gamma_2^2 h_t)\delta_1 h_t P(0,2,0,t)$$
$$+(1-\gamma_1^1 h_t)(1-\gamma_2^2 h_t)P(1,2,0,t)+o(h_t^2). \quad (43)$$

- Consider the state where the buffer level is zero and both V_1 and V_2 are in their down phases 2 and 2 respectively, at time $t+h_t$ such that we have

$$P(2,2,0,t+h_t) = (1-\gamma_1^2 h_t)a_2\gamma_2^1 h_t P(2,1,0,t)+$$
$$a_1\gamma_1^1(1-\gamma_2^2 h_t)P(1,2,0,t)$$
$$+(1-\gamma_1^2 h_t)(1-\gamma_2^2 h_t)P(2,2,0,t)+o(h_t^2). \quad (44)$$

- Note that state $(0,1,0)$ is not reachable from itself since the material level is increased by $c_1 h_t$ in h_t with probability (1 −

*Note that we sligthly modified the state description and included time t in the state vector to be able to write balance equations such as (41).

$\delta_1 h_t)(1-\gamma_2^1 h_t)$.

(b) *Interior and Boundary to Boundary Equations*

- When both V_1 and V_2 are operational and the buffer level is zero, we have

$$P(0,0,0,t+h_t) = (1-\delta_1 h_t)(1-\delta_2 h_t)P(0,0,0,t)+$$
$$\gamma_1^1 h_t(1-a_1)P(1,S,0,t)+\gamma_1^2 h_t P(2,S,0,t)+$$
$$\int_0^{(c_1-c_2)h_z} f_{0,0}(z,t)dz + o(h_t^2). \tag{45}$$

  Note that the last term in (45) is the probability of having the material level in the buffer between 0 and $(c_2-c_1)h_z$ at time t. The above equation is valid only if $c_1 \le c_2$. Otherwise, there will be material accumulation in the buffer at time $(t+h_t)$ regardless of the system state at time t. Consequently, state $(0,0,0)$ becomes transient.

- When V_1 is in down phase 1, V_2 is idle and no material is in the buffer, we have

$$P(1,S,0,t+h_t) = (1-\gamma_1^1 h_t)P(1,S,0,t)$$
$$+(1-\gamma_1^1 h_t)(1-a_2)\gamma_2^1 h_t P(1,1,0,t)$$
$$+\delta_1 h_t(1-\delta_2 h_t)P(0,0,0,t)+(1-\gamma_1^1 h_t)\gamma_2^2 h_t P(1,2,0,t)+$$
$$\int_0^{c_2 h_z} f_{1,0}(z,t)dz + o(h_t^2). \tag{46}$$

- When V_1 is in down phase 2 and V_2 is idle and no material is in the buffer, we have

$$P(2,S,0,t+h_t) = (1-\gamma_1^2 h_t)P(2,S,0,t)+a_1\gamma_1^1 h_t P(1,S,0,t)$$
$$+\int_0^{c_2 h_z} f_{2,0}(z,t)dz + o(h_t^2). \tag{47}$$

(c) *Boundary and Interior to Interior Equations*

- When V_1 is operational and V_2 is in down phase 1, we have

$$\int_0^{c_1 h_z} f_{0,1}(z,t)dz = \gamma_1^2 h_t \int_t^{t+h_t} P(2,1,0,s)ds+$$
$$(1-a_1)\gamma_1^1 h_t \int_t^{t+h_t} P(1,1,0,s)ds$$
$$+\delta_2 h_t \int_t^{t+h_t} P(0,0,0,s)ds + o(h_t^2). \quad (48)$$

- When V_1 is operational and V_2 is in down phase 2, we have

$$\int_0^{c_1 h_z} f_{0,2}(z,t)dz = a_2\gamma_2^1 h_t \int_t^{t+h_t} P(0,1,0,s)ds+$$
$$(1-a_1)\gamma_1^1 h_t \int_t^{t+h_t} P(1,2,0,s)ds$$
$$+\gamma_1^2 h_t \int_t^{t+h_t} P(2,2,0,s)ds + o(h_t^2). \quad (49)$$

Notice that states $(0,1,0)$ and $(0,0,0)$ are not reachable from any of the interior states, and states $(0,0,z)$ and $(1,0,z)$ are not reachable from any of the boundary states.

- Buffer is Full :

(a) *Boundary to Boundary Equations*

- Consider the state where the buffer is full and V_1 and V_2 are both in down phase 1, we have

$$P(1,1,N,t+h_t) = (1-\gamma_1^1 h_t)(1-\gamma_2^1 h_t)P(1,1,N,t)+$$
$$\delta_1 h_t(1-\gamma_2^1 h_t)P(0,1,N,t)$$
$$+(1-\gamma_1^1 h_t)\delta_2 h_t P(1,0,N,t) + o(h_t^2). \quad (50)$$

- Consider the state where the buffer is full and V_1 and V_2 are both in down phases 2 and 1, respectively, we have

$$P(2,1,N,t+h_t) = (1-\gamma_1^2 h_t)(1-\gamma_2^1 h_t)P(2,1,N,t)+$$

$$(1 - \gamma_1^2 h_t)\delta_2 h_t P(2, 0, N, t)$$
$$+a_1\gamma_1^1 h_t(1 - \gamma_2^1 h_t)P(1, 1, N, t) + o(h_t^2). \qquad (51)$$

- Consider the state where the buffer is full and V_1 and V_2 are both in down phases 1 and 2, respectively, we have

$$P(1, 2, N, t + h_t) = (1 - \gamma_1^1 h_t)(1 - \gamma_2^2 h_t)P(1, 2, N, t)+$$
$$\delta_1 h_t(1 - \gamma_2^2 h_t)P(0, 2, N, t)$$
$$+(1 - \gamma_1^1 h_t)a_2\gamma_2^1 h_t P(1, 1, N, t) + o(h_t^2). \qquad (52)$$

- Consider the state where the buffer is full and V_1 and V_2 are both in down phases 2, we have

$$P(2, 2, N, t + h_t) = (1 - \gamma_1^2 h_t)(1 - \gamma_2^2 h_t)P(2, 2, N, t)+$$
$$a_1\gamma_1^1 h_t(1 - \gamma_2^2 h_t)P(1, 2, N, t)$$
$$+(1 - \gamma_1^2 h_t)a_2\gamma_2^1 h_t P(2, 1, N, t) + o(h_t^2). \qquad (53)$$

(b) *Interior and Boundary to Boundary Equations*

- When both V_1 and V_2 are operational and the buffer is full, we have

$$P(0, 0, N, t + h_t) = (1 - \delta_1 h_t)(1 - \delta_2 h_t)P(0, 0, N, t)+$$
$$(1 - a_2)\gamma_2^1 h_t P(B, 1, N, t) + \gamma_2^2 h_t P(B, 2, N, t)+$$
$$\int_{N-(c_1-c_2)h_z}^{N} f_{0,0}(z, t)dz + o(h_t^2) \qquad (54)$$

which is valid only when $c_1 \geq c_2$. Otherwise, the material level would decrease by the amount of $c_2 h_z$ at time $(t+h_t)$. As a consequence, $(1, 1, N)$ becomes a transient state and $P(1, 1, N)$ vanishes.

- When V_1 is blocked and V_2 is in down phase 1, we have

$$P(B, 1, N, t + h_t) = (1 - \gamma_2^1 h_t)P(B, 1, N, t)+$$

$$(1-a_1)\gamma_1^1 h_t(1-\gamma_2^1 h_t)P(1,1,N,t) + (1-\delta_1 h_t)\delta_2 h_t P(0,0,N,t) +$$

$$\gamma_1^2 h_t + (1-\gamma_2^1 h_t)P(2,1,N,t) + \int_{N-c_1 h_z}^{N} f_{0,1}(t,z)dz + o(h_t^2) \quad (55)$$

– When V_1 is blocked and V_2 is in down phase 2, we have

$$P(B,2,N,t+h_t) = (1-\gamma_2^2 h_t)P(B,2,N,t) +$$

$$a_2\gamma_2^1 h_t P(B,1,N,t) + \int_{N-c_1 h_z}^{N} f_{0,2}(t,z)dz + o(h_t^2) \quad (56)$$

(c) *Interior and Boundary to Interior Equations*

– When V_1 is in down phase 1 and V_2 is operational, we have

$$\int_{N-c_2 h_z}^{N} f_{1,0}(z,t)dz = (1-a_2)\gamma_2^1 h_t \int_t^{t+h_t} P(1,1,N,s)ds +$$

$$\gamma_2^2 h_t \int_t^{t+h_t} P(1,2,N,s)ds + \delta_1 h_t \int_t^{t+h_t} P(0,0,N,s)ds \quad (57)$$

– When V_1 is in down phase 2 and V_2 is operational, we have

$$\int_{N-c_2 h_z}^{N} f_{2,0}(z,t)dz = \int_t^{t+h_t} a_1\gamma_1^1 h_t(1-\delta_2 h_t)P(1,0,N,s)ds +$$

$$(1-\gamma_1^2 h_t)\gamma_2^2 h_t \int_t^{t+h_t} P(2,2,N,s)ds +$$

$$(1-\gamma_1^2 h_t)(1-a_2)\gamma_2^1 h_t \int_t^{t+h_t} P(2,1,N,s)ds \quad (58)$$

Notice that the states $(1,0,N),(0,1,N),(0,0,z),(1,0,z)$ are not reachable from any of the boundary states.

As $h \to 0$ and $t \to \infty$, the flow-balance equations (41)-(58) reduce to a set of linear equations which are given in Appendix 1. Therefore, the steady-state probabilities of the boundary states can be found by solving the set of linear equations given in Appendix 1, as functions of the valve parameters and $f_{0,0}(.)$, $f_{0,1}(.)$, $f_{0,2}(.)$, $f_{1,0}(.)$, $f_{2,0}(.)$. However, note that some of the states are transient since the failures are assumed to be operation-dependent. For instance, let us assume that the process is in state $(1,0,z)$ with $z \le c_2 h_z$ at time t. Within the next time interval of length h_t, if V_1 sojourns in down phase 1 with probability $(1 - \gamma_1^1 h_t)$ and V_2 does not fail with probability $(1 - \delta_2 h_t)$, the process enters state $(1,0,0)$. Then, within the next time interval of length h_t, if V_1 remains down in phase 1 with probability $(1 - \gamma_1^1 h_t)$, the process enters state $(1,S,0)$. The process stays in $(1,S,0)$ until either V_1 recovers from its failure or it enters state $(2,S,0)$. The former event occurs with probability $a_1\gamma_1^1 h_t$ and the process enters either state $(0,0,z)$ if $c_1 > c_2$ or state $(0,0,0)$ if $c_1 \leq c_2$. However, if the latter event occurs, with probability $(1 - a_1)\gamma_1^1 h_t$, the process enters in state $(2,S,0)$. Clearly, there is no possibility that the process can sojourn in state $(1,0,0)$ for a long time. Therefore, state $(1,0,0)$ is a transient state. It can be shown through similar arguments that the other states $(0,1,0)$, $(0,2,0)$, $(2,0,0)$, $(0,1,N)$, $(1,0,N)$, $(0,2,N)$ and $(2,0,N)$ are also transient and their corresponding probabilities vanish in the long run. Consequently, the probabilities $P(1,1,0)$, $P(2,1,0)$, $P(1,2,0)$, $P(2,2,0)$, $P(0,0,N)$, $P(1,1,N)$, $P(2,1,N)$, $P(1,2,N)$ in (41)-(58) become zero. Below are the explicit expressions for the remaining system probabilities at boundary states.

$$
\begin{aligned}
P(0,0,0) &= \frac{c_1}{\delta_2} f_{0,1}(0) \\
P(1,S,0) &= \frac{\delta_1}{\gamma_1^1} P(0,0,0) + \frac{c_2}{\gamma_1^1} f_{1,0}(0) \\
P(2,S,0) &= \frac{\gamma_1^1 a_1}{\gamma_1^2} P(1,S,0) + \frac{c_2}{\gamma_1^2} f_{2,0}(0) \\
& c_1 f_{0,2}(0) = 0 \\
P(0,0,N) &= \frac{c_2}{\delta_1} f_{1,0}(N) \\
P(B,1,N) &= \frac{\delta_2}{\gamma_2^1} P(0,0,N) + \frac{c_1}{\gamma_2^1} f_{0,1}(N) \\
& c_2 f_{2,0}(N) = 0 \\
P(B,2,N) &= \frac{a_2\gamma_2^1}{\gamma_2^2} P(B,1,N) + \frac{c_1}{\gamma_2^2} f_{0,2}(N)
\end{aligned} \tag{59}
$$

Notice that the functional values of $f_{2,0}(.)$ and $f_{0,2}(.)$ at $z = N$ and $z = 0$ are 0, respectively. In other words, we obtain conditions of $f_{0,2}(0) = 0$ and $f_{2,0}(N) = 0$ from the boundary analysis of the system behavior. On the other hand, when $c_1 < c_2$, the functional value of $f_{1,0}(.)$ at $z = N$ become 0 since state $(0,0,N)$ becomes transient and the corresponding probability of P(0,0,N) vanishes in the long long run. Similarly, when $c_1 > c_2$, the boundary equations result in the value of zero for $f_{0,1}(.)$, at $z = 0$ since $(0,0,0)$ becomes a transient state. More formally,

Boundary restrictions

(a) When $c_1 < c_2$:

$$f_{0,2}(0) = 0$$
$$f_{1,0}(N) = 0$$
$$f_{2,0}(N) = 0 \tag{60}$$

(b) When $c_1 > c_2$:

$$f_{0,1}(0) = 0$$
$$f_{0,2}(0) = 0$$
$$f_{2,0}(N) = 0 \tag{61}$$

(c) When $c_1 = c_2$:

$$f_{0,2}(0) = 0$$
$$f_{2,0}(N) = 0 \tag{62}$$

The constraint sets given in (60), (61) and (62) on the functional values of $f_{,j}(z)$ obtained from the boundary equations play an important role for finding the unknown parameters of functions $f_{0,0}(z), f_{0,1}(z), f_{0,2}(z), f_{1,0}(z), f_{2,0}(z)$. Once $f_{i,j}$, $i, j = 0, 1, 2$ are explicitly found, the probabilities of the system at the boundary states can be obtained using the respective set of equations given in (59).

3. Solution Approach

We have developed the following approach to solve the set of equations given in (33). We first solve the system of differential equations using the eigenvalue method for homogeneous first order system of equations with constant coefficients. Notice that the system given in (40) is one order less than the one given in (36). Therefore, we concentrate on the system in (36) to implement the approach we use in this study. Since matrix C is a block-diagonalizable (Jordan Form), one can find a diagonal matrix D with entries corresponding to the eigenvalues, such that $D = V^{-1}CV$ where V is the corresponding matrix of eigenvectors, see ([3]). The general solution for the system (36) is a 4-tuple vector with entries corresponding to functions $f_{0,1}(z)$, $f_{0,2}(z)$,

$f_{1,0}(z)$ and $f_{2,0}(z)$. Thus, we have,

$$\begin{pmatrix} f_{0,1}(z) \\ f_{0,2}(z) \\ f_{1,0}(z) \\ f_{2,0}(z) \end{pmatrix} = Vke^{\lambda z} = V \begin{pmatrix} k_1 e^{\lambda_1 z} \\ k_2 e^{\lambda_2 z} \\ k_3 e^{\lambda_3 z} \\ k_4 e^{\lambda_4 z} \end{pmatrix}$$

$$= \begin{pmatrix} v_{11}e^{\lambda_1 z}k_1 + v_{12}e^{\lambda_2 z}k_2 + v_{13}e^{\lambda_3 z}k_3 + v_{14}e^{\lambda_4 z}k_4 \\ v_{21}e^{\lambda_1 z}k_1 + v_{22}e^{\lambda_2 z}k_2 + v_{23}e^{\lambda_3 z}k_3 + v_{24}e^{\lambda_4 z}k_4 \\ v_{31}e^{\lambda_1 z}k_1 + v_{32}e^{\lambda_2 z}k_2 + v_{33}e^{\lambda_3 z}k_3 + v_{34}e^{\lambda_4 z}k_4 \\ v_{41}e^{\lambda_1 z}k_1 + v_{42}e^{\lambda_2 z}k_2 + v_{43}e^{\lambda_3 z}k_3 + v_{44}e^{\lambda_4 z}k_4 \end{pmatrix} \tag{63}$$

where k_i, $i = 1, .., 4$ are real numbers and $0 \le z \le N$. One can find different combinations of $k_i s$, $i = 1, .., 4$ such that the general solution (63) satisfies (36). However, in the present study the problem turns out to be of finding k_is such that

$$0 \le \int_0^N f_{ij}(z)dz \le 1, \quad i = 0, 1, 2 \text{ and } j = 0, 1, 2. \tag{64}$$

$$\text{and} \quad 0 \le P(i, j, 0), P(i, j, N) \le 1, \quad i = B, 0, 1, 2 \text{ and } j = S, 0, 1, 2. \tag{65}$$

Note that the same coefficient matrix C is used to find the eigenvalue and eigenvector pairs for cases $c_1 < c_2$ and $c_1 > c_2$. However, to uniquely obtain respective k_is for cases $c_1 < c_2$ and $c_1 > c_2$, we use (60) and (61) which are obtained from the boundary equations and the constraints (64) and (65). The restrictions in (60) results in the following set of three linear equations in k_i, $i = 1, 2, 3, 4$ when $c_1 < c_2$.

$$\begin{aligned} v_{21}k_1 + v_{22}k_2 + v_{23}k_3 + v_{24}k_4 &= 0 \\ v_{31}e^{\lambda_1 N}k_1 + v_{32}e^{\lambda_2 N}k_2 + v_{33}e^{\lambda_3 N}k_3 + v_{34}e^{\lambda_4 N}k_4 &= 0 \\ v_{41}e^{\lambda_1 N}k_1 + v_{42}e^{\lambda_2 N}k_2 + v_{43}e^{\lambda_3 N}k_3 + v_{44}e^{\lambda_4 N}k_4 &= 0 \end{aligned} \tag{66}$$

On the other hand, the restrictions in (61) for the case $c_1 > c_2$ yield another set of three linear equations in k_i, $i = 1, 2, 3, 4$.

$$\begin{aligned} v_{11}k_1 + v_{12}k_2 + v_{13}k_3 + v_{14}k_4 &= 0 \\ v_{21}k_1 + v_{22}k_2 + v_{23}k_3 + v_{24}k_4 &= 0 \\ v_{41}e^{\lambda_1 N}k_1 + v_{42}e^{\lambda_2 N}k_2 + v_{43}e^{\lambda_3 N}k_3 + v_{44}e^{\lambda_4 N}k_4 &= 0 \end{aligned} \tag{67}$$

Now we have 3 equations in four unknowns for each of the cases. The additional equation that we need can be the following normalization equation.

$$\sum_{\alpha_1=B,0,1,2}\sum_{\alpha_2=S,0,1,2}[\int_0^N f_{\alpha_1,\alpha_2}(z)dz + P(\alpha_1,\alpha_2,0) + P(\alpha_1,\alpha_2,N)] = 1. \quad (68)$$

Consequently, we have the resultant linear system of four equations:

$$\begin{pmatrix} a_{11} & a_{12} & a_{13} & a_{14} \\ a_{21} & a_{22} & a_{23} & a_{24} \\ a_{31} & a_{32} & a_{33} & a_{34} \\ a_{41} & a_{42} & a_{43} & a_{44} \end{pmatrix}\begin{pmatrix} k_1 \\ k_2 \\ k_3 \\ k_4 \end{pmatrix} = \begin{pmatrix} 1 \\ 0 \\ 0 \\ 0 \end{pmatrix} \quad (69)$$

which uniquely determines k_is. Note that, the linear system will be of 3×3 for the case where $c_1 = c_2$.

Coefficients a_{1j}, $j = 1,..,4$ obtained using the normalization equation for cases $c_1 < c_2$ and $c_1 > c_2$ are given in Appendix 3. Furthermore, the coefficients in the 3×3 linear system for case $c_1 = c_2$ are also provided in Appendix 3.

4. Model Verification

In this section, we verify the results obtained from our model with the results obtained from the model proposed by [4] for systems with valves experiencing exponential repair times. We propose to use MGE-2 distributions that are equivalent to exponential distributions as repair times and compute probabilities using our algorithm with the MGE-2 repair times and using [4] with exponential repair times. We expect to obtain exactly the same results.

	$P_2(S)$	$P_1(B)$	$\bar{o}$	$\bar{N}$
		$P_2(Down) = 0.10$		
MGE-2	0.1251	0.0021	63.6171	10617
Expo.	0.1251	0.0021	63.6171	10618
		$P_2(Down) = 0.30$		
MGE-2	0.0591	0.0378	61.4323	53404
Expo.	0.0591	0.0378	61.3423	53404

Table 1. Comparison for the case $c_1 < c_2$

We have considered two cases where $P_2(Down) = 0.10$ and $P_2(Down) = 0.30$ with the expected time to repair for V_2 being 480 time units. The expected time to failure for the corresponding cases is 4320 and 1120 time units, respectively. It is assumed that V_1 is operating 85% of the time with the expected time to failure being 2550 time units. In other words, probability that

	$P_2(S)$	$P_1(B)$	$\bar{o}$	$\bar{N}$
		$P_2(Down) = 0.10$		
MGE-2	0.0055	0.0715	67.1291	129040
Expo.	0.0055	0.0715	67.1291	129043
		$P_2(Down) = 0.30$		
MGE-2	0.0016	0.2450	52.4163	139890
Expo.	0.0016	0.2450	52.4163	139895

Table 2. Comparison for the case $c_1 > c_2$

V_1 being down is 0.15 for each case with the repair rate of $\gamma_1 = 1/450$. Note that we first need to find the MGE-2 representation of the requisite exponential repair time distribution. The parameters (γ_1, γ_2, a) of an MGE-2 distribution that is equivalent to an exponential distribution with rate μ are given below (see [1]).

$$\gamma_1 = 2\mu, \quad \gamma_2 = \mu, \quad a = 0.5$$

Note that the density function of the MGE-2 distribution collapses onto the one of the exponential distribution. We provide results for different rates at which V_1 and V_2 can operate, i.e. $c_1 = 75 < c_2 = 100$, $c_1 = 100 > c_2 = 75$ and $c_1 = c_2 = 100$. As shown in Tables 1, 2 and 3, the results obtained from the model proposed for systems with MGE-2 repair times exactly match the results obtained from the model for systems with exponential repair times and thereby verify the results.

5. Impact of Repair Time Variability

We have also investigated the impact of repair time variability of V_2 on some of the performance metrics such as starvation and blocking probabilities, $P_2(S)$ and $P_1(B)$, the average inventory level in the buffer, $\bar{N}$, and the throughput, $\bar{o}$, of the system.

	$P_2(S)$	$P_1(B)$	$\bar{o}$	$\bar{N}$
		$P_2(Down) = 0.10$		
MGE-2	0.0790	0.0205	83.2572	50173
Expo.	0.0790	0.0205	83.2573	50174
		$P_2(Down) = 0.30$		
MGE-2	0.0131	0.1873	69.0795	119080
Expo.	0.0132	0.1873	69.0795	119085

Table 3. Comparison for the case $c_1 = c_2$

As we have pointed out earlier, the squared coefficient of variation (Cv^2) of the Mixture of Generalized Erlang distribution with two phases can assume any real value greater than or equal to 0.5, (see [1]). In this study, we let $Cv^2 \in [0.5, \ldots, 10.0]$. Recall that the parameters γ_2^2, γ_1^2 and a_2 of the repair time distribution of V_2 for any level of Cv^2 can be obtained using the method of moments. Depending on the Cv^2, one can obtain the above parameters using one of the following expressions ,(see [1]).

- $0.5 \leq Cv^2 < 1$:

$$\gamma_1^2 = \frac{1}{m_2 Cv^2}, \quad \gamma_2^2 = \frac{2}{m_2}, \quad a_2 = 2(1 - Cv^2)$$

- $Cv^2 > 1$:

$$\gamma_1^2 = \frac{2}{m_2}, \quad \gamma_2^2 = \frac{1}{m_2 Cv^2}, \quad a_2 = \frac{0.5}{Cv^2}$$

For this particular example, V_1 is assumed to be operating 85% of the time, with the expected time to failure being 2720 time units. Probability that V_1 is down kept at 0.15 for each case with the repair time parameters $(\gamma_2^1, \gamma_1^1, a_1) = (450, 300, 0.1)$.

In the set of examples, the buffer storage is assumed to hold 150,000 tons of material. Note that the capacity of the buffer has a direct impact on the performance metrics. That is, as the buffer capacity increases (decreases), blocking of V_1 and starvation of V_2 decrease (increase); and also, the throughput and the average material level increase (decrease). Furthermore, the valve capacities c_1 and c_2 also play an important role on the performance metrics. To clarify this point, now assume that V_1 and V_2 are identical except in their capacities. Under these assumptions, it should be clear that $P_1(B)$ and $P_2(S)$ assume fairly significant values when $c_1 < c_2$ and $c_1 > c_2$, respectively. In addition to N, c_1 and c_2, $P_1(Down) < P_2(Down)$ and $P_1(Down) > P_2(Down)$ are the other two conditions affecting $P_1(B)$ and $P_2(S)$. As shown in Tables 4 and 5, $P_1(B)$ is quite significant as opposed to $P_2(S)$, when $P_1(Down) = 0.15 < P_2(Down) = 0.30$. Similarly, $P_2(S)$ is relatively large than $P_1(B)$ when $P_1(Down) = 0.15 > P_2(Down) = 0.10$, as shown in Tables 4 and 6. On the other hand, as also shown in Tables 4 and 5, a decrease in the average material level is the direct consequence of the monotonic decrease in $P_2(Down)$.

We have observed that the impact of the repair time variability on the performance metrics displays exactly the same pattern for $c_1 = c_2$ and $c_1 > c_2$. Although the impact appears to be the same on the $\bar{o}$ and $P_1(B)$, unfortunately, $\bar{N}$ and $P_2(S)$ have shown slightly different behavior when $c_1 < c_2$, particularly for the case $P_2(Down) = 0.10$. By slightly different behavior we mean that monotonic increase and decrease in $P_2(S)$ and $\bar{N}$ is not observed for $Cv^2 <= 2.50$,

			$P_2(Down) = 0.10$		
Cv^2	$P_2(S)$	$P_1(B)$	$P_2(Down)$	$\bar{o}$	$\bar{N}$
0.50	0.0717	0.0174	0.0930	83.5299	5.2651e+04
1.00	0.0753	0.0209	0.0925	83.2215	5.0757e+04
2.50	0.0824	0.0284	0.0918	82.5866	4.5586e+04
5.00	0.0882	0.0345	0.0912	82.0645	4.0620e+04
10.0	0.0928	0.0395	0.0907	81.6454	3.6415e+04
			$P_2(Down) = 0.30$		
0.50	0.0118	0.1865	0.2966	69.1572	1.2153e+05
1.00	0.0138	0.1878	0.2959	69.0347	1.1896e+05
2.50	0.0174	0.1908	0.2948	68.7807	1.1410e+05
5.00	0.0204	0.1933	0.2939	68.5716	1.0994e+05
10.0	0.0229	0.1954	0.2931	68.3948	1.0639e+05

Table 4. Impact of Repair Time Variability for $c_1 = c_2$

			$P_2(Down) = 0.10$		
Cv^2	$P_2(S)$	$P_1(B)$	$P_2(Down)$	$\bar{o}$	$\bar{N}$
0.50	0.0058	0.0690	0.0994	67.1089	1.2900e+05
1.00	0.0060	0.0712	0.0994	67.0928	1.2854e+05
2.50	0.0064	0.0745	0.0994	67.0693	1.2785e+05
5.00	0.0066	0.0768	0.0993	67.0546	1.2740e+05
10.0	0.0067	0.0783	0.0993	67.0446	1.2710e+05
			$P_2(Down) = 0.30$		
0.50	0.0016	0.2416	0.2995	52.4177	1.4014e+05
1.00	0.0018	0.2444	0.2995	52.4067	1.3962e+05
2.50	0.0021	0.2489	0.2994	52.3890	1.3879e+05
5.00	0.0023	0.2519	0.2993	52.3769	1.3823e+05
10.0	0.0025	0.2541	0.2992	52.3680	1.3783e+05

Table 5. Impact of Repair Time Variability for $c_1 > c_2$

respectively. Although, $P_1(B)$ is increasing and $\bar{o}$ is decreasing as a function of the Cv^2, it is also expected that $\bar{N}$ to be decreasing and $P_2(S)$ to be increasing monotonically. Furthermore, $P_2(Down)$ remains almost the same when $c_1 < c_2$ and $Cv^2 <= 2.50$. However, the two performance metrics show the same behavior for $Cv^2 > 2.50$ as they displayed in the other cases.

As shown in Tables 4 and 5, as the variability increases, $P_2(S)$ and $P_1(B)$ increase and $P_2(Down)$ decreases, monotonically. Also, an increase in starvation and blocking causes a reduction on the throughput.

6. Conclusion

A two-valve fluid-flow system with phase-type repair times is modeled using a Markovian approach. Possible transitions between the system's continu-

			$P_2(Down) = 0.10$		
Cv^2	$P_2(S)$	$P_1(B)$	$P_2(Down)$	$\bar{o}$	$\bar{N}$
0.50	0.1265	4.7619e-04	0.0874	63.7196	9.0936e+03
1.00	0.1255	0.0021	0.0875	63.6155	1.0684e+04
2.50	0.1264	0.0102	0.0874	63.0966	1.1951e+04
5.00	0.1286	0.0200	0.0871	62.4734	1.1670e+04
10.0	0.1308	0.0290	0.0869	61.9017	1.0935e+04
			$P_2(Down) = 0.30$		
0.50	0.0575	0.0279	0.2827	61.9734	5.3323e+04
1.00	0.0598	0.0381	0.2821	61.3182	5.3617e+04
2.50	0.0672	0.0613	0.2798	59.8438	5.0827e+04
5.00	0.0750	0.0830	0.2775	58.4579	4.6547e+04
10.0	0.0822	0.1027	0.2753	57.2007	4.2076e+04

Table 6. Impact of Repair Time Variability for $c_1 < c_2$

ous states first led us to cope with a set of homogeneous differential equations. The system is solved using the eigenvalue method for homogeneous first order systems with constant coefficients. On the other hand, possible transitions between the system's discrete and continuous states where the buffer is either empty or full provided us with a set of linear equations. We have obtained empty and full state probabilities in terms of the valve parameters and functions defining the system's behavior when the buffer is neither empty nor full. Given that some of the probabilities are transient, the set of linear equations yielded constraints on the values of the functions defining the system's behavior when the buffer is partially full. Then, these constraints are used to find the particular solution which consists of the set of probability density functions. The approach we used to find the functions defining the system's behavior can be extended for the case where the repair time distribution of valves is of MGE-k where $k \geq 2$. Also, the impact of the repair time variability on the system throughput, blocking and starvation probabilities and the average buffer level is discussed for the cases of $c_1 < c_2$, $c_1 = c_2$ and $c_1 > c_2$. The impact of the repair time variability does not seem to be highly significant. For practical applications, one may simply use the exponential distribution for the repair time as an approximation.

Acknowledgments

This Research has been partially supported by the NSF grants DMI-9812858, DMI-0085659 and Dimacs, Rutgers-The State University of New Jersey, grant STC-91-19999

Appendix

1. Boundary Equations

Below are the set of equations obtained from the analysis of the boundary behavior of the system.

$$(\gamma_1^1 + \gamma_2^1)P(1,1,0) = \delta_1 P(0,1,0) + \delta_2 P(1,0,0)$$

$$(\gamma_1^2 + \gamma_2^1)P(2,1,0) = a_1\gamma_1^1 P(1,1,0) + \delta_2 P(2,0,0)$$

$$(\gamma_1^1 + \gamma_2^2)P(1,2,0) = a_2\gamma_2^1 P(1,1,0) + \delta_1 P(0,2,0)$$

$$(\gamma_1^2 + \gamma_2^2)P(2,2,0) = a_2\gamma_2^1 P(2,1,0) + a_1\gamma_1^1 P(1,2,0)$$

$$(\delta_1 + \delta_2)P(0,0,0) = \gamma_1^1(1 - a_1)P(1,S,0) + \gamma_1^2 P(2,S,0) + (c_2 - c_1)f_{0,0}(0)$$

$$\gamma_1^1 P(1,S,0) = \delta_1 P(0,0,0) + (1 - a_2)\gamma_2^1 P(1,1,0) + \gamma_2^2 P(1,2,0) + c_2 f_{1,0}(0)$$

$$\gamma_1^2 P(2,S,0) = a_1\gamma_1^1 P(1,S,0) + c_2 f_{2,0}(0)$$

$$c_1 f_{0,1}(0) = \gamma_1^2 P(2,1,0) + \delta_2 P(0,0,0) + (1 - a_1)\gamma_1^1 P(1,1,0)$$

$$c_1 f_{0,2}(0) = a_2\gamma_2^1 P(0,1,0) + (1 - a_1)\gamma_1^1 P(1,2,0) + \gamma_1^2 P(2,2,0)$$

$$(\gamma_1^1 + \gamma_2^1)P(1,1,N) = \delta_1 P(0,1,N) + \delta_2 P(1,0,N)$$

$$(\gamma_1^2 + \gamma_2^1)P(2,1,N) = \delta_2 P(2,0,N) + a_1\gamma_1^1 P(1,1,N)$$

$$(\gamma_1^1 + \gamma_2^2)P(1,2,N) = \delta_1 P(0,2,N) + a_2\gamma_2^1 P(1,1,N)$$

$$(\gamma_1^2 + \gamma_2^2)P(2,2,N) = a_1\gamma_1^1 P(1,2,N) + a_2\gamma_2^1 P(2,1,N)$$

$$(\delta_1 + \delta_2)P(0,0,N) = (1 - a_2)\gamma_2^1 P(B,1,N) + \gamma_2^2 P(B,2,N) + (c_1 - c_2)f_{0,0}(N)$$

$$\gamma_2^1 P(B,1,N) = \delta_2 P(0,0,N) + (1 - a_1)\gamma_1^1 P(1,1,N) + \gamma_1^2 P(2,1,N) + c_1 f_{0,1}(N)$$

$$\gamma_2^2 P(B,2,N) = a_2\gamma_2^1 P(B,1,N) + c_1 f_{0,2}(N)$$

$$c_2 f_{1,0}(N) = (1 - a_2)\gamma_2^1 P(1,1,N) + \gamma_2^2 P(1,2,N) + \delta_1 P(0,0,N)$$

$$c_2 f_{2,0}(N) = a_1\gamma_1^1 P(1,0,N) + (1 - a_2)\gamma_2^1 P(2,1,N) + \gamma_2^2 P(2,2,N)$$

2. The Coefficients in the Systems of Differential equations

2.1 Case: $c_1 < c_2$ and $c_1 > c_2$

As we pointed out in Section 3 that coefficient matrix C in the differential equation system (36) is for both cases $c_1 < c_2$ and $c_1 > c_2$. Below, we present the individual entries of matrix C.

$$c_{11} = \frac{1}{c_1}\left[-(\delta_1+\gamma_2^1) + \frac{\gamma_1^1(1-a_1)\delta_1}{(\gamma_1^1+\gamma_2^1)} + \frac{\gamma_1^2 a_1 \gamma_1^1 \delta_1}{(\gamma_1^2+\gamma_2^1)(\gamma_1^1+\gamma_2^1)}\right] - \frac{\delta_2}{(c_1-c_2)}, \quad c_{12} = \frac{-\delta_2}{(c_1-c_2)}$$

$$c_{13} = \frac{1}{c_1}\left[\frac{\gamma_1^1(1-a_1)\delta_2}{(\gamma_1^1+\gamma_2^1)} + \frac{\gamma_1^2 a_1 \gamma_1^1 \delta_2}{(\gamma_1^2+\gamma_2^1)(\gamma_1^1+\gamma_2^1)} + \frac{\delta_2\, ro}{(c_1-c_2)}\right], \quad c_{14} = \frac{1}{c_1}\left[\frac{\gamma_1^2\delta_2}{(\gamma_1^2+\gamma_2^1)} + \frac{\delta_2\, ro}{(c_1-c_2)}\right]$$

$$c_{21} = \frac{1}{c_1}\left[a_2\gamma_2^1 + \frac{(1-a_1)\gamma_1^1 a_2 \gamma_2^1 \delta_1}{(\gamma_1^1+\gamma_2^2)(\gamma_1^1+\gamma_2^1)} + \frac{\gamma_1^2 A \delta_1}{(\gamma_1^2+\gamma_2^2)}\right],$$

$$c_{22} = \frac{1}{c_1}\left[-(\delta_1+\gamma_2^2) + \frac{(1-a_1)\gamma_1^1\delta_1}{(\gamma_1^1+\gamma_2^2)} + \frac{\gamma_1^2 a_1 \gamma_1^1 \delta_1}{(\gamma_1^2+\gamma_2^2)(\gamma_1^1+\gamma_2^2)}\right]$$

$$c_{23} = \frac{1}{c_1}\left[\frac{(1-a_1)\gamma_1^1 a_2\gamma_2^1\delta_2}{(\gamma_1^1+\gamma_2^2)(\gamma_1^1+\gamma_2^1)} + \frac{\gamma_1^2 A\delta_2}{(\gamma_1^2+\gamma_2^2)}\right], \quad c_{24} = \frac{1}{c_1}\left[\frac{\gamma_1^2 a_2\gamma_2^1\delta_2}{(\gamma_1^2+\gamma_2^2)(\gamma_1^2+\gamma_2^1)}\right]$$

$$c_{31} = \frac{-1}{c_2}\left[\frac{(1-a_2)\gamma_2^1\delta_1}{(\gamma_1^1+\gamma_2^1)} + \frac{\gamma_2^2 a_2\gamma_2^1\delta_1}{(\gamma_1^1+\gamma_2^2)(\gamma_1^1+\gamma_2^1)}\right] - \frac{\delta_1\, ri}{(c_1-c_2)}, \quad c_{32} = \frac{-1}{c_2}\left[\frac{\gamma_2^2\delta_1}{(\gamma_1^1+\gamma_2^2)} - \frac{\delta_1\, ri}{(c_1-c_2)}\right]$$

$$c_{33} = \frac{-1}{c_2}\left[-(\delta_2+\gamma_1^1) + \frac{(1-a_2)\gamma_2^1\delta_2}{(\gamma_1^1+\gamma_2^1)} + \frac{\gamma_2^2 a_2\gamma_2^1\delta_2}{(\gamma_1^1+\gamma_2^2)(\gamma_1^1+\gamma_2^1)}\right] - \frac{\delta_1}{(c_1-c_2)}, \quad c_{34} = \frac{-\delta_1}{(c_1-c_2)}$$

$$c_{41} = \frac{-1}{c_2}\left[\frac{(1-a_2)\gamma_2^1 a_1\gamma_1^1\delta_1}{(\gamma_1^2+\gamma_2^1)(\gamma_1^1+\gamma_2^1)} + \frac{\gamma_2^2 A\delta_1}{(\gamma_1^2+\gamma_2^2)}\right], \quad c_{42} = \frac{-1}{c_2}\left[\frac{\gamma_2^2 a_1\gamma_1^1\delta_1}{(\gamma_1^2+\gamma_2^2)(\gamma_1^1+\gamma_2^2)}\right]$$

$$c_{43} = \frac{-1}{c_2}\left[\gamma_1^1 a_1 + \frac{(1-a_2)\gamma_2^1 a_1\gamma_1^1\delta_2}{(\gamma_1^2+\gamma_2^1)(\gamma_1^1+\gamma_2^1)} + \frac{\gamma_2^2 A\delta_2}{(\gamma_1^2+\gamma_2^2)}\right],$$

$$c_{44} = \frac{-1}{c_2}\left[-(\gamma_1^2+\delta_2) + \frac{(1-a_2)\gamma_2^1\delta_2}{(\gamma_1^2+\gamma_2^1)} + \frac{\gamma_2^2 a_2\gamma_2^1\delta_2}{(\gamma_1^2+\gamma_2^2)(\gamma_1^2+\gamma_2^1)}\right]$$

where $A = \frac{a_1\gamma_1^1 a_2\gamma_2^1}{\gamma_1^1+\gamma_2^1}\left[\frac{1}{(\gamma_1^1+\gamma_2^2)} + \frac{1}{(\gamma_1^2+\gamma_2^1)}\right]$

2.2 Case: $c_1 = c_2$

The set of coefficients below are obtained for case $c_1 = c_2$ in the differential equation system in (40).

$$X1 = \frac{1}{c_1}\left[a_2\gamma_2^1 + \frac{(1-a_1)\gamma_1^1 a_2\gamma_2^1\delta_1}{(\gamma_1^1+\gamma_2^2)(\gamma_1^1+\gamma_2^1)} + \frac{\gamma_1^2 A\delta_1}{\gamma_1^2+\gamma_2^2}\right],$$

$$X102 = \frac{1}{c_1}\left[-(\delta_1+\gamma_2^2) + \frac{(1-a_1)\gamma_1^1\delta_1}{\gamma_1^1+\gamma_2^2} + \frac{\gamma_1^2 a_1\gamma_1^1\delta_1}{(\gamma_1^2+\gamma_2^2)(\gamma_1^1+\gamma_2^2)}\right]$$

$$X110 = \frac{1}{c_1}\left[\frac{(1-a_1)\gamma_1^1 a_2\gamma_2^1\delta_2}{(\gamma_1^1+\gamma_2^2)(\gamma_1^1+\gamma_2^1)} + \frac{\gamma_1^2 A\delta_2}{\gamma_1^2+\gamma_2^2}\right], \quad X2 = \frac{-1}{c_2}\left[\frac{(1-a_2)\gamma_2^1\delta_1}{(\gamma_1^1+\gamma_2^1} + \frac{\gamma_2^2 a_2\gamma_2^1\delta_1}{(\gamma_1^1+\gamma_2^2)(\gamma_1^1+\gamma_2^1)}\right],$$

$$X202 = \frac{-1}{c_2}\left[\frac{\gamma_2^2\delta_1}{\gamma_1^1+\gamma_2^2}\right], \quad X210 = \frac{-1}{c_2}\left[-(\delta_2+\gamma_1^1) + \frac{(1-a_2)\gamma_2^1\delta_2}{(\gamma_1^1+\gamma_2^1} + \frac{\gamma_2^2 a_2\gamma_2^1\delta_2}{(\gamma_1^1+\gamma_2^2)(\gamma_1^1+\gamma_2^1)}\right],$$

$$X220 = 0, \quad X120 = \frac{1}{c_1}\left[\frac{\gamma_1^2 a_2\gamma_2^1\delta_2}{(\gamma_1^2+\gamma_2^2)(\gamma_1^2+\gamma_2^1)}\right], \quad X3 = \frac{-1}{c_2}\left[\frac{(1-a_2)\gamma_2^1 a_1\gamma_1^1\delta_1}{(\gamma_1^2+\gamma_2^1)(\gamma_1^1+\gamma_2^1)} + \frac{\gamma_2^2 A\delta_1}{\gamma_1^2+\gamma_2^2}\right],$$

$$X302 = \frac{-1}{c_2}\left[\frac{\gamma_2^2 a_1\gamma_1^1\delta_1}{(\gamma_1^2+\gamma_2^2)(\gamma_1^1+\gamma_2^2)}\right], \quad X310 = \frac{-1}{c_2}\left[\gamma_1^1 a_1 + \frac{(1-a_2)\gamma_2^1 a_1\gamma_1^1\delta_2}{(\gamma_1^2+\gamma_2^1)(\gamma_1^1+\gamma_2^1)} + \frac{\gamma_2^2 A\delta_2}{\gamma_1^2+\gamma_2^2}\right],$$

$$X320 = \frac{-1}{c_2}[-(\gamma_1^2 + \delta_2) + \frac{(1-a_2)\gamma_2^1\delta_2}{\gamma_1^2+\gamma_2^1} + \frac{\gamma_2^2 a_2 \gamma_2^1 \delta_2}{(\gamma_1^2+\gamma_2^2)(\gamma_1^2+\gamma_2^1)}]$$

$$X002 = \frac{\gamma_2^2}{\delta_1+\delta_2} - \frac{(1-a_2)\gamma_2^1}{\delta_1+\delta_2}, \quad X010 = \frac{(1-a_1)\gamma_1^1}{\delta_1+\delta_2} + \frac{(1-a_2)\gamma_2^1}{\delta_1+\delta_2}, \quad X020 = \frac{\gamma_1^2}{\delta_1+\delta_2} + \frac{(1-a_2)\gamma_2^1}{\delta_1+\delta_2}$$

$$C = \begin{pmatrix} X102 - X1 & X110 + \frac{c_2}{c_1}X1 & X120 + \frac{c_2}{c_1}X1 \\ X202 - X2 - \frac{\delta_1}{c_2}X002 & X210 + \frac{c_2}{c_1}X2 - \frac{\delta_1}{c_2}X010 & X220 + \frac{c_2}{c_1}X2 - \frac{\delta_1}{c_2}X020 \\ X302 - X3 & X310 + \frac{c_2}{c_1}X3 & X320 + \frac{c_2}{c_1}X3 \end{pmatrix}$$

3. Coefficients in the Normalization Equation (68)

In this section, the coefficients $a_{1i}, i = 1,2,3,4$ in (69) obtained from (68) are presented. Integrating $f_{00}(z)$, $f_{01}(z)$, $f_{02}(z)$, $f_{10}(z)$, $f_{20}(z)$, $f_{11}(z)$, $f_{12}(z)$, $f_{21}(z)$, $f_{22}(z)$ in the domain they are defined, we obtain the following expressions.

$$b_1 = \frac{-c_2}{(c_1-c_2)} + \frac{\delta_1}{(\gamma_1^1+\gamma_2^1)} + \frac{a_1\,\gamma_1^1\,\delta_1}{(\gamma_1^2+\gamma_2^1)(\gamma_1^1+\gamma_2^1)} + \frac{a_2\,\gamma_2^1\,\delta_1}{(\gamma_1^1+\gamma_2^2)(\gamma_1^1+\gamma_2^1)} + \frac{A\,\delta_1}{(\gamma_1^2+\gamma_2^2)}$$

$$b_2 = \frac{-c_2}{(c_1-c_2)} + \frac{\delta_1}{(\gamma_1^1+\gamma_2^2)} + \frac{a_1\,\gamma_1^1\,\delta_1}{(\gamma_1^2+\gamma_2^2)(\gamma_1^1+\gamma_2^2)}$$

$$b_3 = \frac{c_1}{(c_1-c_2)} + \frac{\delta_2}{(\gamma_1^1+\gamma_2^1)} + \frac{a_1\,\gamma_1^1\,\delta_2}{(\gamma_1^2+\gamma_2^1)(\gamma_1^1+\gamma_2^1)} + \frac{a_2\,\gamma_2^1\,\delta_2}{(\gamma_1^1+\gamma_2^2)(\gamma_1^1+\gamma_2^1)} + \frac{A\,\delta_2}{(\gamma_1^2+\gamma_2^2)}$$

$$b_4 = \frac{c_1}{(c_1-c_2)} + \frac{\delta_2}{(\gamma_1^2+\gamma_2^1)} + \frac{a_2\,\gamma_2^1\,\delta_2}{(\gamma_1^2+\gamma_2^2)(\gamma_1^2+\gamma_2^1)}$$

$$I_{k1} = \frac{e^{\lambda_2 N} - 1}{\lambda_2}[b_1\ v_{11} + b_2\ v_{21} + b_3\ v_{31} + b_4\ v_{41}]$$

$$I_{k2} = \frac{e^{\lambda_3 N} - 1}{\lambda_3}[b_1\ v_{12} + b_2\ v_{22} + b_3\ v_{32} + b_4\ v_{42}]$$

$$I_{k3} = \frac{e^{\lambda_4 N} - 1}{\lambda_4}[b_1\ v_{13} + b_2\ v_{23} + b_3\ v_{33} + b_4\ v_{43}]$$

$$I_{k4} = \frac{e^{\lambda_5 N} - 1}{\lambda_5}[b_1\ v_{14} + b_2\ v_{24} + b_3\ v_{34} + b_4\ v_{44}]$$

However, the sum of the probabilities for the boundary states contributing to the coefficients in (69) depends on the case where $c_1 < c_2$ and $c_1 > c_2$ and $c_1 = c_2$.

Case: $c_1 < c_2$ and $c_1 > c_2$

Using probabilities $P(1,S,0)$, $P(2,S,0)$, $P(0,0,0)$, we obtain the following coefficients.

$$L_{k1} = v_{11}[\frac{c_1}{\delta_2} + \frac{\delta_1 c_1}{(\gamma_1^1\delta_2)} + \frac{a_1\delta_1 c_1}{(\gamma_1^2\delta_2)}] + v_{31}[\frac{c_2}{\gamma_1^1} + \frac{a_1 c_2}{\gamma_1^2}] + \frac{c_2 v_{41}}{\gamma_1^2}$$

$$L_{k2} = v_{12}[\frac{c_1}{\delta_2} + \frac{\delta_1 c_1}{(\gamma_1^1\delta_2)} + \frac{a_1\delta_1 c_1}{(\gamma_1^2\delta_2)}] + v_{32}[\frac{c_2}{\gamma_1^1} + \frac{a_1 c_2}{\gamma_1^2}] + \frac{c_2 v_{42}}{\gamma_1^2}$$

$$L_{k3} = v_{13}[\frac{c_1}{\delta_2} + \frac{\delta_1 c_1}{(\gamma_1^1\delta_2)} + \frac{a_1\delta_1 c_1}{(\gamma_1^2\delta_2)}] + v_{33}[\frac{c_2}{\gamma_1^1} + \frac{a_1 c_2}{\gamma_1^2}] + \frac{c_2 v_{43}}{\gamma_1^2}$$

$$L_{k4} = v_{14}[\frac{c_1}{\delta_2} + \frac{\delta_1 c_1}{(\gamma_1^1\delta_2)} + \frac{a_1\delta_1 c_1}{(\gamma_1^2\delta_2)}] + v_{34}[\frac{c_2}{\gamma_1^1} + \frac{a_1 c_2}{\gamma_1^2}] + \frac{c_2 v_{44}}{\gamma_1^2}$$

Using probabilities $P(B,1,N)$, $P(B,2,N)$, $P(0,0,N)$, we obtain the following coefficients.

$$U_{k1} = v_{31}(e^{\lambda_2 N}[\tfrac{c_2}{\delta_1} + \tfrac{\delta_2 c_2}{\gamma_2^1 \delta_1} + \tfrac{a_2 \delta_2 c_2}{\gamma_2^2 \delta_1}] + \tfrac{c_2}{\gamma_1^1} + \tfrac{a_1 c_2}{\gamma_1^2}) + v_{11} e^{\lambda_2 N}[\tfrac{c_1}{\gamma_2^1} + \tfrac{a_2 c_1}{\gamma_2^2}] + \tfrac{c_1 v_{21}}{\gamma_2^2} e^{\lambda_2 N} + \tfrac{v_{41} c_2}{\gamma_1^2}$$
$$U_{k2} = v_{32}(e^{\lambda_3 N}[\tfrac{c_2}{\delta_1} + \tfrac{\delta_2 c_2}{\gamma_2^1 \delta_1} + \tfrac{a_2 \delta_2 c_2}{\gamma_2^2 \delta_1}] + \tfrac{c_2}{\gamma_1^1} + \tfrac{a_1 c_2}{\gamma_1^2}) + v_{12} e^{\lambda_3 N}[\tfrac{c_1}{\gamma_2^1} + \tfrac{a_2 c_1}{\gamma_2^2}] + \tfrac{c_1 v_{22}}{\gamma_2^2} e^{\lambda_3 N} + \tfrac{v_{41} c_2}{\gamma_1^2}$$
$$U_{k3} = v_{33}(e^{\lambda_4 N}[\tfrac{c_2}{\delta_1} + \tfrac{\delta_2 c_2}{\gamma_2^1 \delta_1} + \tfrac{a_2 \delta_2 c_2}{\gamma_2^2 \delta_1}] + \tfrac{c_2}{\gamma_1^1} + \tfrac{a_1 c_2}{\gamma_1^2}) + v_{13} e^{\lambda_4 N}[\tfrac{c_1}{\gamma_2^1} + \tfrac{a_2 c_1}{\gamma_2^2}] + \tfrac{c_1 v_{23}}{\gamma_2^2} e^{\lambda_4 N} + \tfrac{v_{41} c_2}{\gamma_1^2}$$
$$U_{k4} = v_{34}(e^{\lambda_5 N}[\tfrac{c_2}{\delta_1} + \tfrac{\delta_2 c_2}{\gamma_2^1 \delta_1} + \tfrac{a_2 \delta_2 c_2}{\gamma_2^2 \delta_1}] + \tfrac{c_2}{\gamma_1^1} + \tfrac{a_1 c_2}{\gamma_1^2}) + v_{14} e^{\lambda_5 N}[\tfrac{c_1}{\gamma_2^1} + \tfrac{a_2 c_1}{\gamma_2^2}] + \tfrac{c_1 v_{24}}{\gamma_2^2} e^{\lambda_5 N} + \tfrac{v_{41} c_2}{\gamma_1^2}$$

Therefore,

$$a_{11} = L_{k1} + U_{k1} + I_{k1}$$
$$a_{12} = L_{k2} + U_{k2} + I_{k2}$$
$$a_{13} = L_{k3} + U_{k3} + I_{k3}$$
$$a_{14} = L_{k4} + U_{k4} + I_{k4}$$

Case $c_1 = c_2$

The system to be solved for the unknown parameters k_i, $i = 1, \ldots, 3$ of the density functions is a 3×3 linear system. The coefficients of the system contributed by the normalization equation (68) are given below.

Integrating $f_{00}(z)$, $f_{01}(z)$, $f_{02}(z)$, $f_{10}(z)$, $f_{20}(z)$, $f_{11}(z)$, $f_{12}(z)$, $f_{21}(z)$, $f_{22}(z)$ in the domain they are defined, we obtain the following expressions.

$$X11 = \tfrac{\delta_2}{\gamma_1^1+\gamma_2^1}, \quad X1110 = \tfrac{\delta_1}{\gamma_1^1+\gamma_2^1}, \quad X12 = \tfrac{a_2 \gamma_2^1 \delta_1}{(\gamma_1^1+\gamma_2^2)(\gamma_1^1+\gamma_2^1)}, \quad X1210 = \tfrac{a_2 \gamma_2^1 \delta_2}{(\gamma_1^1+\gamma_2^2)(\gamma_1^1+\gamma_2^1)}$$

$$X21 = \tfrac{a_1 \gamma_1^1 \delta_1}{(\gamma_1^2+\gamma_2^1)(\gamma_1^1+\gamma_2^1)}, \quad X2110 = \tfrac{a_1 \gamma_1^1 \delta_2}{(\gamma_1^2+\gamma_2^1)(\gamma_1^1+\gamma_2^1)}, \quad X2120 = \tfrac{\delta_2}{(\gamma_1^2+\gamma_2^1)},$$

$$X1202 = \tfrac{\delta_1}{(\gamma_1^1+\gamma_2^2)}, X22 = \tfrac{A\delta_1}{(\gamma_1^2+\gamma_2^2)}, \quad X2220 = \tfrac{a2 \gamma_2^1 \delta_2}{(\gamma_1^2+\gamma_2^2)(\gamma_1^2+\gamma_2^1)},$$

$$X2202 = \tfrac{a1 \gamma_1^1 \delta_1}{(\gamma_1^2+\gamma_2^2)(\gamma_1^1+\gamma_2^2)}, \quad X2210 = \tfrac{A\delta_2}{(\gamma_1^2+\gamma_2^2)}$$

$$f1110 = X1110 + X11, \quad f1102 = -X11, \quad f1120 = X11$$

$$f2110 = X2110 + X21, \quad f2102 = -X21, \quad f2120 = X2120 + X21$$

$$f1210 = X1210 + X12, \quad f1202 = X1202 - X12, \quad f1220 = X12$$

$$f2210 = X2210 + X22, \quad f2202 = X2202 - X22, \quad f2220 = X2220 + X22$$

$$f10 = f1110 + f2110 + f1210 + f2210 + X010 + 2$$

$$f02 = f1102 + f2102 + f1202 + f2202 + X002$$

$$f20 = f1120 + f2120 + f1220 + f2220 + X020 + 2$$

$$I_{k1} = \frac{e^{\lambda_3 N} - 1}{\lambda_3}[v_{11} f02 + v_{21} f10 + v_{31} f20]$$
$$I_{k2} = \frac{e^{\lambda_4 N} - 1}{\lambda_4}[v_{12} f02 + v_{22} f10 + v_{32} f20]$$
$$I_{k3} = \frac{e^{\lambda_5 N} - 1}{\lambda_5}[v_{13} f02 + v_{23} f10 + v_{33} f20]$$

On the other hand, using the lower and upper bound equations we obtain the following expressions.

$$B_{k1} = v_{21}(\frac{c_2}{\delta_1} + \frac{c_2}{\gamma_1^1} + \frac{\delta_1 c_2}{\gamma_1^1 \delta_2} + \frac{a_1 c_2}{\gamma_1^2} + \frac{c_2 a_1 \delta_1}{\gamma_1^2 \delta_2} + (\frac{c_2}{\delta_1} + \frac{c_2}{\gamma_2^1} + \frac{\delta_2 c_2}{\gamma_2^1 \delta_1} + \frac{a_2 c_2 \delta_2}{\gamma_2^2 \delta_1} + \frac{c_2 a_2}{\gamma_2^2})e^{\lambda_4 N}) +$$
$$v_{11}[\frac{c_1}{\gamma_2^1} + \frac{c_1}{\gamma_2^2} - \frac{a_2 c_1}{\gamma_2^2}]e^{\lambda_3 N} + v_{31}(\frac{c_2}{\delta_1} + \frac{\delta_1 c_2}{\gamma_1^1 \delta_2} + \frac{c_2}{\gamma_1^2} + \frac{a_1 \delta_1 c_2}{\gamma_1^2 \delta_2})$$
$$B_{k2} = v_{22}(\frac{c_2}{\delta_1} + \frac{c_2}{\gamma_1^1} + \frac{\delta_1 c_2}{\gamma_1^1 \delta_2} + \frac{a_1 c_2}{\gamma_1^2} + \frac{c_2 a_1 \delta_1}{\gamma_1^2 \delta_2} + (\frac{c_2}{\delta_1} + \frac{c_2}{\gamma_2^1} + \frac{\delta_2 c_2}{\gamma_2^1 \delta_1} + \frac{a_2 c_2 \delta_2}{\gamma_2^2 \delta_1} + \frac{c_2 a_2}{\gamma_2^2})e^{\lambda_4 N}) +$$
$$v_{12}[\frac{c_1}{\gamma_2^1} + \frac{c_1}{\gamma_2^2} - \frac{a_2 c_1}{\gamma_2^2}]e^{\lambda_3 N} + v_{32}(\frac{c_2}{\delta_1} + \frac{\delta_1 c_2}{\gamma_1^1 \delta_2} + \frac{c_2}{\gamma_1^2} + \frac{a_1 \delta_1 c_2}{\gamma_1^2 \delta_2})$$
$$B_{k3} = v_{23}(\frac{c_2}{\delta_1} + \frac{c_2}{\gamma_1^1} + \frac{\delta_1 c_2}{\gamma_1^1 \delta_2} + \frac{a_1 c_2}{\gamma_1^2} + \frac{c_2 a_1 \delta_1}{\gamma_1^2 \delta_2} + (\frac{c_2}{\delta_1} + \frac{c_2}{\gamma_2^1} + \frac{\delta_2 c_2}{\gamma_2^1 \delta_1} + \frac{a_2 c_2 \delta_2}{\gamma_2^2 \delta_1} + \frac{c_2 a_2}{\gamma_2^2})e^{\lambda_4 N}) +$$
$$v_{13}[\frac{c_1}{\gamma_2^1} + \frac{c_1}{\gamma_2^2} - \frac{a_2 c_1}{\gamma_2^2}]e^{\lambda_3 N} + v_{33}(\frac{c_2}{\delta_1} + \frac{\delta_1 c_2}{\gamma_1^1 \delta_2} + \frac{c_2}{\gamma_1^2} + \frac{a_1 \delta_1 c_2}{\gamma_1^2 \delta_2})$$

Hence, the system to be solved for k_is when $c_1 = c_2$ is the following.

$$\begin{pmatrix} B_{k1} + I_{k1} & B_{k2} + I_{k2} & B_{k3} + I_{k3} \\ v_{11} & v_{12} & v_{13} \\ v_{31} e^{\lambda_3 N} & v_{32} e^{\lambda_4 N} & v_{33} e^{\lambda_5 N} \end{pmatrix} = \begin{pmatrix} 1 \\ 0 \\ 0 \end{pmatrix}$$

References

[1] T. Altiok. *Performance analysis of Manufacturing Systems.* Springer-Verlag New York, Inc., 175 Fift Avenue, NY 10010, 1997.

[2] H. Le Bihan H. and Y. Dallery. A Robust Decomposition Method for the Analysis of Production Lines with Unreliable Machines and Finite Buffers. *to appear in International Journal of Production Research*, 1999.

[3] C.H. Edwards and Jr. D.E. Penney. *Elementary Differantial Equations with Applications.* volume 5. Prentice Hall, Englewood Cliffs, NJ 07632, second edition, 1989.

[4] S. Gershwin. *Manufacturing Systems Engineering.* Prentice Hall, Inc.,Englewood Cliff, NJ, 1994.

[5] E. Koenigsberg. Production Lines and Internal Storage-A Review. *Management Science*, 5:410-433, 1959.

[6] D. Mitra. Stochastic Theory of a Fluid Model of Producers and Consumers coupled by a buffer. *Adv. Appl. Prob.*, 20:646-676, 1988.

[7] B.A. Sevasyanov. Influence of Storage Bin Capacity on the Average Standstill Time of a Production Line. *Theory of Probability and Its Applications*, 7:429-438, 1962.

[8] J. Wijngaard. The effect of a interstage buffer storage on the output of two unreliable production units in series with different production rates. *AIIE Transactions*,11(1):42-46, 1979.

[9] S. Yeralan and J.E. Muth. A continuous materials flow production line model with station breakdown. *EJOR*, 27:289-300, 1986.

Chapter 12

STOCHASTIC LEAD TIME MODELS FOR SUPPLY CHAIN NETWORKS

N. R. Srinivasa Raghavan
Management Studies
Indian Institute of Science
Bangalore, India 560 012
raghavan@mgmt.iisc.ernet.in

N. Viswanadham
Mechanical and Production Engineering and TLI-Asia Pacific
National University of Singapore
Singapore 119260
mpenv@nus.edu.sg

Abstract

In this paper, we investigate use of stochastic network modeling techniques for analyzing supply chain networks. Supply chains are interconnections of several companies such as suppliers, manufacturers, distributors and retailers with the aim of producing and selling customer desired products. Computing the supply chain lead time, or the order-to-delivery time is an important exercise. In this paper, we present stochastic network models for computing the average lead times of make-to-order supply chains. In particular, we illustrate the use of static probabilistic and generalized queueing networks for computing the lead times.

Keywords: Supply Chain Networks, Performance Analysis, Stochastic Models, Queuing Networks

1. Supply Chain Networks

Manufacturing supply chain networks (SCNs) are formed out of complex interconnections amongst various manufacturing companies and service provi-

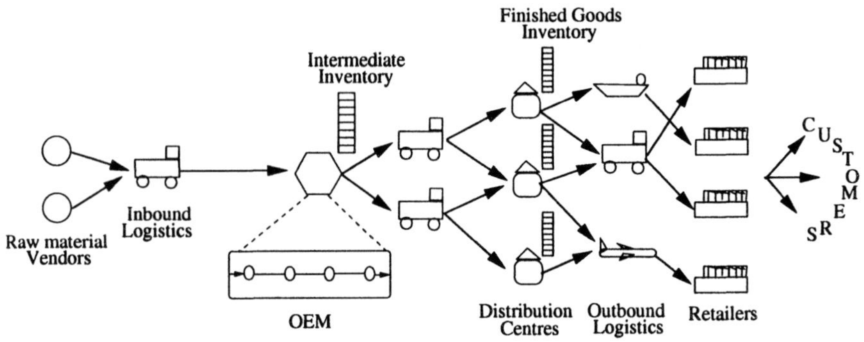

Figure 1. The supply chain network

ders such as raw material vendors, original equipment manufacturers (OEMs), logistics operators, warehouse operators, distributors, retailers and customers (see Figure 1). One can succinctly define supply chain management(SCM) as the coordination or integration of the activities of all the companies involved in procuring, producing, delivering and maintaining products and services to customers located in geographically different places. Traditionally, each company performed marketing, distribution, planning, manufacturing and purchasing activities independently, optimizing their own functional objectives. SCM is a process-oriented approach to coordinating all organizations and all functions involved in the delivery process. Sometimes one can also derive a supply chain within one single enterprise. This typically occurs in a multistage manufacturing system where each downstream machine demands raw parts from its immediate predecessor stages. The methods that we henceforth propose, by default, can also be applied to the above *internal* supply chain.

The product moving through the SCN transits several organizations and each time a transition is made, logistics is involved. Logistics includes all activities associated with movement of material from one enterprise to another. Also since each of the organizations is under independent control, there are interfaces between organizations and material and information flows depend on how these interfaces are managed. We define interfaces as the procedures and vehicles for transporting information and materials across functions or organizations such as negotiations, approvals (so called paper work), decision making, and finally inspection of components/assemblies, etc. For example, the interface between a supplier and manufacturer involves procurement decisions such as price, delivery frequencies and nature of information sharing at the strategic level and the actual order processing and delivery at the operational level. The coordination of the SCN plays a big role in the over-all functioning of the SCP. In most cases, there is an integrator for the network, who could be an original equipment manufacturer, coordinating the flow of orders and materials throughout the network.

Modeling and analysis of such a complex system is crucial for performance evaluation and for comparing competing supply chains. In this paper, we view a supply chain as a probabilistic network and present a modeling approach to compute performance measures such as lead time and work in process inventory. In particular, we investigate the use of static network (PERT like) models and queueing network models for computing the lead time.

1.1 Supply Chain Network Configuration

The configuration of the supply chain defines the interconnection pattern among its facilities. Supplier's manufacturing plants, warehouses, final assembly plants, packaging centers and transshipment points (cross-docking) are examples of facilities. All supply chain networks do not have the same configuration. Depending on the product structure several network configurations are possible. One could identify four major supply chain configurations including serial, converging, diverging, and converging-diverging networks. The supply chains can be operated in different ways: make-to-stock, make-to-order or assemble-to-order. We briefly cover these issues below.

(a) Serial structure: Here, one facility of the network feeds into another and the entire supply chain resembles a single pipeline.

(b) Divergent structure: This structure resembles a cone. At the vertex of the cone is the facility that produces a base product from which are produced the derivatives. The Petroleum industry is a typical case in point. Distribution intensive industry supply chains often have a divergent structure.

(c) Convergent structure: In this case, there is a series of sub-assembly stages finally leading to a finished product as in the case of automobiles and air crafts.

(d) Network structure: This is a tandem combination of convergent and divergent structures as in the computer industry which is both sourcing and distribution intensive.

In this paper, we consider the serial and convergent structures. But the application of the methodology to other structures is straightforward.

1.2 Operational Models

Another important aspect of of the supply chain operation is the supply chain planning and control methodology (SPC). A customer order for a product triggers a series of activities in the supply chain facilities, and these have to be synchronized so that the end customer order is satisfied. The SPC specifies the business model and hence determines the paths for the information and material flow in the supply chain. There are three broad models followed in practice:

(a) **Make-to-stock**(MTS): Here, the end customer products are satisfied from stocks of inventory of finished goods that are kept at various retail points of the SCN.

(b) **Make-to-order**(ATO): In this SPC technique, a *confirmed* customer order triggers the flow of materials and information in the supply chain. Each customer order is unique in terms of manufacturing, procuring, packaging or logistics requirements. There is very little or no inventory maintained of the finished goods or component materials.

(c) **Assemble-to-order**(ATO): Here, a variety of products are assembled to order from components and sub-assemblies, which are either manufactured-to-stock or outsourced. One crucial issue in such a business model is the location of the customer order decoupling point, a point in the supply chain until which subassemblies are made to forecast and beyond which products are built to order.

The SPC technique defines the inventory control rules followed by the supply chain. Each facility of the supply chain could follow its own inventory control policies and ways of handling the sourcing and delivery functions. It is possible that not all the facilities produce to stock. Some facilities might assemble goods to order. The crucial issues of when to order and how much to order define these policies. For instance in base stock policies, one unit (alternatively, a stock keeping unit, SKU) of inventory is replenished as soon as a unit of goods held at the facility is depleted. On the other hand, if the facility is following a reorder point based policy, it replenishes items as soon as a preset reorder level is reached, ordering each time such that a targeted level of inventory is reached. For a detailed discussion, see [Vollman et al., 1998].

1.3 Literature Survey

In this section, we briefly survey the literature on mathematical models for supply chains. The analytical modeling of supply chain networks can broadly be classified into two areas: network design, and performance analysis methods. The network design models help in strategic and tactical decision making. They are basically mixed-integer programming models and are used to decide what products to produce and for what markets, where and how to produce them, and using what resources. There is a large amount literature on this subject and a number of survey papers also appeared on this subject (see [Erenguc et al., 1999] and the references cited therein). There is also a large body of literature in the development of multi-echelon inventory control models. A comprehensive review of these models can be found in the book by Silver et.al. [A.E. et al., 1998]. Other studies include location of production and inventory facilities [Arntzen et al., 1995, Sankaran and Raghavan, 1997], choosing make to order or make to stock policies [D. Connors et al., 1995], etc.

Performance analysis of SCNs is basically conducted to determine the lead time, variation, cost, reliability and flexibility. SCNs are discrete event dynamical systems (DEDS) in which the evolution of the system depends on the complex interaction of the timing of various discrete events such as the arrival of components at the supplier, the departure of the truck from the supplier, the start of an assembly at the manufacturer, the arrival of the finished goods at the customer, payment approval by the seller, etc. The state of the system changes only at discrete events in time. Over the last two decades, there has been a tremendous amount of research interest in this area.

Very attractive higher-level general-purpose simulation packages are now available that can faithfully model the value delivery processes of a manufacturing enterprise. These include SIMPROCESS, PROMODEL, and TAYLOR II, to name a few. The simulation of a SCN involves developing a simulation model, coding it, validating it, designing the experiments, and finally conducting a statistical analysis to obtain the performance measures.

1.4 Analytical Methods

Our aim here is to summarize five analytical techniques useful for modeling SCNs including series-parallel graphs, Markov chains, queueing networks, Petri nets and system dynamics models.

1.4.1 Series parallel graphs. Series parallel graphs can model an SCN by assigning probability distributions to the lead time of the activities in the graphs. These are graphs, showing the precedence and concurrency of the activities of the material and information flow. Assuming that all the activities are statistically independent, one can determine the mean and variance of the lead times.

1.4.2 Markov chains. The use of Markov models in the study of performance of manufacturing systems [Viswanadham and Narahari, 1992] is well known. Direct modeling of an SCN as a Markov chain would be very difficult and expensive. Higher level models based on Petri nets and queueing networks are ultimately solved as Markov chains using software packages such as SPNP.

1.4.3 Petri nets. Faithful modeling of iteration, synchronization, forks, and joins that arise in SCNs is possible using Petri nets. If too detailed models are developed, numerical solution, however, may turn out to be a nightmare. Hierarchical modeling discussed in this paper provides a tractable way of handling largeness here.

1.4.4 Queueing networks. The most general SCN can be modeled as a fork-join queueing network model with iteration or reentrancy. An analytical solution of these general models is not available, and approximations are available in only special cases. Some solutions can be found in [Raghavan, 1998]. This is an area of active research.

1.4.5 System dynamics models. Here the SCN is modeled using differential equations. Forrester first explained the bullwhip effect using these models [J.W, 1961]. In this paper, we consider the series-parallel graph and queueing models.

2. Lead Time Models

The total average lead time for an order entering the supply chain is a crucial performance measure. We can compute the same in the static case as well as the dynamic case at the aggregate level, using some of the models that we present in this paper. In the static case, we require information on the processing time (or, its probability distribution) of all the orders that are to be scheduled across the supply chain. Whereas, in the dynamic and stochastic setting, orders for end products arrive in a random fashion, and demand random processing times at various facilities. In this paper, we don't present the models for the dynamic and static case.

The static lead time models presented in Section 3 are based on an earlier work by Wilhem [Wilhem and Marandi, 1982]. The author provides approximation schemes for computing the lead time for jobs in a manufacturing system which have to undergo complex assembly operations. Such models can be used to arrive at the total expected supply chain lead time and its variance. This can provide the vital link between the order processing agents and the customer, especially during the order acceptance. For instance, an idea on the mean and variance of the supply chain lead time can help one in quoting the delivery time reliably for a given customer order.

We then cast the dynamic lead time computation problem for a class of make-to-order supply chains as multi-class open generalized queueing networks and utilize the existing efficient approximation algorithms for computing the expected supply chain lead time and variance. This is the subject matter of Section 5. We conclude this paper in Section 6.

3. Lead Time Models for Performance Analysis

Here, we present several numerical examples illustrating the computation of lead times in supply chains. We assume that the lead times are random variables and present methods of computing the distribution of the lead time. The analysis is similar to computing lead time distributions in PERT networks and for acyclic graphs ([Martin, 1965], [Sculli, 1983] and [Wilhem and Marandi, 1982]).

3.1 Lead Time Computation

3.1.1 Serial Networks. Consider, for example a pipeline supply chain with a single supplier, single manufacturer and a single distributor with logistical operations in between. Let X_i be the random variable representing the corresponding lead times. Assume that they are independently and identically distributed. Then the distribution of supply chain lead time $F_{\text{LT}}(t)$ is given by

$$F_{\text{LT}}(t) = F_{X_1}(t) \otimes F_{X_2}(t) \otimes F_{X_3}(t) \otimes F_{X_4}(t) \otimes F_{X_5}(t) \quad (1)$$

where $\otimes$ is the convolution operation. In the above equation, $F_{X_i}(t)$ is the probability that the random variable X_i doesn't exceed t. For a given distribution of X_i, the mean and variance of LT can be numerically computed. If we assume that all X_is are normally distributed, then the mean and variance of the SCN can be easily obtained by simple addition. While the above is true for make-to-order supply chains, we can use the method for computing the lead time in the make-to-stock supply chains too. For instance, if the probability of stock outs at the various facilities in the serial network is known apriori, then one can easily compute the time taken for serving a customer order (see numerical example at the end of this section).

3.1.2 Convergent Networks. Consider for example the case where a manufacturer sources three subassemblies from three independent suppliers, who deliver directly to the factory floor. Let X_i be the random variable denoting the delivery time from supplier i, and let $F_{X_i}(t)$ be its distribution function. Let T be the random variable denoting the max (X_1, X_2, X_3). Then T is the time after which all the subassemblies are available at the manufacturer. Also,

$$F_T(t) = F_{X_1}(t) * F_{X_2}(t) * F_{X_3}(t)$$

If A is the random variable signifying the assembly time at the manufacturer, let $F_A(t)$ be its distribution function. Then the distribution of the supply chain lead time LT is given by

$$F_{\text{LT}}(t) = F_T(t) \otimes F_A(t)$$

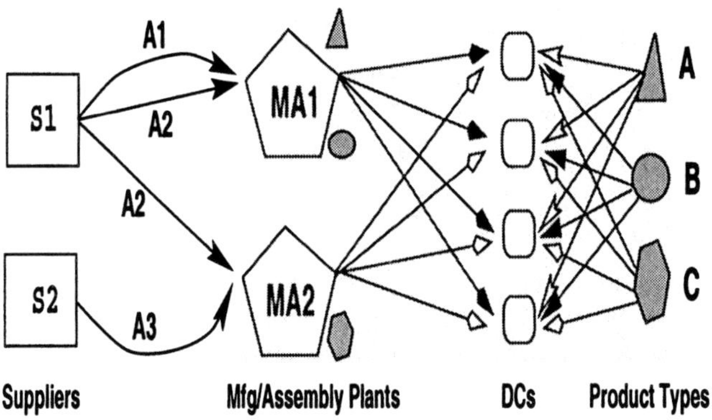

Figure 2. The supply chain network considered for analysis

where ⊗ represents the convolution operator. The mean and variance of LT can easily be computed. For given distributions of X_i and A, one can numerically compute $F_{\text{LT}}(t)$. Also , the probability of delivering an order within a specified time limit can be easily computed.

3.1.3 General Networks. A general supply chain network can be modelled as a graph with the facilities as nodes and the arcs model the interconnection between the nodes. Suppose the probability distributions of the processing times at each of the nodes is known, then it is straightforward to write down the expression for the supply chain lead time as in the above two cases. The expression would involve multiplications, convolutions, etc of the individual facility distributions. In principle it is possible to numerically solve for the lead time. But it is computationally expensive.

In the following, we assume that the distributions are normal and describe a method of computing the approximate lead time using a result of Clarke [Clarke, 1961].

3.2 Lead Time with Normal Distributions

In this section, we discuss an interesting application of Clarke's results to compute the lead time for supply chains with assembly, i.e., the convergent structure. Consider the supply chain network shown in Figure 1, with facilities such as manufacturers, suppliers, etc., inter-connected by logistics and various interorganizational interfaces. All the facilities operate in make-to-order mode. We assume that all processing times are normally distributed with mean μ_i and standard deviation σ_i at facility i. Also we assume that $\mu_i \geq 3\sigma_i$ so that the probability of processing time going negative is less than 0.05. In the following section, we consider a convergent supply chain and compute the lead time.

3.2.1 Lead Time in Convergent Supply Chains. Consider a supply chain network consisting of three product lines, A, B and C, which are assembled in two plants MA1 and MA2, and distributed using dedicated logistics operators (See Figure 2). The Bill-of-subassemblies (BOS) for each batch of the products is: A needs one each of A1 and A2; B, one each of A1, A2 and A3; and C, one each of A2 and A3. For simplicity, we model only one hierarchy for the BOS, albeit with subassembly commonality. Supplier S1 provides sub-assemblies A1 and A2; and S2 supplies A3. For instance, the three products could be different varieties of desk-jet printers, with the sub-assemblies representing the front panels (A1), cartridges (A3) and a base product (here, A2). We assume that the organization has strategic relationships with the two suppliers. The distribution is handled by two independent third party logistics providers: one for the sub-assemblies and the other for the outbound finished goods. Batch processing is assumed everywhere. We are not concerned here with optimal batch sizes and the like. For our purposes, a job represents a batch of sub-assemblies or finished products.

Our approach is the following. First, we focus on each facility and find the mean and variance of the lead time (assumed Normal) individually for each job type produced by the facility. For example, supplier S1 produces sub-assemblies A1 and A2 for both plants MA1 and MA2. We find the lead times for the three possible combinations: A1 for MA1, A2 for MA1 and finally A2 for MA2. Next, we decompose the supply chain network into K subnetworks, one for each job type, where K is the number of job types. Finally, we use the algebra of probability distributions to determine the supply chain lead time. Now we illustrate this procedure for the above example.

Let us consider product type C. The reduced network for C would be as in Figure 3. In this figure, S_1 and S_2 are the suppliers, I_3 is the interface between the suppliers and the manufacturing plant MA_2, while O_3 is the outbound logistics facility. Let the various lead times be denoted as T_1 ... T_7, which are computed as discussed in the earlier paragraph. We assume that these times are normally distributed with means greater than thrice respective standard deviations. We obtain the average supply chain lead time as $\max\{(T_1+T_2),(T_3+T_4)\}+T_5+T_6+T_7$. We know that the sum of any number of independent normal random variables is normal with additive means and variances. Also, to compute the maximum of two normal random variables, we use Clarke's results [Clarke, 1961], by assuming that the flow times at the various individual facilities are independent of each other. Thus, the total lead time is easily computed .

We cite below Clarke's results, for the maximum of two Gaussian random variables D_1 and D_2. The maximum is again assumed to be Gaussian (D) [Clarke, 1961]. The maximum of any number of such random variables is

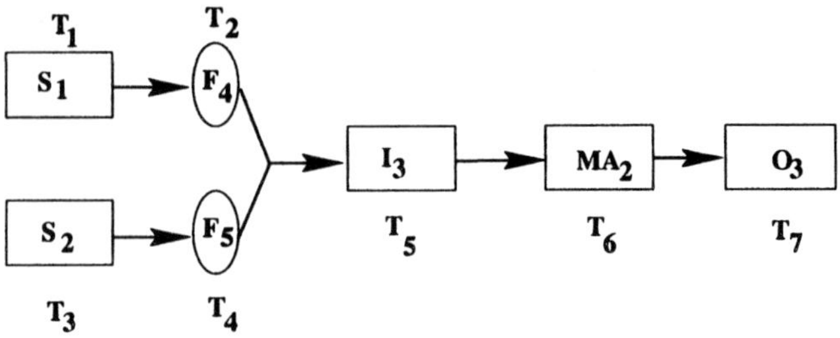

Figure 3. The reduced supply chain network for product C

easily derivable, thus resulting in a quick and effective method for performance modeling of even large supply chains.

$$\begin{aligned}
\mathbb{E}[D] &= \mathbb{E}[D_1]\Phi(\alpha) + \mathbb{E}[D_2]\Phi(-\alpha) + a\phi(\alpha) &(2)\\
\mathbb{E}\left[D^2\right] &= (\mathbb{E}^2[D_1] + \sigma^2_{D_1})\Phi(\alpha) + (\mathbb{E}^2[D_2] + \sigma^2_{D_2})\Phi(-\alpha) &(3)\\
&\quad + (\mathbb{E}[D_1] + \mathbb{E}[D_2])\, a\, \phi(\alpha) &\\
\alpha &= \frac{1}{a}(\mathbb{E}[D_1] - \mathbb{E}[D_2]) &(4)\\
a^2 &= \sigma^2_{D_1} + \sigma^2_{D_2} - 2\sigma_{D_1}\sigma_{D_2}\varrho &(5)
\end{aligned}$$

In the above equations, ϕ is the standard normal density function and Φ is the corresponding cumulative distribution function. Also, ϱ is the coefficient of correlation between the variables D_1 and D_2, which is assumed to be 0 in our case, owing to the independence assumption.

The above result easily extends to the case where we have to evaluate the maximum of more than two normal random variables by observing that $\max\{A, B, C\} = \max\{\max\{A, B\}, C\}$.

3.2.2 Numerical Examples. In this section we illustrate the use of the above methods for computing the lead time in various supply chains.

(a) Serial Network MTO case: Consider a tandem supply chain which is operating in a make-to-order fashion. We assume that the lead times at the individual facilities of the supply chain are i.i.d. normal random variables with given mean and variance. We know that the convolution of several normal distributions results in a normal distribution. Let the number of facilities be five, including the raw material vendor, the inbound logistics, the OEM plant, the outbound logistics and a customizing plant. Let X_i be the respective lead time random variables as follows: $X_1 \sim N(10, 2)$, $X_2 \sim N(2, 1)$, $X_3 \sim N(5, 1)$, $X_4 \sim N(2, 1)$, $X_5 \sim N(1, 0.5)$, where $N(\mu, \sigma^2)$ is the Gaussian with mean μ and standard deviation σ. The unit of time is days. Thus the mean total supply chain lead time in the above case would be 20 days with a standard deviation of 2.35 days. We can now easily determine the probability of on-time delivery. Let us assume that the customer has requested for a due date of 18 days. We thus need to compute $P(X \leq 18)$ where X is the total supply chain lead time. In this case, this probability turns out to be 0.1967.

(b) Serial Network MTS case: Let us assume that the supply chain now operates in a make-to-stock mode with the same parameters as above, for the lead times at the individual facilities. We need more information on the probability of stock out at the above facilities. Let us assume that these values are available apriori. (Usually, firms form an alliance to fix a value of the probability of stock out at their respective facilities.) Let us assume p_i is the probability of stock out given to us as follows: $p_1 = 0.0$, $p_2 = 0.01$, $p_3 = 0.10$, $p_4 = 0.10$, $p_5 = 0.01$. We assume that the facility that is left most in the supply chain (for e.g., raw material vendor) cannot go out of stock.

Our aim here is to compute the customer order delivery time by its first two moments. We observe that the order delivery is immediate if the last stage of the supply chain (the retailer) is not out of stock. Otherwise the order has to wait until it is sourced from the immediate supplier. Applying this recursively, we obtain the expression for the total supply chain lead time as:

$$X = X_n + p_n X_{n-1} + p_n p_{n-1} X_{n-2} + \ldots + p_n p_{n-1} p_{n-2} \cdot\cdot p_2 X_1.$$

We can now easily compute the mean and variance. For the above input values the mean lead time turns out to be 1.82 days with a standard deviation of 0.715 days.

(c) Computing probability of stock out: Continuing example 2 above, we note that the equation for the total lead time in the MTS case can be used to determine the allowable probability of stock out at a given facility of the supply chain. Let us assume that we are given the following: the probability that the customer order lead time will over shoot 2 days is 0.4. Mathematically, $P(X \geq 2) = 0.4$. We also assume that we are given the lead time and the stock out probabilities at all other facilities. Assuming that the various times are normally distributed we proceed as follows. Let us be given to determine p_5, given all other probabilities and lead times as in example 2 above. We get the following equation:

$\Phi^{-1}(0.4) = \frac{2.0-(1.8+2p_5)}{0.7148+p_5^2}$, where Φ is the standard Normal c.d.f.
Solving the above equation we get $p_5 = 0.011$.

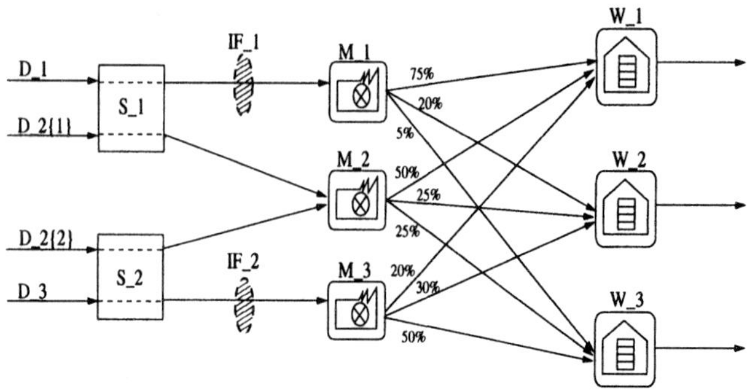

Figure 4. A supply chain Example

(d) Convergent network using Clarke's results: We now compute the lead time for the supply chain shown in Figure 2. We are given the following cycle times at the individual facilities: $T_1 \sim N(10, 2.0)$, $T_2 \sim N(1, 0.5)$, $T_3 \sim N(15, 4.0)$, $T_4 \sim N(2, 1.0)$, $T_5 \sim N(1, 0.5)$, $T_6 \sim N(3, 2.0)$, $T_7 \sim N(1, 0.5)$. Using Clarke's results (shown in equations 1-4), the mean and variance of the total supply chain lead time is computed. We assume that the network operates in make-to-order mode. Thus $\mathbb{E}[T] = 22.013$ days, while $\sigma(T) = 3.586$ days. Assuming that the resulting distribution is also Normal, we can now compute the probability of delivering a product in specified lower and upper limits. For instance, the probability that the above supply chain will deliver the goods within 20 days but not after 25 days is computed as $P(20 \leq T \leq 25) = \Phi(\frac{25-22.013}{3.586}) - \Phi(\frac{20-22.013}{3.586}) = 0.511$.

4. Generalized Queueing Network Models

In this section, we model make-to-order supply chains using general queueing networks, and evaluate their performance. The orders for end products arrive at random points of time, and demand a random processing time at each one of the facilities. Refer [Buzacott, 1997] for instances where business processes of an enterprise are modeled using queueing networks.

In this paper, we analyse the situation wherein three *different* manufacturers supply a common end product (for e.g. three different television producers) to a common market consisting of three distribution centres (DCs). We bring out the significance of strategic alliances which when absent, appear as *interfaces* in our models.

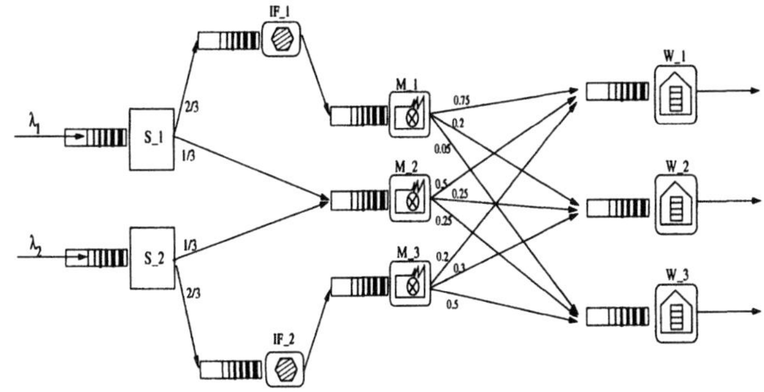

Figure 5. The generalized open queueing network model

4.1 The Single Product Multiple Competitors Case

Consider the network shown in Figure 4. Let us assume that there is a steady demand for products of a type, at the three warehouses W_1, W_2, and W_3. There are three different manufacturers M_1, M_2, and M_3 for this product, who supply to the same three markets. There are two common suppliers S_1 and S_2. The alliances between the suppliers and the manufacturers is as shown in the Figure 4. We assume that M_2 has good alliance (For eg. vendor managed inventories) in place with both S_1 and S_2. The other manufacturers have no such arrangements. Thus, we have interfaces between M_1 and S_1, and M_3 and S_2, which essentially add delay to the process of supplying raw materials/sub-assemblies to the manufacturers. Each manufacturer shares the production with the warehouses as shown in Figure 4. The demand for the end products translate into demand for the required raw materials to the suppliers. We model the physics of the above supply chain as a generalized open queueing network of the single class type, for purposes of approximate analysis. Figure 5 shows the generalized queueing network (GQN) model for the example supply chain in Figure 4 with input parameters as in Table 1.

We model various facilities (suppliers, manufacturers, etc) as servers of the queueing network. Since we consider the single product type case, we *aggregate* the demand for all the three manufacturers as two independent arrival streams to the suppliers. The probability of routing the raw parts/sub-assemblies to the manufacturers (which is computed as a percentage of demand outsourced by the manufacturers to the suppliers) is depicted in the Figure 5. The performance analysis of the queueing models is conducted using a queueing network package developed locally [Kumar, 1994], based on the principles of the Queueing Network Analyzer of Ward Whitt [Whitt, 1983].

$D_1 = D_2 = D_3 = D = 20$ units per day; SCV=2.0 each. with $\lambda_1 = \lambda_2 = 30$

Server Name	Service Rate in units/day	
	Mean	SCV
S_1	40	1.5
S_2	40	1.0
IF_1	25	0.5
IF_2	25	0.5
M_1	45	1.5
M_2	50	1.0
M_3	45	0.5
W_1	60	0
W_2	60	0
W_3	60	0

Table 1. The input parameters for the supply chain of Figure 4

Manufacturer	Arrival Rates		
	D+D/2, D+D/2	D+0.05D, D+0.95D	D+D/4, D+3D/4
M_1	0.3629	1.1195	0.3509
M_2	0.1535	0.9078	0.1984
M_3	0.3313	0.3342	0.3882

Table 2. Effect of order release on average cycle times (in days): base case

Manufacturer	(SCV_{IF_1}, SCV_{IF_2})		
	(1.5, 0.5)	(1.5, 1.0)	(0.0, 0.0)
M_1	0.4509	0.4510	0.3188
M_2	0.1540	0.1541	0.1530
M_3	0.3316	0.3754	0.2886

Table 3. Effect of SCV of interface times on average cycle times, with arrival rates 3D/2 at both the suppliers

Manufacturer	Arrival Rates		
	D+D/2, D+D/2	D+0.05D, D+0.95D	D+D/4, D+3D/4
M_1	0.3110	0.8676	0.2909
M_2	0.1270	0.6670	0.1553
M_3	0.2810	0.2686	0.3267

Table 4. Effect of SCV of *arrivals* on average cycle times, with $SCV_1 = 1.0, SCV_2 = 1.0$

4.2 Discussion of Results

The performance measures of interest to us are the average cycle time and average total WIP at the three manufacturing plants. In the context of supply chain management, a manufacturer would usually be interested in knowing how much of his requirements he has to procure from a given supplier. We studied three different scenarios for the above issue, with M_2 procuring different amounts of its demand from S_1 and S_2. This is modeled as different splitting rates of demand, thus influencing the arrival rates to the two suppliers. In the tables from Table 2, we considered three different such *procurement* options: one with equal splitting of M_2's demand among the suppliers, second with very little going to S_1, and the third is a case in between. The values for the average waiting time and average queue lengths have been tabulated for all the three cases. On critically observing the tables, we conclude as follows:

- By splitting demand equally, M_2 gains a lot in terms of reduced lead times and queue lengths (which turn out to be WIP in a manufacturing setting).

- Before issuing procurement orders to its suppliers, a manufacturer must keep in view the orders from *all other customers of* S_1 *and* S_2, *who could very well turn out to be his own competitors.*

  If M_1 and M_3 decide on the amount to be procured from S_1 and S_2 without paying attention to what M_2 decides, it would prove sub-optimal to them. For example, M_1's optimal value occurs when M_2 decides to procure $3/4^{th}$ of its demand from S_2. But this won't happen, since, M_2 has decided to split demand equally! (Here, by optimal, we mean being *economical* in terms of lead times and in-process inventories and not the absolute minimum value for these two metrics.)

- The effect of interface time variance on the performance of M_1 and M_3 can be seen clearly from Table 3. As the SCV (and hence, variance) of the interface IF_2 decreases from 1.0 to 0.0, (keeping the mean service rates at their base values), both the average lead time and the average inventory decrease. It is also seen, (see Tables 2, 3) that if the SCV of IF_1 is also reduced, in conjunction with IF_2, it will reduce the performance measures for M_1 and M_3. The best performance is obtained when the SCV's are zero.

- Let us now compare Table 2 with Table 4. For the former, the SCV of inter-arrival times are each 2.0, while for the latter, we decreased it to 1.0, each, fixing the mean inter-arrival times. We see that irrespective of the splitting pattern, the performance measures have improved. The average cycle times and inventories have reduced when we reduced the SCV of the arrival rate.

4.3 Implications

As discussed in the earlier section, our simple analysis (though approximated at a very aggregate level) brings out some very relevant issues in managing supply chains, unequivocally. They are:

(a) Decisions by a manufacturer (retailer) regarding how much to be procured from which supplier (manufacturer), are not to be made in isolation. A thorough survey of the number, demand patterns and suppliers of the competing manufacturers (retailers) has to be made at the outset.

(b) Although interfaces exist in the event of alliances not being proper, efforts must be put towards reducing the variance (and mean, if possible) of the interface process cycle times. Also, it would be better for M_1 and M_3 to get into alliance with the suppliers, like M_2 has got into, so that their interfaces are smoothened out.

(c) The variance of arrival rate should be properly adjusted and controlled to yield better performance. Efforts aimed at reducing SCV of arrival rate (keeping its mean at the same value) yields better results.

5. Conclusions

In this paper, we have presented graph theoretic and queueing network based lead time models for analysing the supply chain networks. The graph models are based on the assumption that the facility processing times are described by a probability distribution function. We presented a general method of obtaining the distribution of the supply chain lead time. For the special case when the distributions are normally distributed, we use Clarke's approximation to get estimates of the mean and variance of the total supply chain lead time. This can be used in determining the delivery reliability for end customer orders. We have also proposed generalized open queueing network models for the dynamic performance evaluation of supply chain process. These are basically meant for make-to-order situations. Through use of simple examples, we have illustrated the advantages of communications among various players in the supply chain.

Supply chain modeling and analysis has very rich potential in terms of improving the industry productivity. For meaningful results and conclusions it is essential to model real world supply chain, collect data and estimate the distributions and their parameters.

References

[A.E. et al., 1998] A.E., S., Pyke, D., and R.Peterson (1998). *Inventory Management and Production Planning and Scheduling*. Wiley, New York.

[Arntzen et al., 1995] Arntzen, B. C., Brown, G. G., and Harrison, T. P. (Jan-Feb 1995). Global supply chain management at Digital Equipment Corporation . *Interfaces*, 25(1):69–93.

[Buzacott, 1997] Buzacott, J. (1997). Commonalities in reengineered business processes: Models and issues. *Management Science*, 42(5):768–782.

[Clarke, 1961] Clarke, C. E. (1961). The greatest of a finite set of random variables. *Operations Research*, pages 145–161.

[D. Connors et al., 1995] D. Connors, C. An, S. Buckley, G. Feigin, A. Levas, N. Nayak, R. Petrakian, and R. Srinivasan (1995). Dynamic modeling of re-engineering supply chains. Technical Report RC-19944, IBM Research Center, Almaden.

[Erenguc et al., 1999] Erenguc, S., Simpson, N., and A.J.Vakharia (1999). Integrated production /distribution planning in supply chains: An invited review. *European Journal of Operations Research*, 115:219–236.

[J.W, 1961] J.W, F. (1961). *Industrial Dynamics*. MIT Press, Cambridge, MA.

[Kumar, 1994] Kumar, V. K. (1994). A package for performance analysis of generalized queueing networks. Technical report, Department of Computer Science and Automation, Indian Institute of Science, Bangalore.

[Martin, 1965] Martin, J. J. (1965). Distribution of the time through a directed, acyclic network. *Operations Research*, 13:46–74.

[Raghavan, 1998] Raghavan, N. R. S. (1998). *Performance Analysis and Scheduling of Manufacturing Supply Chain Networks*. Ph.D Thesis, Indian Institute of Science, Bangalore.

[Sankaran and Raghavan, 1997] Sankaran, J. K. and Raghavan, N. R. S. (Nov-Dec 1997). Locating and sizing plants to bottle Propane in South India. *Interfaces*, 27(6):1–15.

[Sculli, 1983] Sculli, D. (1983). The completion time of PERT networks. *Journal of the Operational Research Society*, 34(2):155–158.

[Viswanadham and Narahari, 1992] Viswanadham, N. and Narahari, Y. (1992). *Performance Modeling of Automated Manufacturing Systems*. Prentice Hall, Englewood Cliffs, NJ.

[Vollman et al., 1998] Vollman, T. E., Berry, W. L., and Whybark, D. C. (Fourth Edition 1998). *Manufacturing Planning and Control Systems*. The Dow Jones-Irwin/APICS Series in Production Management.

[Whitt, 1983] Whitt, W. (1983). The queueing network analyzer. *Bell Systems Technical Journal*, 62(9):2779–2815.

[Wilhem and Marandi, 1982] Wilhem, W. E. and Marandi, S. A. (1982). A methodology to describe operating characteristics of assembly systems. *IIE Transactions*, 14(3):204–213.

Chapter 13

MODELING AND PERFORMANCE EVALUATION OF BASE-STOCK CONTROLLED ASSEMBLY SYSTEMS

N. Sbiti [1], M. Di Mascolo [2], T. Bennani [3] and M. Amghar [4]
[1] [3] [4] Ecole Mohammadia d'Ingénieurs. Université Mohammed V. Maroc
Tel: (212) 37 77 19 05/06
e-mails: sbiti@emi.ac.ma, bennani@emi.ac.ma, amghar@emi.ac.ma
[2] Laboratoire d'Automatique de Grenoble , CNRS-INPG-UJF. France
Tel : (33) 4 76 82 64 16, e-mail: Maria.Di-Mascolo@inpg.fr

Abstract In this work, we are interested in base-stock controlled assembly systems. We consider production systems, realizing an assembly operation between components. In a previous work, we defined precisely how the base-stock mechanism can be applied in the case of assembly systems, then we proposed a queueing network model for base-stock controlled assembly systems and we established some properties for this model. In this paper, we propose an analytical technique for performance evaluation of base-stock controlled assembly systems, based on a decomposition method. This analytical method provides steady-state performance measures, like the probability of immediately satisfying a demand, the average number of backordered demands, the average number of finished products, etc. It can be applied for systems assembling any number of components and containing any number of stages in series after the assembly operation.

Keywords Pull systems, Base-stock, Assembly, Analysis, Markov chain, Matrix-geometric, Performance evaluation, Queueing networks, Decomposition method.

1. INTRODUCTION

In this work, we are interested in pull controlled production systems [3][4][10], with a make to stock production. Generally such systems are decomposed into stages, each one consisting of a production subsystem producing semi-finished parts. We consider the special case of base-stock controlled systems. The particularity of these systems is their great reactivity: when an external demand arrives, it is instantaneously transmitted to each stage. These systems have one adjustment parameter per stage, that represents the initial level of finished products for the considered stage. The interest of the base-stock control policy has already been shown in other works [6][7], and the analysis of such systems was performed, particularly by developing a decomposition technique, but only in the case of serial stages. On the other hand, in [16], the author analyses production systems, in which individual components are made to stock, using a base-stock policy, but final products are assembled to customer orders. Note that in [16], it is assumed that no operation can be performed after the assembly operation, so the assembled product is the final product. In the present work, we are interested in the analysis of base-stock controlled assembly systems, in which assembly operation can be followed by other operations.

In this paper, we present first an example of a base-stock controlled system, realizing an assembly operation between two components, and containing one stage after the assembly. Then, we propose an analytical technique for performance evaluation, based on the decomposition method we mentioned before. This analytical technique provides steady-state performance measures, like the probability of immediately satisfying a demand, the average number of backordered demands, the average number of finished products, etc. Then, this analytical method is extended for systems assembling any number of components and containing any number of stages in series after the assembly. Finally, we present some numerical results for the method we proposed and we compare them with simulation results.

2. BASE-STOCK CONTROLLED ASSEMBLY SYSTEMS

Let us present the principle of base-stock controlled assembly systems, by an example. We consider a production system with 3 machines m_1, m_2 and m_3 (see Figure 1). This system produces one type of finished parts, by assembling 2 semi-finished parts that are provided by machines m_1 and m_2, and then by processing the assembled product on machine m_3 to obtain the finished part. Each of machines m_1 and m_2 is supplied by one type of raw parts. In our example, this production system is decomposed into 3 stages.

Each stage i (i=1,2,3) consists of a single machine and an output buffer O_i. In each machine, a part can be waiting for or receiving service. When a part finishes its processing on machine m_i (i=1,2,3), it enters buffer O_i. In the following, it is assumed that raw parts are always available at the inputs of stages 1 and 2.

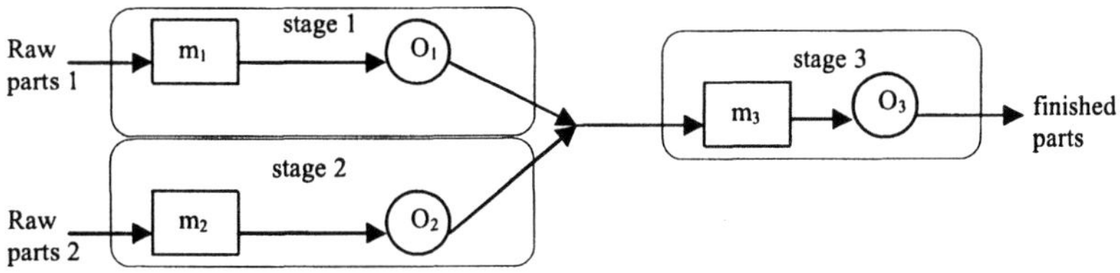

Figure 1: Example of an assembly production system

Each buffer O_i (i=1,2,3) contains initially S_i finished products of stage i. All the machines are initially empty.

When an external demand arrives in the system, it simultaneously causes the three following events:

- The departure of one finished product from the system, if there is at least one finished part in buffer O_3. If buffer O_3 is empty, then this external demand is backordered until a finished part becomes available.

- The instantaneous assembly of one finished part from stage 1 and one finished part from stage 2 (if O_1 and O_2 are both not empty, otherwise, this demand is backordered and will be satisfied later). The obtained product is then processed in stage 3.

- The starting of a new production in stages 1 and 2.

We can notice, that as in the case of serial stages, the advantage of the base-stock policy, if we compare it with the other pull policies, is its great reactivity. Indeed, when an external demand arrives, it is *simultaneously* transmitted to all the stages of the system.

3. MODELING

The base-stock system described above can be modeled as a queueing network with synchronization mechanisms (see Figure 2). Stations M_1, M_2 and M_3 model the machines m_1, m_2 and m_3 respectively. Queues P_i (i=1,2,3) model the output buffers O_i. Queues A_3 and A_4 contain the authorizations for transferring semi-finished parts from stages 1 and 2 into stage 3, and finished parts towards the consumer, respectively.

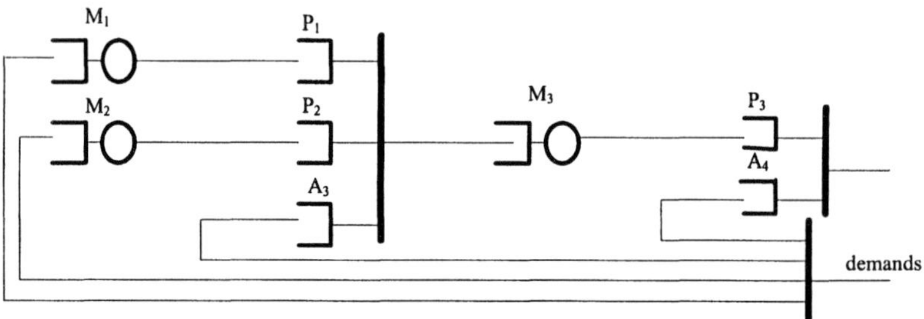

Figure 2: The queueing network model of the production system of Figure 1.

Let us now study the behavior of this system. When an external demand arrives in the system, it is instantaneously transmitted to all the stages, by supplying A_3 and A_4, and starting production in stages 1 and 2. The supplying of A_4 causes the departure of one entity from P_3 (if P_3 is not empty) and the supplying of A_3 authorizes the simultaneous transfer of one entity from P_1 and one entity from P_2 (if P_1 and P_2 are both not empty), into M_3.

We assume that queue A_3 has a *finite capacity* a_3, which can be chosen as large as we like. So, when an external demand arrives, it is taken into account if the number of customers in queue A_3 is less than a_3, otherwise this demand is rejected as a whole, i.e. it will not supply queues A_3 and A_4, and will not start the production in stations M_1 and M_2. On the other hand, we suppose that all the other queues of the system have an infinite capacity. Note that the assumption of finite capacity for A_3 is necessary to obtain an *exact* analysis of stages 1 and 2 (see section 4).

In a previous work [15], we have seen that base-stock controlled assembly systems are a particular case of generalized kanban controlled assembly systems. The queueing network corresponding to these systems is composed of several closed sub-networks [15]. The number of customers present in these closed sub-networks is thus constant and equal to the number of customers that are present at the initial state. We defined then the invariants corresponding to these sub-networks and then, in the particular case of base-stock controlled assembly systems, we obtained the following invariants (the bold characters denote the state of a queue, i.e. the number of customers present in this queue; for example, $\mathbf{M_i}$ denotes the number of customers in station M_i):

$$\mathbf{M_i} + \mathbf{P_i} = \mathbf{A_3} + S_i \qquad i=1,2 \tag{1}$$

$$\mathbf{A_3} + \mathbf{M_3} + \mathbf{P_3} = \mathbf{A_4} + S_3 \tag{2}$$

Since $\mathbf{A_3}$ is bounded by a_3, considering invariants (1) and the synchronization station (P_1, P_2, A_3), we can determine the following bounds:

$$0 \leq \mathbf{M_i} \leq a_3 + S_i \qquad i=1,2 \tag{3}$$
$$0 \leq \mathbf{P_i} \leq a_3 + S_i \qquad i=1,2 \tag{4}$$

And now, considering invariant (2) and the synchronization station (P_3, A_4), we have:

$$0 \leq \mathbf{P_3} \leq S_3 \tag{5}$$

$\mathbf{M_3}$ and $\mathbf{A_4}$ are not bounded and can theoretically reach infinite values.

Note that the assembly structure of the first part of the system (see Figure 2) has some similarities with the so-called fork-join queue problem [8][9]. This is known to be a hard problem due to the strong correlation between the parallel queues, caused by the fact that jobs for the queues arrive simultaneously. In some special and simple cases (only two paths), the fork-join queue problem can be solved exactly in order to calculate some performance (see [8][9]). However, for more complex systems [5][11][12][13], the authors make approximations in order to analyze the problem. For example, in [13], the average waiting time is obtained by assuming that the component manufacturing processes are not correlated. In our study, we consider systems that realize the assembly between two or more components resulting in three or more parallel paths. The difference with fork-join queues problems is due mainly to the fact that one of the parallel paths does not contain a server and consists of a single queue. In our work, for systems realizing the assembly of two components (three parallel paths), the problem is solved in section 4 of this paper by assuming a finite capacity for queue A_3. For systems realizing the assembly of more than two components (more than three parallel paths with infinite capacity queues), we propose in section 8 an approximate solution based on the ignorance of the correlation existing between the component manufacturing processes.

4. ANALYSIS

In our considered system, we are particularly interested in the following steady-state performance measures: the proportion of demands that are immediately satisfied, the proportion of backordered demands, the proportion of rejected demands, the average number of backordered demands, the average number of finished products for each stage, the average work in process for each stage and the average waiting time for a demand. These performance measures depend on the characterization of each manufacturing process, the characterization of the arrival process of demands, and the number of finished parts present at the initial state, in each stage.

In this work, we are making the following assumptions: demands arrive according to a Poisson process with rate λ, the processing times for the manufacturing process i are i.i.d. exponentially distributed variables, with rate μ_i (i = 1,2,3) and the average arrival rate of demands is less than the production capacity of the system.

In order to calculate the steady-state performance measures, we use a decomposition method, which is described in [7] for systems composed of serial stages. The idea is to decompose the whole system into several subsystems. Our considered system, that we denote by Γ2, can be decomposed into 2 subsystems (see Figure 3.1 and Figure 3.2). The model of Figure 3.1 (subsystem 1) is exactly the subsystem composed of stages 1 and 2 of Γ2. For the second subsystem (see Figure 3.2), we can notice that this model includes the whole stage 3 of the system. Furthermore, we have added the modeling of the arrival process of entities at station M3 by introducing a routing at the entry of stage 3, as well as a station DD3 that contains the deferred demands.

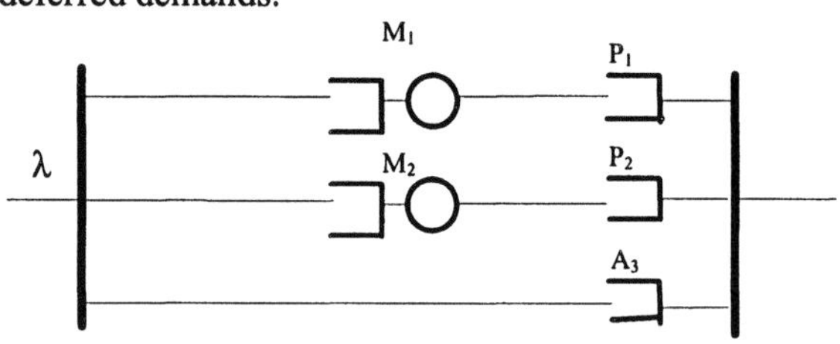

Figure 3.1: Subsystem 1

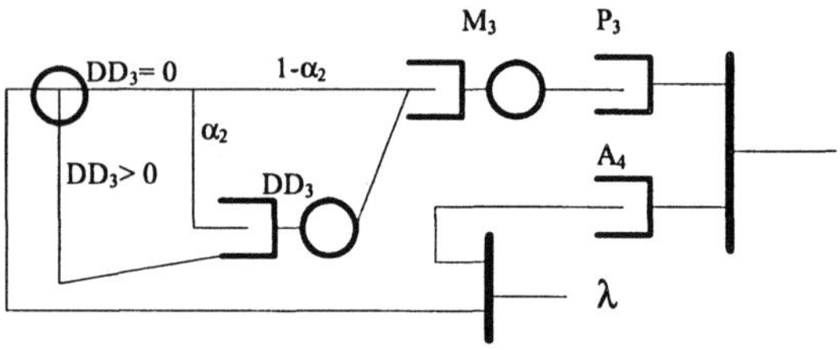

Figure 3.2: Subsystem 2

- the case $\mathbf{DD_3} > 0$ reflects the situation where an arriving request for components finds other requests waiting (hence a backlog) and hence joins the queue of waiting component requests.

- the case < $\mathbf{DD_3} = 0, \alpha_2$ > reflects the situation that an arriving request for components finds no other requests waiting but still finds that at least one component stock is empty, with probability α_2.
- the case < $\mathbf{DD_3} = 0, 1 - \alpha_2$ > reflects the case where a request for the two components can be fulfilled immediately, i.e. assembly may start.

In reality, subsystems 1 and 2 are correlated because they have the same arrival process of demands. Taking this correlation into account, the probability α_2 depends on the number k of jobs at station M_3; in fact we should write $\alpha_2(k)$. This probability is the steady-state probability that an arriving demand in station DD_3 has to wait, given that it finds no other waiting demands in front of it and $\mathbf{M_3} = k$. The difficulty in this description is of course in the dependence of $\alpha_2(k)$ on k (see [1] for more details). The method we propose in this paper is based on the ignorance of this dependence. So in the following, variables $\mathbf{M_3}$ and $\mathbf{A_3}$ will be considered as independent variables.

It is important to notice that station DD_3 is used to model the waiting process of demands at the input of station M_3. In fact, the content of station DD_3 is exactly equal to the content of queue A_3 of Γ_2. So, thanks to the steady-state probabilities determined for subsystem 1, we can determine the probability $1-\alpha_2$ (and then α_2):

$$1-\alpha_2 = p(\mathbf{P_1} \neq 0 \text{ and } \mathbf{P_2} \neq 0 / \mathbf{DD_3} = 0) = p(\mathbf{P_1} \neq 0 \text{ and } \mathbf{P_2} \neq 0 / \mathbf{A_3} = 0) \quad (6)$$

4.1 Analysis of stations M_1 and M_2:

Now, we are interested in the study of the first subsystem (see Figure 3.1). The aim is to calculate, for each queue of this subsystem, the probabilities for being in a given state. Thanks to invariants (1) for i=1,2 we just need to consider the evolution of the contents of stations M_1 and M_2. Indeed, if we consider invariants (1) for i=1,2 and the synchronization station (P_1, P_2, A_3), then we have:

$$p(\mathbf{P_1} = i_1 \text{ and } \mathbf{P_2} = i_2 \text{ and } \mathbf{A_3} = k) = p(\mathbf{M_1} = k+S_1-i_1 \text{ and } \mathbf{M_2} = k+S_2-i_2) \quad (7)$$

$$\text{With} \quad \begin{cases} i_1 \leq k+S_1 \\ i_2 \leq k+S_2 \\ k.i_1.i_2 = 0 \\ i_1 \geq 0,\ i_2 \geq 0,\ k \geq 0 \end{cases}$$

As we will see in section 5.1, expression (7) allows to calculate the steady-state probabilities relative to queues P_1, P_2 and A_3. $(\mathbf{M_1}, \mathbf{M_2})$ is a continuous-time Markov chain with finite state-space $\{(n_1,n_2) \mid 0 \leq n_1 \leq N_1$ and $0 \leq n_2 \leq N_2\}$ where $N_1 = S_1 + a_3$ and $N_2 = S_2 + a_3$. A state transition can only occur if a demand arrives or if one entity in M_1 or M_2 is released. This Markov chain is represented in Figure 4. Specifically,

- with transition rate λ, the system evolves from state (n_1, n_2) towards state (n_1+1, n_2+1), with $n_1 < N_1$ and $n_2 < N_2$.
- with transition rate μ_1, the system evolves from state (n_1, n_2) towards state (n_1-1, n_2), with $n_1 > 0$.
- with transition rate μ_2, the system evolves from state (n_1, n_2) towards state (n_1, n_2-1), with $n_2 > 0$.

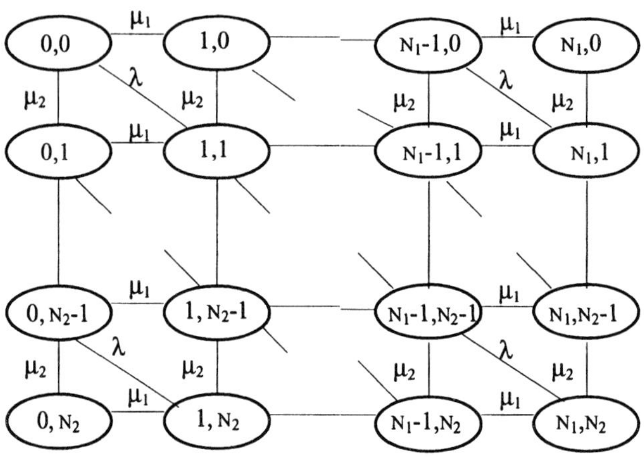

Figure 4: Markov chain for $(\mathbf{M_1}, \mathbf{M_2})$

We can decompose this Markov chain into sets of states corresponding to the same value of $\mathbf{M_1}$. There are state transitions between neighboring sets corresponding either to the processing of a part on M_1 or to the arrival of a part in M_1. Thus, $(\mathbf{M_1}, \mathbf{M_2})$ can be regarded as a quasi birth and death process [14]. Then we build the infinitesimal generator Q_1 by ordering the state space of $(\mathbf{M_1}, \mathbf{M_2})$ relatively to $\mathbf{M_1}$ values as follows: $\{[(0,0),(0,1),\ldots,(0,N_2)], [(1,0),(1,1),\ldots,(1, N_2)], \ldots, [(N_1,0),(N_1,1)\ldots (N_1,N_2)]\}$ and we obtain:

$$Q_l = \begin{pmatrix} B_{01} & A_{01} & 0 & 0 & \dots & 0 & 0 \\ A_{21} & A_{11} & A_{01} & 0 & \dots & 0 & 0 \\ 0 & A_{21} & A_{11} & A_{01} & \dots & 0 & 0 \\ \vdots & \vdots & \vdots & \vdots & \dots & \vdots & \vdots \\ 0 & 0 & 0 & 0 & \dots & A_{01} & 0 \\ 0 & 0 & 0 & 0 & \dots & A_{11} & A_{01} \\ 0 & 0 & 0 & 0 & \dots & A_{21} & B_{11} \end{pmatrix}$$

We denote by B_{01}, B_{11}, A_{01}, A_{11} and A_{21} the following square matrices of order N_2+1 :

$$B_{01} = \begin{pmatrix} -\lambda & 0 & \cdots & 0 & 0 \\ \mu_2 & -(\lambda+\mu_2) & \cdots & 0 & 0 \\ 0 & \mu_2 & \cdots & 0 & 0 \\ \vdots & \vdots & \vdots & \vdots & \vdots \\ 0 & 0 & \cdots & -(\lambda+\mu_2) & 0 \\ 0 & 0 & \cdots & \mu_2 & -\mu_2 \end{pmatrix}$$

$$B_{11} = \begin{pmatrix} -\mu_1 & 0 & \cdots & 0 & 0 \\ \mu_2 & -(\mu_1+\mu_2) & \cdots & 0 & 0 \\ 0 & \mu_2 & \cdots & 0 & 0 \\ \vdots & \vdots & \vdots & \vdots & \vdots \\ 0 & 0 & \cdots & -(\mu_1+\mu_2) & 0 \\ 0 & 0 & \cdots & \mu_2 & -(\mu_1+\mu_2) \end{pmatrix}$$

$$A_{01} = \begin{pmatrix} 0 & \lambda & 0 & \cdots & 0 \\ 0 & 0 & \lambda & \cdots & 0 \\ \vdots & \vdots & \vdots & \cdots & \vdots \\ 0 & 0 & 0 & \cdots & \lambda \\ 0 & 0 & 0 & \cdots & 0 \end{pmatrix}$$

$$A_{11} = B_{01} - \mu_1 I$$

$$A_{21} = \mu_1 I$$

Where I is the identity square matrix of order N_2+1.

We can notice that Q_l has the same repetitive form (A_{01} A_{11} A_{21}) as in [16]. Let $\mathbf{p}_{lk}$ be the stationary probability of states [(k,0),(k,1),…,(k, N_2)]

where $k = 0, \ldots, N_1$. Then the stationary distribution of $(\mathbf{M_1}, \mathbf{M_2})$ is represented by the following vector:

$$\mathbf{p_1} = (\mathbf{p_{10}}, \mathbf{p_{11}}, \ldots, \mathbf{p_{1N_1}})$$

Therefore, we can express the steady-state balance equations in the matrix form:

$$\begin{cases} \mathbf{p_{10}}B_{01} + \mathbf{p_{11}}A_{21} = \mathbf{0} & (8) \\ \mathbf{p_{1k-1}}A_{01} + \mathbf{p_{1k}}A_{11} + \mathbf{p_{1k+1}}A_{21} = \mathbf{0}, \quad 1 \leq k < N_1 & (9) \\ \mathbf{p_{1N_1-1}}A_{01} + \mathbf{p_{1N_1}}B_{11} = \mathbf{0} & (10) \end{cases}$$

Thus, we can write:

$$\mathbf{p_{1N_1}} = -\,\mathbf{p_{1N_1-1}}A_{01}B_{11}^{-1} \equiv \mathbf{p_{1N_1-1}}R_{N_1} \quad (11)$$

$$\mathbf{p_{1k}} = \mathbf{p_{1k-1}}R_k, \quad 1 \leq k < N_1 \quad (12)$$

Where R_k can be found recursively by

$$R_{N_1} = -\,A_{01}B_{11}^{-1} \quad (13)$$

$$R_k = -\,A_{01}[\,A_{11} + R_{k+1}A_{21}\,]^{-1}, \quad k=N_1-1, \ldots, 1 \quad (14)$$

Now $\mathbf{p_{10}}$ satisfies

$$\mathbf{p_{10}}(B_{01} + R_1A_{21}) = \mathbf{0} \quad (15)$$

and satisfies also the normalization equation

$$\sum_{k=0}^{N_1} \mathbf{p_{1k}}\mathbf{e} = \mathbf{p_{10}} \sum_{k=0}^{N_1} \prod_{i=0}^{k} R_i = 1 \quad (16)$$

where $R_0 = I$ and $\mathbf{e}$ is the column vector of ones.

Therefore we can determine vector $\mathbf{p_{10}}$ and then, vectors $\mathbf{p_{11}}, \mathbf{p_{12}}, \ldots, \mathbf{p_{1N_1}}$, by using (12).

Thus we obtain the steady-state probabilities for the system being in one of the states represented in Figure 4, i.e. $p(\mathbf{M_1} = n_1$ and $\mathbf{M_2} = n_2, \forall\, n_1, n_2)$.

4.2 Analysis of station M_3

Now we are interested in the study of subsystem 2, and particularly, we consider the evolution of the content of station M_3. Also, the aim is to calculate, for each queue of subsystem 2, the probabilities for being in a given state. It is easy to see that states of queues P_3 and A_4 could be determined if states of M_3 and A_3 are given. Indeed, if we consider invariant (2), and if we notice that (P_3, A_4) is a synchronization station, we have:

$$p(\mathbf{P_3} = i_3 \text{ and } \mathbf{A_4} = k_4) = \sum_{k_3=0}^{\min(k_4+S_3-i_3, a_3)} p(\mathbf{M_3} = k_4 + S_3 - k_3 - i_3 \text{ and } \mathbf{A_3} = k_3) \quad (17)$$

With $\begin{cases} i_3 \leq S_3 \\ k_4.i_3 = 0 \\ i_3 \geq 0,\ k_4 \geq 0 \end{cases}$

As we will see in section 5.2, expression (17) allows to calculate the steady-state probabilities relative to queues P_3 and A_4.

We must notice that for the analysis of station M_3, we do not consider only subsystem 2, but we have also to take into account stages 1 and 2 of the system.

First, let us define several probabilities which are used for the construction of the Markov chain corresponding to the state of M_3:

$$\beta_{11} = p(\mathbf{DD_3} > 1 \text{ and } \mathbf{P_2}=1 \,/\, (\mathbf{DD_3} \neq 0 \text{ and } \mathbf{P_1}=0 \text{ and } \mathbf{P_2}\neq 0)) \quad (18)$$
$$\beta_{12} = p(\mathbf{DD_3} > 1 \text{ and } \mathbf{P_2}>1 \,/\, (\mathbf{DD_3} \neq 0 \text{ and } \mathbf{P_1}=0 \text{ and } \mathbf{P_2}\neq 0)) \quad (19)$$
$$\beta_{13} = p(\mathbf{DD_3} = 1 \,/\, (\mathbf{DD_3} \neq 0 \text{ and } \mathbf{P_1}=0 \text{ and } \mathbf{P_2}\neq 0)) \quad (20)$$
$$\beta_{21} = p(\mathbf{DD_3} > 1 \text{ and } \mathbf{P_1}=1 \,/\, (\mathbf{DD_3} \neq 0 \text{ and } \mathbf{P_1}\neq 0 \text{ and } \mathbf{P_2}=0)) \quad (21)$$
$$\beta_{22} = p(\mathbf{DD_3} > 1 \text{ and } \mathbf{P_1}>1 \,/\, (\mathbf{DD_3} \neq 0 \text{ and } \mathbf{P_1}\neq 0 \text{ and } \mathbf{P_2}=0)) \quad (22)$$
$$\beta_{23} = p(\mathbf{DD_3} = 1 \,/\, (\mathbf{DD_3} \neq 0 \text{ and } \mathbf{P_1}\neq 0 \text{ and } \mathbf{P_2}=0)) \quad (23)$$
$$\gamma_1 = p(\mathbf{P_1}=0 \text{ and } \mathbf{P_2}=0 / \mathbf{DD_3} = 0) \quad (24)$$
$$\gamma_2 = p(\mathbf{P_1}=0 \text{ and } \mathbf{P_2}>0 / \mathbf{DD_3} = 0) \quad (25)$$
$$\gamma_3 = p(\mathbf{P_1}>0 \text{ and } \mathbf{P_2}=0 / \mathbf{DD_3} = 0) \quad (26)$$

and we have:

$$\gamma_1 + \gamma_2 + \gamma_3 = \alpha_2 \quad (27)$$
$$\beta_{11} + \beta_{12} + \beta_{13} = 1 \quad (28)$$
$$\beta_{21} + \beta_{22} + \beta_{23} = 1 \quad (29)$$

All these probabilities can be calculated by using the results obtained in the analysis of subsystem 1. Indeed, we have,

$$\beta_{11} = p(\mathbf{DD_3} > 1 \text{ and } \mathbf{P_1}=0 \text{ and } \mathbf{P_2}=1) / D_1$$
$$= \left\{ \sum_{k=2}^{a3} p(\mathbf{M_1}=S_1+k \text{ and } \mathbf{M_2}=S_2+k-1 \right\} / D_1$$

$$\beta_{12} = p(\mathbf{DD_3} > 1 \text{ and } \mathbf{P_1}=0 \text{ and } \mathbf{P_2}>1) / D_1$$

$$= \left\{ \sum_{k=2}^{a_3} \sum_{i_2=2}^{S_2+k} p(\mathbf{M_1}=S_1+k \text{ and } \mathbf{M_2}=S_2+k-i_2) \right\} / D_1$$

$$\beta_{13} = 1-(\beta_{11}+\beta_{12})$$

where

$$D_1 = p(\mathbf{DD_3} \neq 0 \text{ and } \mathbf{P_1}=0 \text{ and } \mathbf{P_2}\neq 0)$$

$$= \sum_{k=1}^{a_3} \sum_{i_2=1}^{S_2+k} p(\mathbf{M_1}=S_1+k \text{ and } \mathbf{M_2}=S_2+k-i_2)$$

and

$$\beta_{21} = p(\mathbf{DD_3} > 1 \text{ and } \mathbf{P_1}=1 \text{ and } \mathbf{P_2}=0) / D_2$$

$$= \left\{ \sum_{k=2}^{a_3} p(\mathbf{M_1}=S_1+k-1 \text{ and } \mathbf{M_2}=S_2+k) \right\} / D_2$$

$$\beta_{22} = p(\mathbf{DD_3} > 1 \text{ and } \mathbf{P_1}>1 \text{ and } \mathbf{P_2}=0) / D_2$$

$$= \left\{ \sum_{k=2}^{a_3} \sum_{i_1=2}^{S_1+k} p(\mathbf{M_1}=S_1+k-i_1 \text{ and } \mathbf{M_2}=S_2+k) \right\} / D_2$$

$$\beta_{23} = 1-(\beta_{21}+\beta_{22})$$

where

$$D_2 = p(\mathbf{DD_3} \neq 0 \text{ and } \mathbf{P_1}\neq 0 \text{ and } \mathbf{P_2}=0)$$

$$= \sum_{k=1}^{a_3} \sum_{i_1=1}^{S_1+k} p(\mathbf{M_1}=S_1+k-i_1 \text{ and } \mathbf{M_2}=S_2+k)$$

We have

$$\gamma_1 = p(\mathbf{P_1}=0 \text{ and } \mathbf{P_2}=0 \text{ and } \mathbf{DD_3}=0) / p(\mathbf{DD_3}=0)$$

$$= p(\mathbf{M_1}=S_1 \text{ and } \mathbf{M_2}=S_2) / p(\mathbf{DD_3}=0)$$

$$\gamma_2 = p(\mathbf{P_1}=0 \text{ and } \mathbf{P_2}>0 \text{ and } \mathbf{DD_3}=0) / p(\mathbf{DD_3}=0)$$

$$= \left\{ \sum_{i_2=1}^{S_2} p(\mathbf{M_1}=S_1 \text{ and } \mathbf{M_2}=S_2-i_2) \right\} / p(\mathbf{DD_3}=0)$$

$$\gamma_3 = p(\mathbf{P}_1>0 \text{ and } \mathbf{P}_2=0 \text{ and } \mathbf{DD}_3=0) / p(\mathbf{DD}_3=0)$$

$$= \left\{ \sum_{i_1=1}^{S_1} p(\mathbf{M}_1=S_1-i_1 \text{ and } \mathbf{M}_2=S_2) \right\} / p(\mathbf{DD}_3=0)$$

where $$p(\mathbf{DD}_3=0) = \sum_{i_1=0}^{S_1} \sum_{i_2=0}^{S_2} p(\mathbf{M}_1=S_1-i_1 \text{ and } \mathbf{M}_2=S_2-i_2)$$

Now, let us consider the evolution of the content of station M_3: the departure rate of entities from station M_3 is constant and equal to μ_3, but the arrival rate in station M_3 depends on the states of DD_3, P_1 and P_2 (see Figures 2 and 3.2). That is why we represent the state of M_3 (see Figure 5) by the state vector $(\mathbf{M}_3,X)$, where X describes the states of queues DD_3, P_1 and P_2. We have:

X = x00	if $\mathbf{DD}_3\neq0$ and $\mathbf{P}_1=0$ and $\mathbf{P}_2=0$
X = x0x	if $\mathbf{DD}_3\neq0$ and $\mathbf{P}_1=0$ and $\mathbf{P}_2\neq0$
X = xx0	if $\mathbf{DD}_3\neq0$ and $\mathbf{P}_1\neq0$ and $\mathbf{P}_2=0$
X = 0	if $\mathbf{DD}_3=0$ $(\forall\ \mathbf{P}_1, \mathbf{P}_2)$

In Figure 5, we decompose the Markov chain corresponding to station M_3 into sets of states corresponding to the same values of $\mathbf{M}_3$ (***Set$_0$***, ***Set$_1$***, ...). There are state transitions between neighboring sets corresponding either to the processing of a part on M_3 or to the arrival of a part in M_3. Thus we have a quasi birth and death process with an infinite number of states, because $\mathbf{M}_3$ is not bounded.

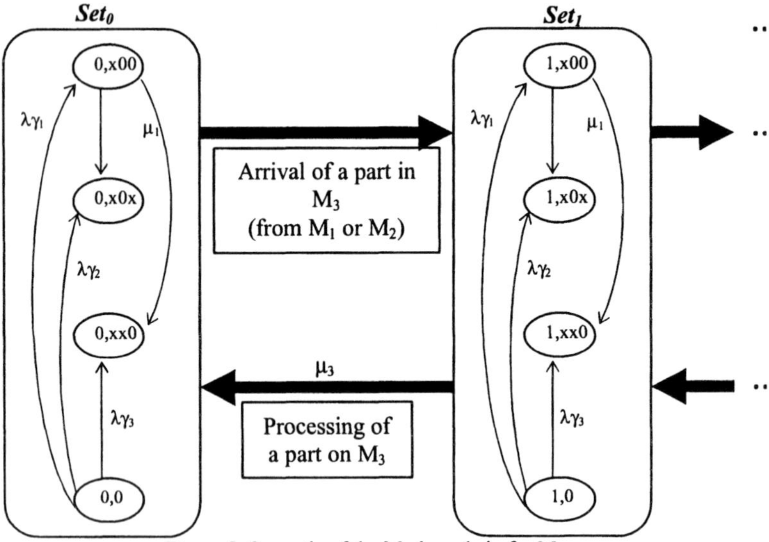

Figure 5: Synoptic of the Markov chain for M_3

Let us explain how transitions are done inside a set of states $\boldsymbol{Set_k}$ corresponding to $\mathbf{M_3}$=k, k≥0:

- From state (k,x00), corresponding to the case where DD_3 is not empty and P_1 and P_2 are both empty, there are 2 possible transitions inside $\boldsymbol{Set_k}$: if M_1 finishes production (rate μ_1), the content of queue P_1 will be incremented and the system will be in state (k,xx0) and if M_2 finishes production (rate μ_2), the content of queue P_2 will be incremented and the system will be in state (k,x0x).

- From state (k,0), corresponding to the case where DD_3 is empty, there are 3 possible transitions inside the set of states $\boldsymbol{Set_k}$ when a demand occurs (rate λ):
    - if P_1 and P_2 were both empty (probability γ_1) the system will be in state (k,x00).
    - if P_1 was empty and P_2 not empty (probability γ_2) the system will be in state (k,x0x).
    - if P_1 was not empty and P_2 empty (probability γ_3) the system will be in state (k,xx0).

- From state (k,x0x) the only possible transition inside $\boldsymbol{Set_k}$ is towards this state itself, due to an arrival of a new request. Also for state (k,xx0) the only possible transition inside $\boldsymbol{Set_k}$ is towards this state itself, due to an arrival of a new request. For clarity, these transitions are not represented in Figures 5 and 6, since they cancel when setting up the steady-state equations.

The complete Markov chain for M_3 is represented in Figure 6.

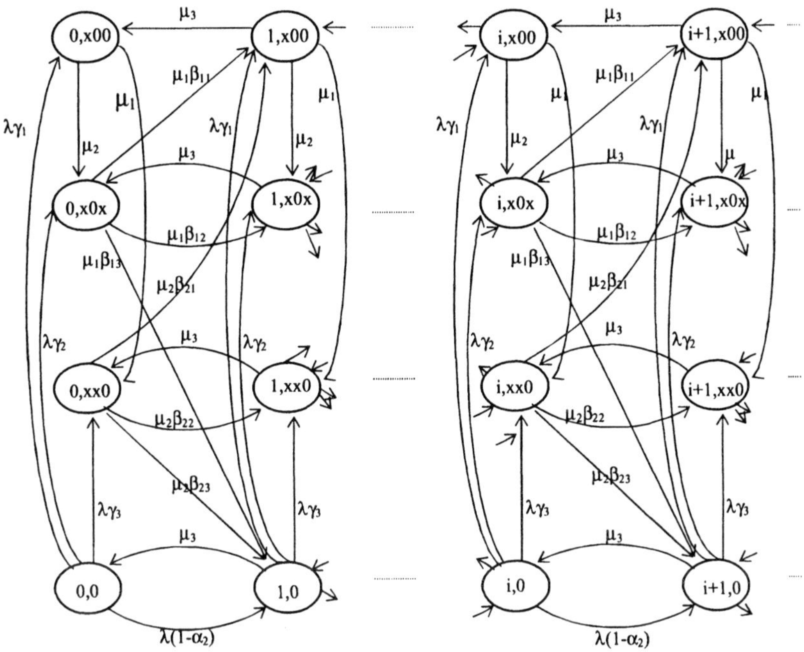

Figure 6: Markov chain for M_3

The only possible transitions from a set of states $\boldsymbol{Set_k}$ towards the set of states $\boldsymbol{Set_{k-1}}$ (k≥1) are the transitions of rate μ_3 occurring from each state (k,X) towards state (k-1,X).

From a set of states $\boldsymbol{Set_k}$ towards the set of states $\boldsymbol{Set_{k+1}}$ (k≥0), we have to distinguish different types of transitions:

- From state (k,x0x) corresponding to the case where DD_3 and P_2 are not empty and P_1 is empty:
  - If DD_3 contains more than one demand and P_2 contains exactly one entity (probability β_{11}), then, when M_1 finishes processing (with rate μ_1), the system will evolve towards state (k+1,x00).
  - If DD_3 contains more than one demand and P_2 contains more than one item (probability β_{12}), then, when M_1 finishes

processing (with rate μ_1), the system will evolve towards state (k+1,x0x).

- If DD_3 contains exactly one demand (probability β_{13}), then, when M_1 finishes processing (with rate μ_1), the system will evolve towards state (k+1,0).

- From state (k,xx0) corresponding to the case where DD_3 and P_1 are both not empty and P_2 is empty:
  - If DD_3 contains more than one demand and P_1 contains exactly one entity (probability β_{21}), then, when M_2 finishes processing (with rate μ_2), the system will evolve towards state (k+1,x00).
  - If DD_3 contains more than one demand and P_1 contains more than one item (probability β_{22}), then, when M_2 finishes processing (with rate μ_2), the system will evolve towards state (k+1,xx0).
  - If DD_3 contains exactly one demand (probability β_{23}), then, when M_2 finishes processing (with rate μ_2), the system will evolve towards state (k+1,0).

- From state (k,0) corresponding to station DD_3 empty, when a demand arrives (rate λ), if P_1 and P_2 are both not empty (probability $1-\alpha_2$), then the system will be in state (k+1,0).

- From state (k,x00) corresponding to station DD_3 not empty, and P_1 and P_2 both empty, there is no possible transition towards a state of $\boldsymbol{Set_{k+1}}$.

Therefore, we can deduce the corresponding infinitesimal generator Q_2, by ordering the states of the system as follows: ((0,x00), (0,x0x), (0,xx0), (0,0), (1,x00), (1,x0x), (1,xx0), (1,0), ...).

$$Q_2 = \begin{pmatrix} B_{02} & A_{02} & 0 & 0 & \cdots \\ A_{22} & A_{12} & A_{02} & 0 & \cdots \\ 0 & A_{22} & A_{12} & A_{02} & \cdots \\ 0 & 0 & A_{22} & A_{12} & \cdots \\ 0 & 0 & 0 & A_{22} & \cdots \\ 0 & 0 & 0 & 0 & \cdots \\ \vdots & \vdots & \vdots & \vdots & \vdots \end{pmatrix}$$

We denote by B_{02}, A_{02}, A_{12}, and A_{22} the following square matrices of order 4 :

$$B_{02}=\begin{pmatrix} -(\mu_1+\mu_2) & \mu_2 & \mu_1 & 0 \\ 0 & -\mu_1 & 0 & 0 \\ 0 & 0 & -\mu_2 & 0 \\ \lambda\gamma_1 & \lambda\gamma_2 & \lambda\gamma_3 & -\lambda \end{pmatrix}$$

$$A_{02}=\begin{pmatrix} 0 & 0 & 0 & 0 \\ \mu_1\beta_{11} & \mu_1\beta_{12} & 0 & \mu_1\beta_{13} \\ \mu_2\beta_{21} & 0 & \mu_2\beta_{22} & \mu_2\beta_{23} \\ 0 & 0 & 0 & \lambda(1-\alpha_2) \end{pmatrix}$$

$$A_{12}=\begin{pmatrix} -(\mu_1+\mu_2+\mu_3) & \mu_2 & \mu_1 & 0 \\ 0 & -(\mu_1+\mu_3) & 0 & 0 \\ 0 & 0 & -(\mu_2+\mu_3) & 0 \\ \lambda\gamma_1 & \lambda\gamma_2 & \lambda\gamma_3 & -(\lambda+\mu_3) \end{pmatrix}$$

and $A_{22} = \mu_3 I$

where I is the identity matrix of order 4.

Q_2 has the same repetitive form as in [6]; however we must notice that the block matrices constituting Q_2 are bigger (order 4, instead of order 2 in the case of serial stages). Then we can solve the steady-state equations by using also the matrix-geometric solution [14].

We denote by:
$p_2(k,x00)$ the probability for being in state (k,x00),
$p_2(k,x0x)$ the probability for being in state (k,x0x),
$p_2(k,xx0)$ the probability for being in state (k,xx0),
$p_2(k,0)$ the probability for being in state (k,0),
$\mathbf{p}_{2k} = (p_2(k,x00), p_2(k,x0x), p_2(k,xx0), p_2(k,0))$

Therefore, we can express the steady-state balance equations in the matrix form:

$$\begin{cases} \mathbf{p}_{20} B_{02} + \mathbf{p}_{21} A_{22} = \mathbf{0} & (30) \\ \mathbf{p}_{2(k-1)} A_{02} + \mathbf{p}_{2k} A_{12} + \mathbf{p}_{2(k+1)} A_{22} = \mathbf{0}, \quad k\geq 1 & (31) \end{cases}$$

We can define a constant matrix R of order 4, as follows:

$$\mathbf{p}_{2k} = \mathbf{p}_{20}R^k \tag{32}$$

R is a solution of equation $A_{02}+RA_{12}+R^2A_{22} = \mathbf{0}$. Then matrix R can be computed using an iterative procedure until convergence.

Thus, using equation (30) and the normalization equation, we can determine $\mathbf{p}_{20}$. Therefore, using expression (32), we can calculate the vectors of probabilities $\mathbf{p}_{2k}$.

Then we obtain the steady-state probabilities for the system being in one of the states represented in Figure 6:

$$p(\mathbf{M}_3 = k \text{ and } X = (x00 \text{ or } x0x \text{ or } xx0 \text{ or } 0)), \text{ with } k = 0,1,2, \ldots$$

5. PERFORMANCE EVALUATION

Now, $p(\mathbf{M}_1 = n_1, \mathbf{M}_2 = n_2)$ and $p(\mathbf{M}_3 = n_3)$ are perfectly defined for all values of n_1, n_2 and n_3. Therefore, we can determine all the steady-state probabilities relative to queues M_1, M_2, M_3, P_1, P_2, P_3, A_3 and A_4.

5.1 Steady-state probabilities relative to subsystem 1

♦ *Steady-state probabilities for M_1:*

$$\underset{0 \le x \le N_1}{p(\mathbf{M}_1 = x)} = \sum_{j=0}^{N_2} p(\mathbf{M}_1 = x, \mathbf{M}_2 = j) \tag{33}$$

♦ *Steady-state probabilities for M_2:*

$$\underset{0 \le x \le N_2}{p(\mathbf{M}_2 = x)} = \sum_{i=0}^{N_1} p(\mathbf{M}_1 = i, \mathbf{M}_2 = x) \tag{34}$$

♦ *Steady-state probabilities for A_3:*

Using expression (7) one can write:

$$\underset{0 < x \le a_3}{p(\mathbf{A}_3 = x)} = \sum_{i_1=0}^{x+S_1} \sum_{i_2=0}^{x+S_2} p(\mathbf{M}_1 = x+S_1-i_1, \mathbf{M}_2 = x+S_2-i_2)$$
$$- \sum_{i_1=1}^{x+S_1} \sum_{i_2=1}^{x+S_2} p(\mathbf{M}_1 = x+S_1-i_1, \mathbf{M}_2 = x+S_2-i_2) \tag{35}$$

and then, $$p(\mathbf{A_3}=0) = 1 - \sum_{x=1}^{a_3} p(\mathbf{A_3}=x) \quad (36)$$

♦ *Steady-state probabilities for P_1:*

Also from expression (7) we obtain:

$$\underset{0<x\leq S_1}{p(\mathbf{P_1}=x)} = \sum_{k=0}^{a_3}\sum_{i_2=0}^{S_2+k} p(\mathbf{M_1}=k+S_1-x, \mathbf{M_2}=k+S_2-i_2)$$
$$- \sum_{k=1}^{a_3}\sum_{i_2=1}^{S_2+k} p(\mathbf{M_1}=k+S_1-x, \mathbf{M_2}=k+S_2-i_2) \quad (37)$$

and

$$\underset{S_1<x\leq S_1+a_3}{p(\mathbf{P_1}=x)} = \sum_{k=x-S_1}^{a_3} p(\mathbf{M_1}=k+S_1-x, \mathbf{M_2}=k+S_2) \quad (38)$$

Then we can deduce the expression of $p(\mathbf{P_1}=0)$:

$$p(\mathbf{P_1}=0) = 1 - \sum_{x=1}^{S_1} p(\mathbf{P_1}=x) - \sum_{x=S_1+1}^{S_1+a_3} p(\mathbf{P_1}=x) \quad (39)$$

The expressions of *steady-state probabilities for P_2* are similar to those of $p(\mathbf{P_1}=x)$, by replacing indexes 1 and 2 by indexes 2 and 1, respectively.

5.2 Steady-state probabilities relative to subsystem 2

♦ *Steady-state probabilities for M_3:*

$$p(\mathbf{M_3}=k) = p(\mathbf{M_3}=k \text{ and } X=x00) + p(\mathbf{M_3}=k \text{ and } X=xx0)$$
$$+ p(\mathbf{M_3}=k \text{ and } X=x0x) + p(\mathbf{M_3}=k \text{ and } X=0) \quad (40)$$

♦ *Steady-state probabilities for queues P_3 and A_4:*

Since we have already determined the steady-state probabilities corresponding to station M_3 and queue A_3, we can calculate the steady-state probabilities corresponding to queues P_3 and A_4, by using expression (17). Therefore, we have,

$$p(\mathbf{P_3} = x) \underset{0 < x \leq S_3}{} = \sum_{k_3=0}^{\min(S_3-x,a_3)} p(\mathbf{M_3} = S_3 - k_3 - x \text{ and } \mathbf{A_3} = k_3)$$

$$p(\mathbf{A_4} = x) \underset{x>0}{} = \sum_{k_3=0}^{\min(S_3+x,a_3)} p(\mathbf{M_3} = S_3 - k_3 + x \text{ and } \mathbf{A_3} = k_3)$$

and since variables $\mathbf{M_3}$ and $\mathbf{A_3}$ are independent, we can write:

$$p(\mathbf{P_3} = x) \underset{0 < x \leq S_3}{} = \sum_{k_3=0}^{\min(S_3-x,a_3)} p(\mathbf{M_3} = S_3 - k_3 - x)\, p(\mathbf{A_3} = k_3) \quad (41)$$

and $$p(\mathbf{A_4} = x) \underset{x>0}{} = \sum_{k_3=0}^{\min(S_3+x,a_3)} p(\mathbf{M_3} = S_3 - k_3 + x)\, p(\mathbf{A_3} = k_3) \quad (42)$$

And then, we can easily deduce,

$$p(\mathbf{P_3} = 0) = 1 - \sum_{x=1}^{S_3} p(\mathbf{P_3} = x) \quad (43)$$

$$p(\mathbf{A_4} = 0) = 1 - \sum_{x=1}^{+\infty} p(\mathbf{A_4} = x) \quad (44)$$

5.3 Some performance measures

From the steady-state probabilities determined above, we can evaluate the following performance measures.

♦*Proportion of Immediately Satisfied Demands (ID), Backordered Demands (BD) and Rejected Demands (RD):*

- A demand will be immediately satisfied if there is at least one finished part in P_3 and if queue A_3 is not full (i.e. $\mathbf{A_3} < a_3$), then,

$$p(\mathbf{P_3} = x \text{ and } \mathbf{A_3} < a_3) \underset{0<x \leq S_3}{} = p(\mathbf{M_3} = S_3 - \mathbf{A_3} - x \text{ and } \mathbf{A_3} < a_3)$$

$$= \sum_{k_3=0}^{\min(S_3-x,a_3-1)} p(\mathbf{M_3}=S_3-k_3-x \text{ and } \mathbf{A_3}=k_3)$$

Since $\mathbf{M_3}$ and $\mathbf{A_3}$ are independent, then

$$p(\mathbf{P_3} = x \text{ and } \mathbf{A_3} < a_3) \underset{0<x\le S_3}{} = \sum_{k_3=0}^{\min(S_3-x,a_3-1)} p(\mathbf{M_3}=S_3-k_3-x)\, p(\mathbf{A_3}=k_3)$$

Thus,

$$ID = \sum_{x=1}^{S_3} \sum_{k_3=0}^{\min(S_3-x,a_3-1)} p(\mathbf{M_3}=S_3-k_3-x)\, p(\mathbf{A_3}=k_3) \tag{45}$$

- A demand will be backordered if P_3 is empty and if A_3 is not full (i.e. $\mathbf{A_3} < a_3$). We have:

$$p(\mathbf{A_3}<a_3) = p(\mathbf{A_3}<a_3 \text{ and } \mathbf{P_3} = 0) + p(\mathbf{A_3}<a_3 \text{ and } \mathbf{P_3} \neq 0)$$

and then

$$BD = p(\mathbf{A_3}<a_3) - ID = \sum_{x=0}^{a_3-1} p(\mathbf{A_3} = x) - ID \tag{46}$$

- A demand will be rejected if A_3 is full, then

$$RD = p(\mathbf{A_3} = a_3) \tag{47}$$

Note that $ID + BD + RD = 1$.

♦*Average number of Backordered demands (AB):*

AB is the average number of entities in queue A_4. So we have :

$$AB = E(\mathbf{A_4}) = \sum_{x=1}^{+\infty} x\, p(\mathbf{A_4} = x) \tag{48}$$

♦*Average number of Finished products for each stage (AF_i):*

The average number of finished products in P_i (i=1,2,3), is

$$AF_i = E(\mathbf{P_i}) = \sum_{x=1}^{Ni} x\, p(\mathbf{P_i} = x) \quad \text{for } i=1,2 \tag{49}$$

$$AF_3 = E(\mathbf{P_3}) = \sum_{x=1}^{S3} x\, p(\mathbf{P_3} = x) \tag{50}$$

♦ *Average Work in process for each stage (AW_i):*

AW_i is the average number of parts in station M_i (i=1,2,3). So we have:

$$AW_i = \sum_{x=1}^{Ni} x\, p(\mathbf{M_i} = x) \qquad \text{for } i=1,2 \qquad (51)$$

and

$$AW_3 = \sum_{x=1}^{+\infty} x\, p(\mathbf{M_3} = x) \qquad (52)$$

♦ *Average Waiting Time for a demand (WT):*

When a new demand arrives in the system, the following events can occur:

→ If queue P_3 is not empty, i.e. finished parts are available, then this demand will be immediately satisfied and there is no waiting time for this demand.

→ If queue P_3 is empty, i.e. there is no available finished part in the system, then this demand will wait during a certain time. If queue A_4 contains k_4 entities, then the arriving demand will be added to A_4 and will be the (k_4+1)st waiting demand in A_4. This demand will be processed after the k_4 demands that are already waiting. Two cases are possible:

- If station M_3 contains more items than queue A_4 ($\mathbf{M_3} > \mathbf{A_4}$) when the new demand arrives, then the average waiting time for each demand in A_4 is the average processing time of station M_3: $1/\mu_3$.
- Otherwise, if $\mathbf{M_3} \leq \mathbf{A_4}$ when the new demand arrives, and if we denote by **n** the number of parts in station M_3, then the average waiting time for each of the first **n** demands in A_4 is $1/\mu_3$. The average waiting time for each of the other $(k_4 + 1 - \mathbf{n})$ demands in A_4 depends also on the first subsystem and can be approximated by: $1/\mu_3 + \max(1/\mu_1, 1/\mu_2)$.

In the case where $a_3 \geq S_3$, using expression (17) and the independence of variables $\mathbf{M_3}$ and $\mathbf{A_3}$, an approximation for the average waiting time for a demand is given by :

$$WT = \sum_{k_4=0}^{+\infty} \left\{ \left[\sum_{k_3=0}^{S_3-1} p(\mathbf{M_3}=k_4+S_3-k_3)\, p(\mathbf{A_3}=k_3) \right] \left[(k_4+1)(1/\mu_3) \right] + \right.$$

$$\left[\sum_{k_3=S_3}^{\min(a_3,k_4+S_3)} p(\mathbf{M}_3=k_4+S_3-k_3)\, p(\mathbf{A}_3=k_3)\{(k_4+1)(1/\mu_3)+(k_3-S_3+1)(\max(1/\mu_1, 1/\mu_2))\}\right]\Bigg\} \quad (53)$$

And in the case where $a_3 < S_3$, we have:

$$WT = \sum_{k_4=0}^{+\infty}\left\{\left[\sum_{k_3=0}^{a_3} p(\mathbf{M}_3=k_4+S_3-k_3)\, p(\mathbf{A}_3=k_3)\right]\left[(k_4+1)(1/\mu_3)\right]\right\} \quad (54)$$

6. GENERALIZATION TO MORE THAN 3 STAGES

In this part, we generalize the proposed analysis method to systems containing more than one stage after the assembly operation. For example, if there are two serial stages after the assembly operation, then the queueing network of the obtained system Γ_3 is represented in Figure 7.

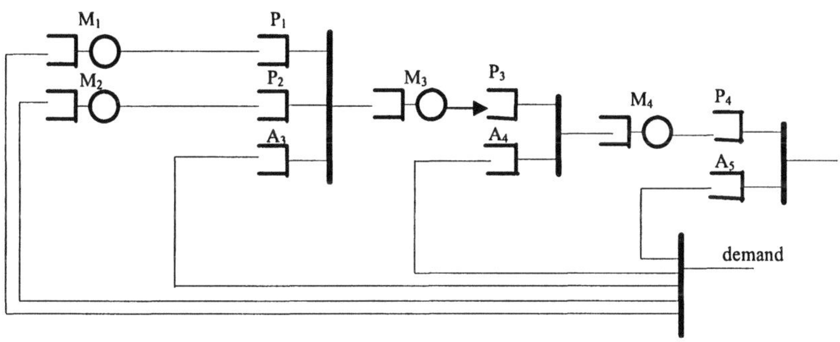

Figure 7: The queueing network model corresponding to Γ_3

System Γ_3 is decomposed into 3 subsystems (see Figures 8.1, 8.2 and 8.3). We notice that subsystems 1 and 2 (Figures 8.1 and 8.2) are similar to those of Figures 3.1 and 3.2, and thus the analysis is identical. Now let us consider subsystem 3 (Figure 8.3). The principle of modeling is the same as in the case of subsystem 2: the model of Figure 8.3 includes the whole stage 4 of the system and the modeling of the arrival process of entities in station M_4. Similarly, station DD_4 contains the deferred demands.

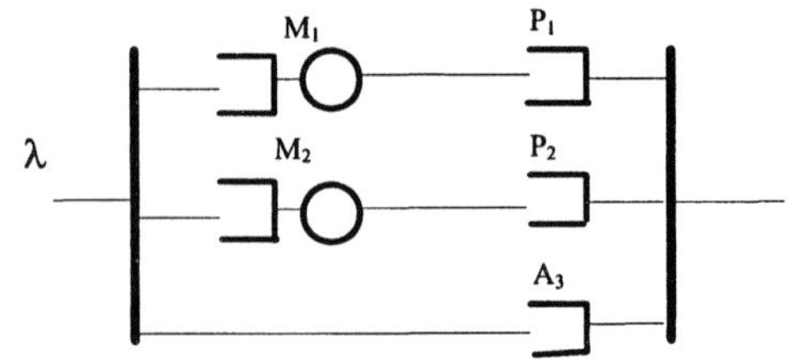

Figure 8.1: Subsystem 1

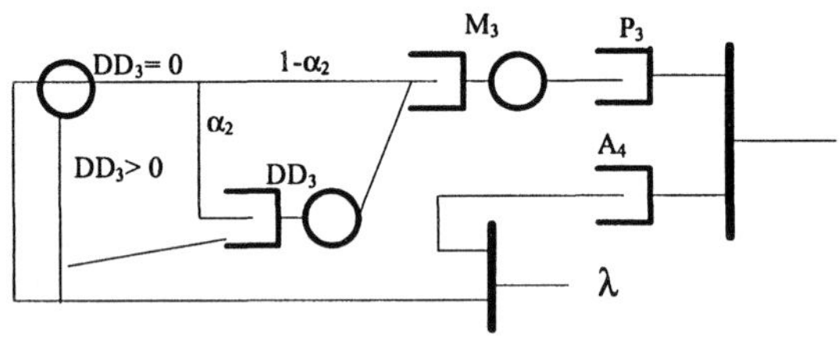

Figure 8.2: Subsystem 2

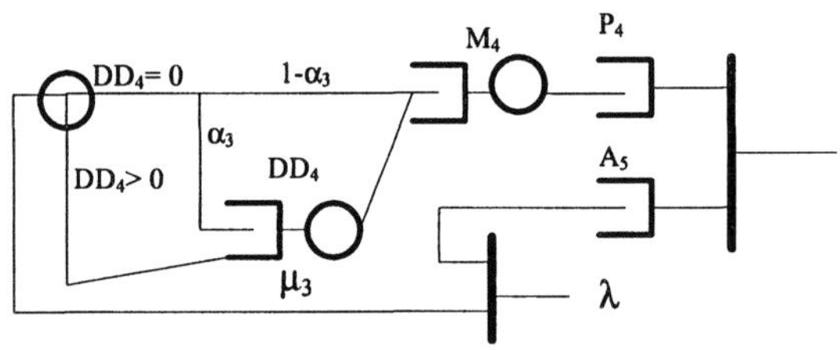

Figure 8.3: Subsystem 3

When a demand arrives in the system, two cases are possible:

- If station DD_4 was empty, then we consider the probability that P_3 is empty:

$$\alpha_3 = p(P_3 = 0 / DD_4 = 0)$$

- With probability α_3, the entity will enter station DD_4 in order to wait until at least one finished part from stage 3 becomes available. This waiting time is the processing time of station DD_4.
  With probability $1-\alpha_3$ (queue P_3 is not empty), the entity will immediately enter M_4.

- If station DD_4 was not empty, then there are already waiting demands in station DD_4, so the arriving demand will be added to the input buffer of station DD_4, and will be processed after the demands that were already waiting (FIFO).

Now, if we consider a system Γ_N, constituted by 2 stages being assembled, and N-1 serial stages after the assembly operation, then Γ_N can be decomposed into N subsystems. Obviously, subsystems 1 and 2 of Γ_N are similar to those of Γ_2 and Γ_3, and thus their analysis is identical. Subsystem i-1 (i=4,..., N+1) corresponds to the modeling of stage i of Γ_N and is represented in Figure 9. The principle of modeling is the same as for subsystem 3; station DD_i contains the deferred demands, and $\alpha_{i-1} = p(P_{i-1} = 0 / DD_i = 0)$.

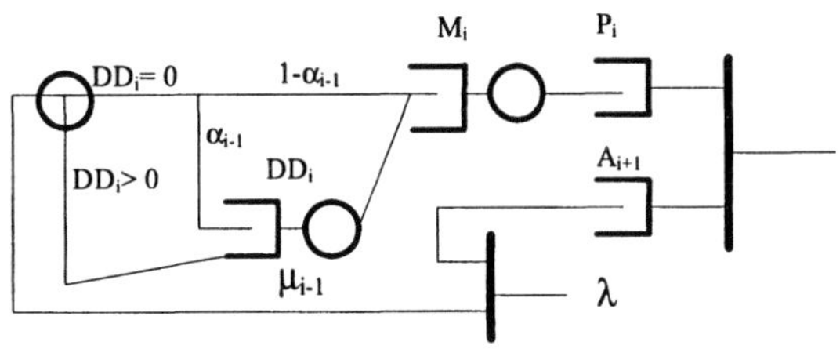

Figure 9: Modeling of stage i (subsystem i-1)

If we consider subsystem 3, an entity in station DD_4 is waiting for a finished part coming from station M_3. But in this case, station M_3 could be empty because P_1 or P_2 (or both) is (are) also empty when the demand arrived; so one must wait for an assembled product coming from subsystem 1. Consequently, as we have to take into account all precedent subsystems, the study of station DD_4 is more complex than the study of DD_3.

Therefore, as in [7], we do the following approximation: if there are entities in DD_4, then station M_3 is not empty. Thus, station DD_4 is modeled as an exponential machine with rate μ_3. We do the same approximation for DD_5,..., DD_{N+1}, that are modeled as exponential machines with rates μ_4, ..., μ_N, respectively.

However, we must notice that because we do this approximation, now station DD_i (i=4, ..., N+1) does not exactly represent queue A_i.

The analysis of subsystem i-1 necessitates the analysis of DD_i and M_i (i=4,..., N+1). The resolution is identical to that proposed in the case of serial stages [6] and performance measures, like those calculated for Γ_2, can be determined.

7. NUMERICAL RESULTS

In this part, we present some numerical results, in order to test the accuracy of the decomposition method we presented, by comparing its results to simulation results. For examples 1 and 2, we consider a base-stock controlled assembly system Γ_2, with one stage after the assembly operation (see Figure 2), and for example 3, we have a base-stock controlled assembly system Γ_3, with two stages after the assembly operation (see Figure 7).

7.1 Example 1

In this example, the average arrival rate of demands, λ, is 0.8. The 3 machines have the same average processing rate $\mu_1=\mu_2=\mu_3=1$. We consider that all the stocks of finished products have the same initial value S. The capacity a_3 of queue A_3 is equal to 50. Then, for different values of parameter S, we can see, on Figures 10 and 11, the evolution of the following performance measures: the proportion of demands that are immediately satisfied, ID, and the average number of finished products in the system, AF_3.

We can note that the approximation results are close to the simulation results, especially for large values of S. Indeed, for ID, the maximum value of the relative error is 9.0%, obtained for the smallest value of S: 4. But generally, for the other points this error is less than 2.5%. And for AF_3, the maximum value of the relative error is 9.1%, obtained for the smallest value of S: 4. But generally, for the other points this error is less than 4.8%.

As expected, these two performance measures increase when S, the initial value of finished products in the stocks, increases.

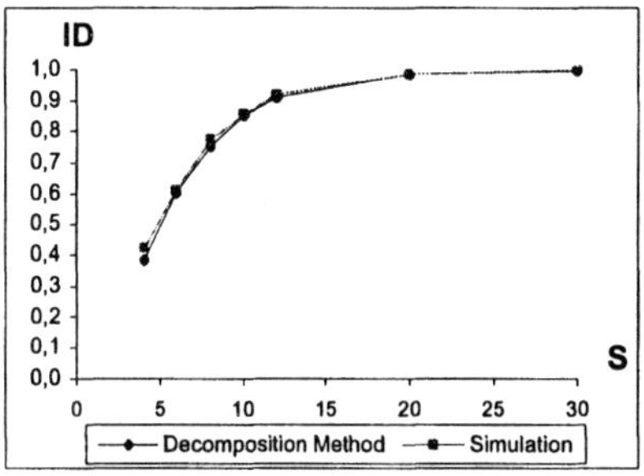

Figure 10: Proportion of demands that are immediately satisfied - Example 1.

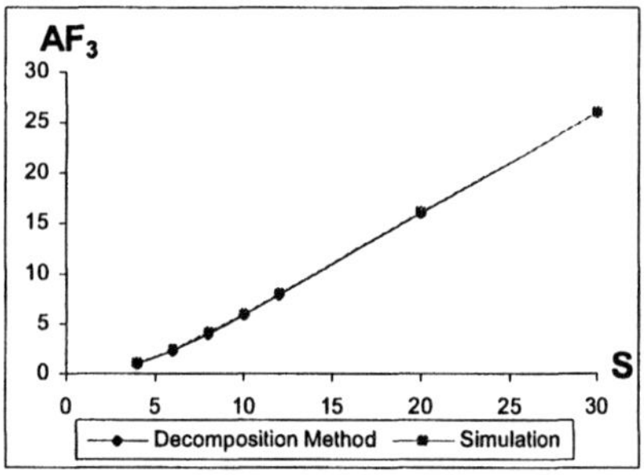

Figure 11: Average number of finished products in stage 3 – Example 1.

7.2 Example 2

In this example, the 3 machines have the same average processing rate $\mu_1=\mu_2=\mu_3=1$. We consider that all the stocks of finished products have the same initial value S=30. The capacity a_3 of queue A_3 is equal to 50. Then, for different values of parameter λ, we can see, on Figures 12 and 13, the evolution of the following performance measures: the proportion of demands that are immediately satisfied, ID, and the average number of finished products in the system, AF_3.

We can note that the approximation results are close to the simulation results, especially for values of λ far enough from saturation. Indeed, for ID, the relative error in this case has a maximum value of 1.0%, obtained for the largest value of λ: 0.9. But generally, for the other points, this error is less than 0.1%. And the relative error for AF_3 has a maximum value of 0.5%, obtained for the largest value of λ: 0.9. But generally, for the other points, this error is less than 0.1%.

As expected these two performance measures decrease when λ, the average arrival rate of demands, increases.

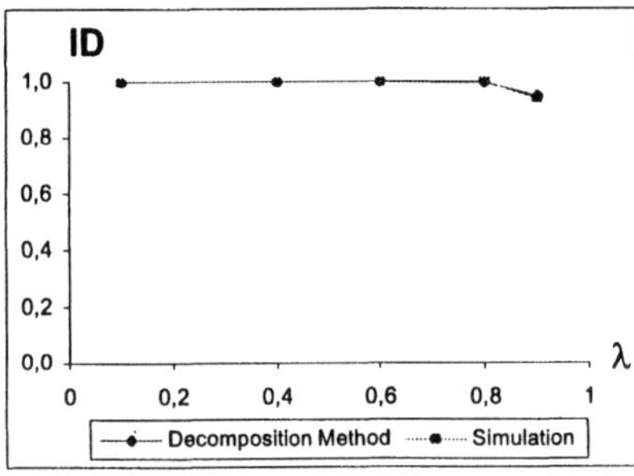

Figure 12: Proportion of demands that are immediately satisfied – Example 2.

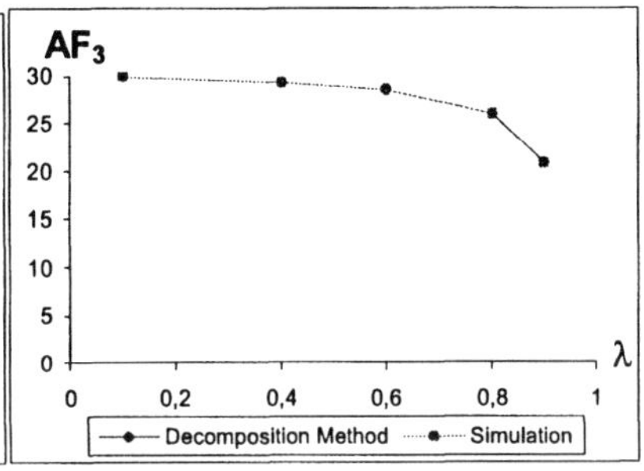

Figure 13: Average number of finished products in stage 3 – Example 2.

7.3 Example 3

In this example, the system contains two serial stages after the assembly operation. The average arrival rate of demands λ is 0.8. Stations M_1, M_2, M_3 and M_4 have the same service rate $\mu_1=\mu_2=\mu_3=\mu_4=1$. We consider that all the stocks of finished products have the same initial value S. The capacity a_3 of queue A_3 is equal to 50. Then, for different values of parameter S, we can see, in Figures 14 and 15, the evolution of AW_4 (the average work in process in the last stage) and WT (the average waiting time for a demand), relatively to parameter S.

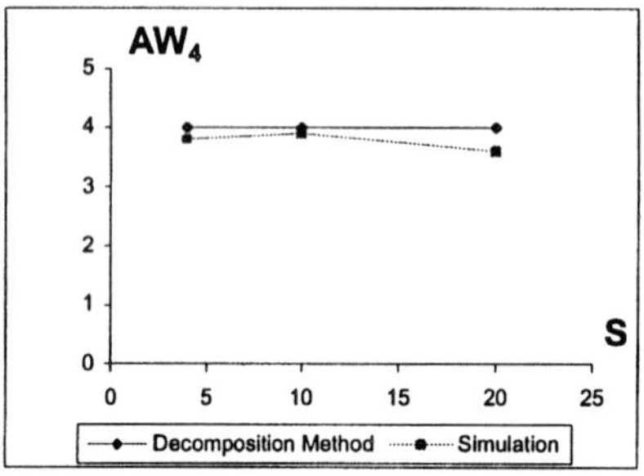

Figure 14: Average work in process in the last stage – Example 3.

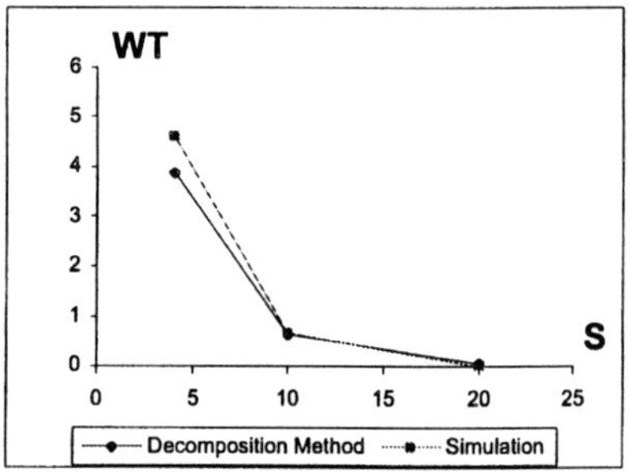

Figure 15: Average waiting time for a demand – Example 3.

Note that in Figure 14 the value of $\mathbf{AW_4}$ is constant according to S, and equal to 4. This value corresponds to the average number of customers in an M/M/1 station, with $\mu = 1$ and $\lambda = 0.8$. Hence, the arrival process of demands in station M_4 is approximately a Poisson process with rate λ.

We can note that the approximation results are close to the simulation results. Indeed, for AW_4, the relative error in this case is around 6.0%. For S=4, the relative error for WT is equal to 15.7%. But, for large values of S, the difference between simulation and decomposition method values for WT, is around zero.

Generally, in base-stock controlled systems, all the queues are supposed to have infinite capacities. For our considered system, queue A_3 has a finite capacity a_3. However, with the simulation method, for the examples we presented above, the content of queue A_3 never reaches its capacity a_3, which is equal to 50 (the proportion of rejected demands RD is equal to zero). Therefore, our analytical method can be used even for systems where queue A_3 is not bounded, by taking large values for a_3.

8. GENERALIZATION TO THE ASSEMBLY OF MORE THAN TWO PRODUCTS

In this part, we consider a system producing one type of finished parts, by assembling n semi-finished parts that are provided by n machines, and then by processing the assembled product on a (n+1)st machine. We can see the queueing network corresponding to this system in Figure 16. We assume that all the queues of the system have *infinite* capacities.

In order to calculate the steady-state performance measures, the system of Figure 16 is decomposed into 2 subsystems (see Figures 17.1 and 17.2).

The model of Figure 17.1 (subsystem 1) is exactly the subsystem composed of stages 1 to *n* of our considered system. The model of Figure 17.2 (subsystem 2) is analog to the model of Figure 3.2. Similarly, station DD_{n+1} is used to model the waiting process of demands at the input of station M_{n+1}, and the content of station DD_{n+1} is exactly equal to the content of queue A_{n+1}. So, thanks to the steady-state probabilities determined for subsystem 1, we can determine the probability $1\text{-}\alpha_2$ (and then α_2):

$$\begin{aligned} 1-\alpha_2 &= p(\mathbf{P_1} \neq 0, \mathbf{P_2} \neq 0, \ldots, \mathbf{P_n} \neq 0 \,/\, \mathbf{DD_{n+1}} = 0) \\ &= p(\mathbf{P_1} \neq 0, \mathbf{P_2} \neq 0, \ldots, \mathbf{P_n} \neq 0 \,/\, \mathbf{A_{n+1}} = 0) \end{aligned} \quad (55)$$

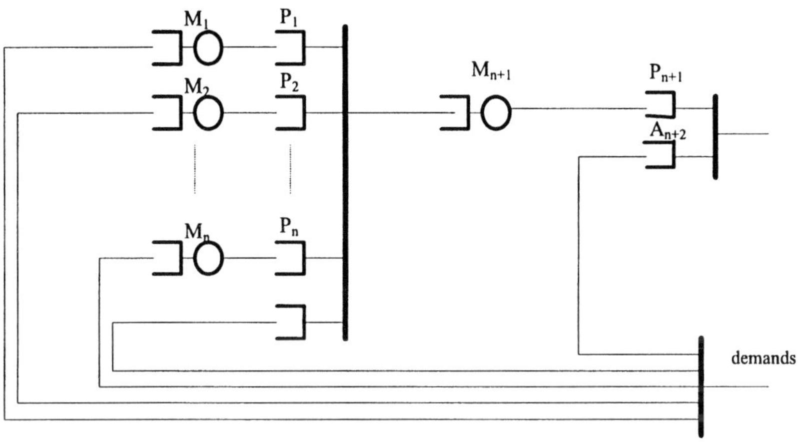

Figure 16: Queueing network corresponding to the assembly of *n* products.

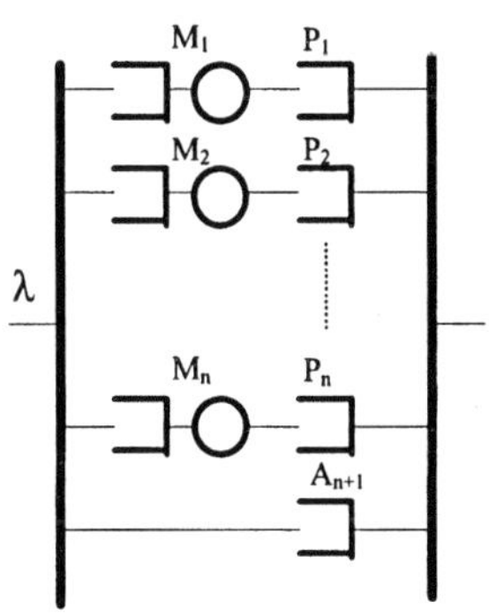

Figure 17.1: Subsystem 1

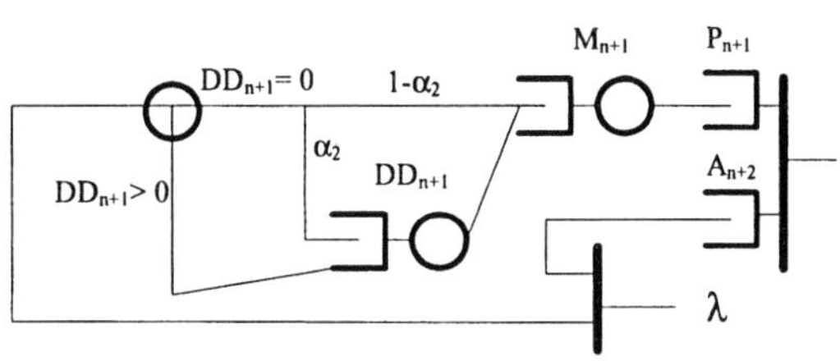

Figure 17.2 : Subsystem 2

8.1 Analysis of stations $\mathbf{M}_1$ to $\mathbf{M}_n$

We are interested here in the study of subsystem 1 (Figure 17.1). In Section 4.1, the analysis of the first subsystem was an *exact* analysis, which is not easy to apply in the case of the assembly of an important number of products. Thus, in the following, we do the approximation that variables $\mathbf{M}_1$ to $\mathbf{M}_n$ are *independent*. We can see, in Table 1, that this approximation is correct in the case of the assembly of 2 products.

Table 1. Comparison between the results of the decomposition method (MD), the results of this method when we consider that $\mathbf{M}_1$ and $\mathbf{M}_2$ are independent (IMD) and the results of the simulation method (Simu) – ($\mu_1=\mu_2=\mu_3=1$, $\lambda=0.8$).

S	ID-MD	ID-IMD	ID-Simu	AF_3-MD	AF_3-IMD	AF_3-Simu
8	0.755	0.744	0.775	4.0	3.9	4.2
12	0.911	0.909	0.919	7.9	7.9	8.0

Therefore, stations M_1 to M_n can be analyzed separately as M/M/1 stations, and the expressions of their steady-state probabilities are easily determined:

$$p(\mathbf{M}_m = x) = \rho_m{}^x (1- \rho_m) \qquad m\in \{1,2,...,n\} \tag{56}$$

where $\rho_m = \lambda / \mu_m$

8.2 Steady-state probabilities of the other queues of subsystem 1

Subsystem 1 (Figure 17.1) has n invariants defined as follows:

$$\begin{cases} \mathbf{M}_1 + \mathbf{P}_1 & = & \mathbf{A}_{n+1} + S_1 \\ \mathbf{M}_2 + \mathbf{P}_2 & = & \mathbf{A}_{n+1} + S_2 \\ & \vdots & \\ \mathbf{M}_n + \mathbf{P}_n & = & \mathbf{A}_{n+1} + S_n \end{cases}$$

Thanks to these invariants and if we consider the synchronization station (P_1,..., P_n, A_{n+1}), we have:

$$p(\mathbf{P}_1=i_1, ..., \mathbf{P}_n=i_n, \mathbf{A}_{n+1}=k) \quad = \quad p(\mathbf{M}_1=k+S_1-i_1, ..., \mathbf{M}_n=k+S_n-i_n)$$

$$\text{with} \quad \begin{cases} i_j \le k + S_j & \text{for } j=1,\dots,n \\ k.\prod_{j=1}^{n} i_j = 0 \end{cases}$$

And since $\mathbf{M_1}, \dots, \mathbf{M_n}$ are independent,

$$p(\mathbf{P_1}=i_1, \dots, \mathbf{P_n}=i_n, \mathbf{A_{n+1}}=k) = \prod_{j=1}^{n} p(\mathbf{M_j}=k+S_j-i_j) \quad (57)$$

$$\text{with} \quad \begin{cases} i_j \le k + S_j & \text{for } j=1,\dots,n \\ k.\prod_{j=1}^{n} i_j = 0 \end{cases}$$

♦ *Steady-state probabilities for P_m (m=1,2,...,n):*

$$\underset{0<x\le S_m}{p(\mathbf{P_m}=x)} = p(\mathbf{P_1}\ge 0, \mathbf{P_2}\ge 0,\dots, \mathbf{P_m}=x, \dots, \mathbf{P_n}\ge 0, \mathbf{A_{n+1}}\ge 0) - p(\mathbf{P_1}>0, \mathbf{P_2}>0,\dots, \mathbf{P_m}=x, \dots, \mathbf{P_n}>0, \mathbf{A_{n+1}}>0)$$

$$= \sum_{k=0}^{+\infty} p(\mathbf{M_m}=k+S_m-x) \prod_{j\in\{1,\dots,n\}-\{m\}} \left(\sum_{i_j=0}^{S_j+k} p(\mathbf{M_j}=k+S_j-i_j)\right)$$

$$- \sum_{k=1}^{+\infty} p(\mathbf{M_m}=k+S_m-x) \prod_{j\in\{1,\dots,n\}-\{m\}} \left(\sum_{i_j=1}^{S_j+k} p(\mathbf{M_j}=k+S_j-i_j)\right) \quad (58)$$

$$\underset{x>S_m}{p(\mathbf{P_m}=x)} = p(\mathbf{P_1}\ge 0, \mathbf{P_2}\ge 0,\dots, \mathbf{P_m}=x, \dots, \mathbf{P_n}\ge 0, \mathbf{A_{n+1}}\ge x-S_m) - p(\mathbf{P_1}>0, \mathbf{P_2}>0,\dots, \mathbf{P_m}=x, \dots, \mathbf{P_n}>0, \mathbf{A_{n+1}}\ge x-S_m)$$

$$= \sum_{k=x-S_m}^{+\infty} p(\mathbf{M_m}=k+S_m-x) \left\{ \prod_{j\in\{1,\dots,n\}-\{m\}} \left(\sum_{i_j=0}^{S_j+k} p(\mathbf{M_j}=k+S_j-i_j)\right) - \prod_{j\in\{1,\dots,n\}-\{m\}} \left(\sum_{i_j=1}^{S_j+k} p(\mathbf{M_j}=k+S_j-i_j)\right) \right\} \quad (59)$$

and then,

$$p(\mathbf{P_m}=0) = 1 - \sum_{x=1}^{+\infty} p(\mathbf{P_m}=x) \quad (60)$$

♦ *Steady-state probabilities for A_{n+1}:*

$$p(\mathbf{A_{n+1}}=x) = p(\mathbf{P_1}\geq 0, \mathbf{P_2}\geq 0, \ldots, \mathbf{P_n}\geq 0, \mathbf{A_{n+1}}=x) - p(\mathbf{P_1}>0, \mathbf{P_2}>0, \ldots, \mathbf{P_n}>0, \mathbf{A_{n+1}}=x)$$
$x>0$

$$= \prod_{j=1}^{n} \sum_{i_j=0}^{S_j+x} p(\mathbf{M_j}=x+S_j-i_j) - \prod_{j=1}^{n} \sum_{i_j=1}^{S_j+x} p(\mathbf{M_j}=x+S_j-i_j) \quad (61)$$

and then,

$$p(\mathbf{A_{n+1}}=0) = 1 - \sum_{x=1}^{+\infty} p(\mathbf{A_{n+1}}=x) \quad (62)$$

8.3 Analysis of station M_{n+1}

In this part, we are interested in the study of subsystem 2 (see Figure 17.2), and particularly, we consider the evolution of the content of station M_{n+1}. Also here, as previously in section 4.2, the aim is to calculate the steady-state probabilities relative to all the queues of subsystem 2.

As in the case of the assembly of 2 products, we define similarly several probabilities which are used for the construction of the infinitesimal generator Q_2:

β_{ij} for $i=1, \ldots, n$ and for $j=1, \ldots, 2^{n-1}+1$

γ_k for $k=1, \ldots, 2^n-1$

And we have:

$$\sum_{k=1}^{2^n-1} \gamma_k = \alpha_2 \quad (63)$$

$$\sum_{j=1}^{2^{n-1}+1} \beta_{ij} = 1 \qquad \text{for } i=1,\ldots,n \quad (64)$$

All these probabilities can be calculated by using the results obtained in the analysis of subsystem 1.

Now, let us consider the evolution of the content of station M_{n+1}: the departure rate from station M_{n+1} is constant and equal to μ_{n+1}, but the arrival rate in this station depends on the states of queues DD_{n+1}, P_1, ..., P_n. So, the state of M_{n+1} is represented by the state vector $(\mathbf{M_{n+1}},X)$, where X describes the states of queues DD_{n+1}, P_1, ..., P_n.

The state vectors are ordered as follows: ((0,x0...00),(0,x0...0x),...,(0,xx...x0),(0,0),(1,x0...00),(1,x0...0x),..., (1,xx...x0),(1,0),...)

The corresponding infinitesimal generator Q_2 has exactly the same form as in section 4.2:

$$Q_2 = \begin{pmatrix} B_{02} & A_{02} & 0 & 0 & \cdots \\ A_{22} & A_{12} & A_{02} & 0 & \cdots \\ 0 & A_{22} & A_{12} & A_{02} & \cdots \\ 0 & 0 & A_{22} & A_{12} & \cdots \\ 0 & 0 & 0 & A_{22} & \cdots \\ 0 & 0 & 0 & 0 & \cdots \\ \vdots & \vdots & \vdots & \vdots & \vdots \end{pmatrix}$$

where B_{02}, A_{02}, A_{12} and A_{22} are square matrices of order 2^n.

We have: $A_{12} = B_{02} - \mu_3 I$ and $A_{22} = \mu_3 I$, where I is the identity square matrix of order 2^n.

Therefore, in order to define completely Q_2, we only have to determine matrices B_{02} and A_{02}. These matrices have 2^n lines corresponding to states (0,x0...00),(0,x0...0x),...,(0,xx...x0),(0,0) respectively. The following algorithms allow the construction of these matrices, for all values of n.

Algorithm 1 - Construction of B_{02}

Step 0 Initialization $B_{02} = \mathbf{0}$

Step 1

1.1 $B_{02}(2^n,2^n) = -\lambda$

1.2 For j=1 to 2^{n-1} do:

$B_{02}(2^n,j) = \lambda\,\gamma_j$

-Return to **1.2**

Step 2 For i= 1 to 2^n - 1 do:

- Calculate set SI of indexes k corresponding to $\mathbf{P_k}$=0 for the state corresponding to line i of matrix B_{02}, where $k \in \{1,...,n\}$.
- If SI contains *exactly 1 element* k_0 then do:

$B_{02}(i,i) = B_{02}(i,i) - \mu_{k_0}$

Endif

• If SI contains *at least 2 elements* then

2.a For each element $k \in SI$ do:

$B_{02}(i,i) = B_{02}(i,i) - \mu_k$

$B_{02}(i,i+2^{n-k}) = \mu_k$

-Return to **2.a**

Endif

-Return to **Step 2**

Algorithm 2 - Construction of A_{02}

Step 0 Initialization:

$A_{02} = \mathbf{0}$

$A_{02}(2^n,2^n) = \lambda\ (1 - \alpha_2)$

Step 1 For k=1 to n do:

1.1 m=1

1.2 For r=1 to 2^{k-1} do:

1.2.1 For $j= (2^r-2)2^{n-k}+1$ to $(2^r-1)2^{n-k}$ do:

$\bullet A_{02}(2^n-2^{n-k},j) = \mu_k\beta_{km}$

$\bullet m = m+1$

-Return to **1.2.1**

-Return to **1.2**

1.3 $A_{02}(i,2^n) = \mu_k\beta_{km}$

-Return to **Step 1**

Therefore, the resolution of the steady-state equations here is similar to the resolution in the case of the assembly of 2 products (see section 4.2), using also the matrix-geometric solution [14]. Then the same performance measures can be calculated.

9. CONCLUSION

We have presented a model of base-stock controlled assembly systems that includes stochastic processing times. We have considered that the assembly operation was not necessarily the last operation, so it could be followed by any other operation(s). We proposed an approximate analysis method for which procedures were developed in order to evaluate several performance measures. Thus, we have compared the obtained numerical results with simulation results, and we can conclude that our proposed method provides a good evaluation of the considered performance measures. Finally, we have extended this method to the case of the assembly of more than two products.

In this work, for simplicity, we have made the assumption that the fabrication stages and the assembly stage each consist of a single machine.

This simplified assumption is necessary to begin with ; now, we are extending the developed method to a more general system where stages consist in a network of machines.

REFERENCES

[1] AVSAR Z.M., ZIJM W.H.M., Resource-constrained two-echelon inventory models for repairable item systems, *Memorandum no. 1521, University of Twente, Faculty of Mathematical Sciences*, July 2000.

[2] BUZACOTT J.A., SHANTHIKUMAR J.G., "Stochastic Models of Manufacturing Systems", *Prentice Hall*, 1993.

[3] CHAOUIYA C., LIBEROPOULOS G., DALLERY Y., "The Extended Kanban control systems for production coordination of assembly systems", *IIE Transactions on Design and Manufacturing, Special Issue on Decentralized Control of Manufacturing Systmes*, Vol. 32, No. 10, pp. 999-1012, 2000..

[4] DI MASCOLO M., DALLERY Y., "Performance Evaluation of Kanban Controlled Assembly Systems", *CESA'96 IMACS Multiconference on Computational Engineering in Systems Applications, Symposium Discrete Events and Manufacturing Systems*, Lille, July 96.

[5] DIETZ D.C., JENKINS R.C., "Analysis of Aircraft Sortie Generation with the Use of a Fork-Join Queueing Network Model", *Nav. Res. Logist. 44, N°.2, 153-164,* 1997.

[6] DURI C., "Etude Comparative de Gestions à Flux Tirés", *Thèse de Doctorat (PhD) de l'INPG, Laboratoire d'Automatique de Grenoble*, January 1996.

[7] DURI C., FREIN Y., DI MASCOLO M., "Analysis of Base Stock Systems Using a Decomposition Method", *CESA'96 IMACS Multiconference on Computational Engineering in Systems Applications, Symposium Discrete Events and Manufacturing Systems*, Lille, July 96.

[8] FLATTO L., HAHN S., "Two Parallel Queues Created by Arrivals with Two Demands, I, *SIAM J. Applied Math, vol. 44, pp. 1041-1053*, 1984.

[9] FLATTO L., HAHN S., "Two Parallel Queues Created by Arrivals with Two Demands, II, *SIAM J. Applied Math, vol. 45, pp. 861-878*, 1985.

[10] FREIN Y., DI MASCOLO M., DALLERY Y., "On the Design of Generalized Kanban Control Systems", *International Journal of Operations & Production Management*, Vol. 15, N°9, 1995, pp. 158.184.

[11] KIM C., AGRAWALA A.K., "Analysis of the Fork-Join Queue", *IEEE Trans. Computers, Vol. 38(2), pp. 250-255*, 1989.

[12] LIU Y.C., PERROS H.G., "A Decomposition Procedure for the Analysis of a Closed Fork/Join Queueing System", *IEEE Trans. Computers, Vol. 40(3), pp. 365-370*, 1991.

[13] NELSON R., TANTAWI A.N., "Approximate Analysis of Fork/Join Synchronization in Parallel Queues", *IEEE Trans. Computers, Vol. 37(6), pp. 739-743*, 1988.

[14] NEUTS M.F. , "Matrix Geometric Solutions in Stochastic Models", *The John Hopkins University Press*, Baltimore, London, (1981).

[15] SBITI N., DI MASCOLO M., AMGHAR M., "Modélisation et Propriétés du Pilotage Kanban Généralisé pour les Systèmes de Production avec Assemblage", *MOSIM'99, Deuxième Conférence Francophone de Modélisation et Simulation, MOSIM'99*, Annecy, October 1999.

[16] SONG J.S., XU S.H., LIU B., "Order-Fulfillment Performance Measures in an Assemble-to-Order System with stochastic Leadtimes", *Operations Research, 47 (1), 1999.*

Chapter 14

DESIGNING MANUFACTURING CELLS USING A TABU SEARCH APPROACH

K. Spiliopoulos and S. Sofianopoulou
Department of Industrial Management, University of Piraeus, Greece

Abstract: *We present a Tabu Search scheme for the solution of the NP-hard part of the manufacturing cells formation problem, on a simplified yet concise model which adequately fits to practice. The design can then be finalized using simple methods to incorporate additional parameters. This scheme integrates in a systematic way proper short and long-term memory structures and an overall search strategy for their use. At the code development stage, special care was taken to enhance the explorative capability of the algorithm by correlating hash information with the values of the search parameters. The resulting algorithm is robust and produces more than promising results for problems with up to 30 machines, with negligible computational effort. In the absence of known optimal solutions for larger dimensions and therefore of a testing benchmark, fine-tuning may be required. This could be achieved by incorporating dynamic parameters in the place of the static ones.*

Key words: *Tabu Search, Cellular Manufacturing.*

1. INTRODUCTION

The design of manufacturing cells is a central issue in the area of Group Technology. The goal is to decentralize production by first grouping the machines into clusters and the various part types into part families, and then allocating the processing of each part family to a single machine cluster. The benefits from such an arrangement include the reduction of transport, queuing and processing times, the elimination of the need for frequent set-ups and tool changes in the machines, the reduction of inventories and the simplification of the production plan. These benefits lead naturally to better delivery times, quality improvements, more efficient management and customer satisfaction. The application of cellular manufacturing is also an appropriate first step towards a JIT production scheme.

In practice however, the creation of completely autonomous manufacturing cells is very seldom feasible [Wemmerlov and Hyer, 1989] and the aim becomes to decrease as much as possible the between-cells traffic i.e. the traffic generated by some parts which visit machines in different clusters. Optimization methods tackle the NP-hard part of the design, that is the grouping of machines into clusters. The parts grouping is an almost trivial task, based on locating part similarities.

The manufacturing systems Cell (or Cluster) Formation Problem (CFP) addresses this problem of creating cells with minimum interactions. The formulation examined in this paper was first presented in [Sofianopoulou, 1994] and takes into account the most important parameters, namely the sequencing order of the parts in the routing through the various machines, the associated «distances» between the various machines, and the maximum allowed size for the cells. The importance of considering the sequencing order of the parts in the routing, a somewhat neglected factor, is discussed in [Harhalakis, Nagi and Proth, 1990]. The upper limit on the number of machines in each cell is appropriately imposed to face practical restrictions, arising for example from location considerations or available space limitations. This formulation combines simplicity and adequate fit to realistic situations for a preliminary design.

This is in line with the staged approach suggested by [Burbidge,1993] for clustering problems in Group Technology. The second stage of the design would then be to modify the solution found to fit even more to practical situations, by considering options to eliminate remaining intercellular traffic. These options include machine duplication, part subcontracting and changes in the parts routings, and can be incorporated in the design with very simple methods such as the ones suggested by [Kern and Wei, 1991] and [Seifoddini, 1989]. Note that, by changes in the parts routings, we do not refer to alternative routings being available a priori, e.g. through multiple process plans and/or replicate machines. Such cases fall in the area of Generalized Group Technology (GGT) and are best solved with special models, see for example [Sofianopoulou, 1996], [Sofianopoulou, 1999], and [Lozano et al., 1999a]. The proposed model is summarized in Figure 1.

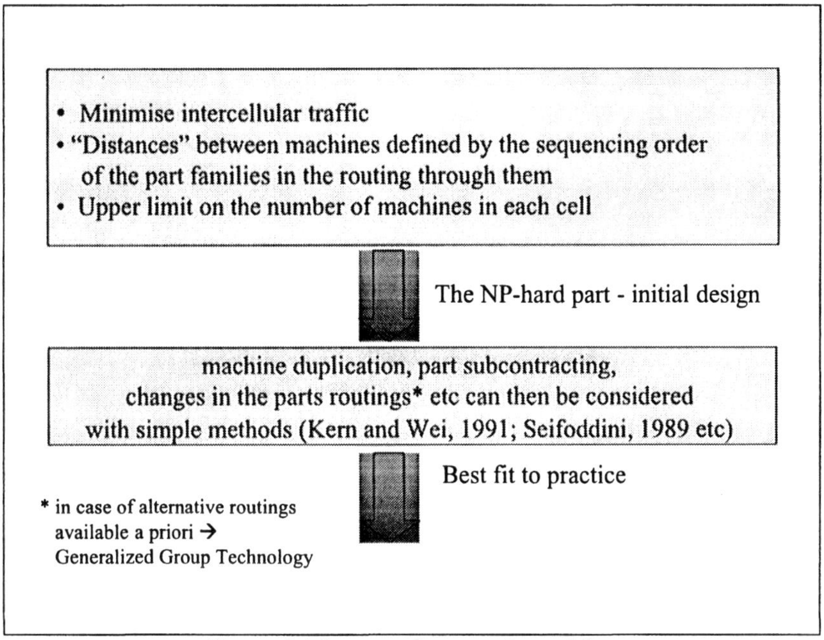

Figure 1. The staged approach model.

Most of the existing algorithms for the CFP are based on clustering methods and carry the weaknesses of the statistical clustering theory with the plethora of different criteria and the inherent inability to find guaranteed good solutions. Exact algorithms have also been proposed, mostly based on mathematical programming approaches. These latter however, do not follow such a staged approach and as a consequence they are very hard to solve due to the large number of variables involved and the difficulty of obtaining tight bounds in branch-and-bound techniques. In general, existing algorithms for the CFP either oversimplify the problem or cannot deal with instances of practical size using reasonable computing resources. A brief description and evaluation of methods proposed for the CFP is given in [Cheng, 1992], [Kandiller, 1994], [Sofianopoulou, 1997] and [Cheng, Goh and Lee, 1996], among others. A recently proposed branch and bound method in [Spiliopoulos and Sofianopoulou, 1998] was proved capable of dealing with medium-sized instances with over 20 machines (25 with some unpublished improvements), but more is needed to tackle realistic situations.

In this work, we turn our attention to the most successful Tabu Search (TS) paradigm. TS is a meta-heuristic and as such, open to strategic adaptation of the master ideas. We present a proper framework for the short and long-term memory structures and the overall search strategy, so as to reach the benefits of intelligent search, namely adaptive memory and responsive exploration. At the development stage, special care was taken to enhance the explorative capability of the algorithm by correlating hash function information with the values of the various parameters. This way, a robust code was designed which produced most promising results with negligible computational effort. Other TS applications in the area are either based on different approaches or have a different focus. For example, [Lozano et al., 1999b] propose a quadratic assignment model with side constraints in an attempt to group both machines and parts simultaneously. In the [Skorin-Kapov and Vakharia, 1993] application cited in [Glover and Laguna, 1997], the problem addressed is the one of scheduling of jobs within a single manufacturing cell under varying parameter conditions.

The problem statement is presented in section 2, including the Integer Programming formulation and the calculation of machines "distances" published in [Sofianopoulou, 1997]. Section 3 presents some necessary definitions and the basic elements of the TS framework proposed, namely the short and long-term memory structures and aspects. Section 4 integrates these basic elements into a general strategy and outlines the search course. Computational results and conclusions are given in sections 5 and 6, respectively.

2. PROBLEM STATEMENT

CFP can be described in graph-theoretic terms, if the machines are considered as vertices of a weighted graph where the edge weights represent the cost of parts traffic between the machines. The following problem statement describes in a compact form our objective, that is the formation of cells of bounded size, with minimum interactions:

> Problem CFP: "Given a graph G = (V,E), where V is the set of n nodes and E the set of edges, and weights W_i, $i=1,\dots,|E|$ on the edges, partition V into disjoint subsets of maximum size M, so as to minimize the total weight of edges connecting nodes in different subsets".

CFP is known to be NP-hard, see [Garey and Johnson, 1979]. In fact, it can be equivalently formulated (with fixed number of cells and cell sizes) as a Quadratic Assignment, a Quadratic Transportation or a Matrix Norm Minimization problem.

The following 0-1 IP formulation for the manufacturing systems Cell Formation Problem is given in [Sofianopoulou, 1997]:

$$\min z = \sum_{i=1}^{n-1} \sum_{j=i+1}^{n} c_{ij}\left(1 - x_{ij}\right) \tag{2.1}$$

$$\text{s.t.} \sum_{i=1}^{k-1} x_{ik} + \sum_{j=k+1}^{n} x_{kj} \leq M - 1 \qquad k = 1,\ldots,n \tag{2.2}$$

$$\begin{aligned} x_{ij} + x_{ik} - x_{jk} &\leq 1 \\ x_{ij} - x_{ik} + x_{jk} &\leq 1 \\ -x_{ij} + x_{ik} + x_{jk} &\leq 1 \\ i = 1,\ldots,n-2, \quad j = i+1,\ldots,n-1, &\quad k = j+1,\ldots,n \end{aligned} \tag{2.3}$$

$$x_{ij} = \begin{cases} 1, & \text{if machines } i, j \text{ are in the same cell,} \\ 0, & \text{otherwise} \end{cases} \qquad i = 1,\ldots,n-1 \quad j = i+1,\ldots,n \tag{2.4}$$

In the above formulation, n machines are allocated to cells of maximum size M. Two machines are said to be "connected" if they belong to the same cell. Constraint (2.2) ensures that machine k is connected with at most M-1 other machines, and constraints (2.3), the "triangle" constraints, ensure that within each cell all machines are connected to each other.

The cost elements c_{ij} are calculated from data on the traffic of P parts between the n machines. Given a Pxn part-machine incidence matrix $\mathbf{A}=(a_{pi})$ p=1,...,P, i=1,...,n which describes the sequence of machine operations on each part:

$$a_{pi} = \begin{cases} t, \text{ if part } p \text{ visits machine } i \text{ at step } t \ (1 \le t \le n), \\ 0, \text{ otherwise} \end{cases} \tag{2.5}$$

$$\text{we set } c_{ij} = \sum_{p=1}^{P} \delta_{ijp} \qquad i = 1,\dots,n-1, \qquad j = i+1,\dots,n \tag{2.6}$$

$$\text{where } \delta_{ijp} = \begin{cases} 1, \text{ if } a_{pi}a_{pj} \neq 0 \text{ and } \left|a_{pi} - a_{pj}\right| = 1 \\ 0, \text{ otherwise} \end{cases} \tag{2.7}$$

Thus, d_{ijp} in (2.7) shows whether part p visits machines i,j in two successive steps or not, and c_{ij} in (2.6) collects the contributions from all parts, to represent the traffic between machines i and j. The part-machine incidence matrix used above is not the one commonly seen in the literature where the important factor of sequencing order of the parts in part routings is largely neglected.

The modifications for the case of parts being processed more than once in some machine are easy. The part-machine incidence matrix becomes multi-valued, i.e. each element a_{pi} p=1,...,P, i=1,...,n is taken to mean a set which includes the steps at which part p visits machine i. Equation (2.7) is accordingly adjusted to consider all pairs of values, one from set a_{pi} and the other from set a_{pj}.

3. DEFINITIONS - THE BASIC ELEMENTS OF THE TABU SEARCH FRAMEWORK

3.1 The Shape of Solutions

The *shape* of a solution, that is of an allocation of machines into cells, is the number of cells and their sizes. For example, the shape with 4 cells and sizes {4,5,5,6} is one of the *feasible* shapes for problems with n=20 machines and M=6 maximum cell size. During the search, *infeasible* shapes are also encountered. These violate the implicit constraint that in an optimal solution, no two cells can be merged. Extensive experimentation has led to the conclusion that the most possible shape of the optimal solution is the one with the maximum number of cells of the maximum allowed size M, e.g.

{2,6,6,6) in the previous example. One can easily construct however counter-examples to show that optimality is not guaranteed.

3.2 Definition of Moves

Each solution visited during the search has an associated neighborhood. The basic operation to reach one of the neighboring solutions is called a *move.* The two types of moves used in the proposed algorithm are:

- *simple move* of a machine from one cell to another
- *swap* of two machines which are allocated to different cells.

The simple move changes the shape of the solution, a highly desirable feature because the search can potentially lead to any point in the feasible space. In fact, simple moves can temporarily guide the search out of the feasibility boundary, to an infeasible shape, therefore serving the overall diversification strategy of the method. This happens because an infeasible shape is quite inferior, is met for very "local" reasons and cannot be preserved for long. When such a shape is met, the search automatically returns to the feasible area in a few iterations. We note here that diversification is the general TS target of guiding the search towards unexplored areas of the solution space. "Oscillation" between feasible and infeasible areas greatly enhances the chances of achieving this goal, and in fact much research is devoted to exploiting almost only this idea, see for example [Glover and Laguna, 1997].

3.3 Short-Term Memory: Tabu Moves

As a result of a move (simple move or swap), some connections are removed and new ones are created. For example, after a simple move of a machine i from cell c_1 to cell c_2:

- for every machine j remaining in cell c_1, all connections (j,i) are removed
- for every machine k already in cell c_2, new connections (i,k) are created.

as shown in the figure below.

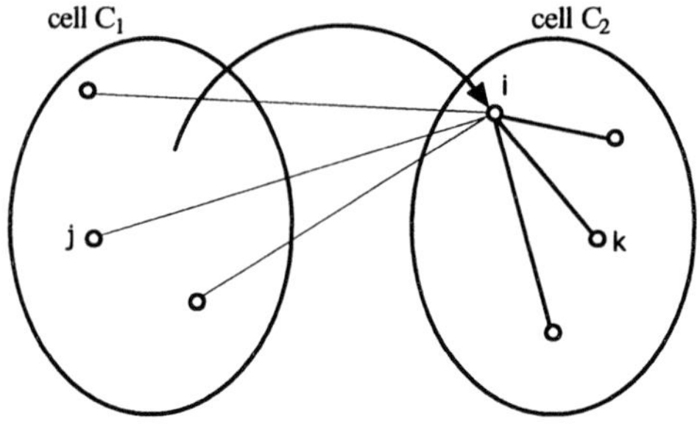

Figure 2. Simple move – removal and creation of connections.

The short-term memory employed is the history of recent moves (recency memory) and aims primarily at avoiding cycling by imposing restrictions on candidate moves. To this end, we use a symmetric tabu matrix $\mathbf{T}=(t_{ij})$ $i,j=1,\ldots,n$, initially with all elements zeros. In the case of a simple move of machine i from cell c_1 to cell c_2, a constant positive value (the tabu tenure) is given to all removed connections, that is to all elements t_{ji}, where j is any machine remaining in cell c_1 (see Figure 2). At the same time, the non-zero values in all the other elements of matrix **T** are reduced by one.

On the basis of the tabu values in matrix **T**, a simple move of a machine i to a cell c is prohibited (considered tabu) if:

- either all new connections involved have positive tabu values, that is $t_{ik} > 0$ for each machine k in cell c,
- or the mean value of the non-zero tabu values corresponding to new connections is greater than a predefined lower limit (which is set to half the tabu tenure value).

The first condition is the standard approach for the application of this short-term memory. The second condition is an improvement for the particular problem. The standard approach alone is quite restrictive and also leads to frequent returns to solutions already visited and eventual re-appearance of whole solutions sequences. An example would be to permit a simple move which involves new connections that have all tabu values close to the maximum (the tabu tenure) except from one which is zero. This is allowed in the standard approach, although it is an "almost reversal" and thus most probably would result to cycling.

Analogous rules are applied in the case of swaps. A swap is considered tabu if either all new connections involved have positive tabu values or the mean value of the non-zero tabu values is greater than the lower threshold.

3.4 Short-Term Memory: Aspiration Criteria

When certain conditions are satisfied, it is desirable to override the tabu status and allow tabu moves to be candidates, an obvious example being a move which improves the best value found so far in the search. Using the standard TS terminology, these conditions are called aspiration criteria. We emphasize that in the present algorithm, overriding the tabu status of a move does not necessarily mean that this move is finally selected, as it only becomes candidate again. The final selection for the next move depends on the general search strategy which is described in section 4.

The first aspiration criterion resembles the typical and customarily used global form of aspiration by objective where the tabu status of a move is overridden when the move improves the best value found so far, as in the example mentioned. For the class of problems to which CFP belongs this criterion is not sufficient, hence we modify it, by monitoring the best value after a global improvement. With the exception of early stages in the search, the sequence between two successive global improvements can be very lengthy. To avoid extreme deterioration within this sequence (from which a deep "drop" to a global improvement would be difficult), we also remove the tabu status of moves that improve on this intermediate value. Again using standard TS terminology, this rule can be considered a variation of aspiration by objective/ regional form.

The second criterion is the standard aspiration by search direction (one-way), where the tabu status is overridden if:

- the move improves the current solution, and
- all connections that characterize this move as tabu received their tabu status, similarly during a locally improving iteration.

The third criterion serves the aspiration by influence target. The notion of influence is related to the introduction of some perturbation in the search, to uncover solutions which are structurally different from those that would be visited normally. Influential moves are also carried out as part of the long-term memory application (see section 3.5) but we prefer to enhance this element by introducing it also in the aspiration process.

To this end, we define as strong machines those that have the largest total connection cost to all others and therefore bear more influence when moved, and in particular the top n/4 machines. The conditions for the application of this criterion is then:

- the tabu move is locally improving,
- the machine involved in a simple move is a strong one (in the case of swaps at least one machine is strong),
- there is a certain delay from the last global improvement, and
- the mean value of the non-zero tabu values that correspond to new connections is lower or equal to an upper threshold (which is in fact again half of the tabu tenure, as the threshold used to characterize a move as tabu).

3.5 Long -Term Memory

The long-term memory introduces the following strategic elements that complement each other:

- diversification of the search i.e. guidance towards unexplored areas of the solution space
- intensification of the search, that is, focusing to "good" and/or promising areas, through the introduction of attributes of already visited "elite" solutions and also through the selection of influential moves.

The first long-term memory used is a residence frequency one, based on the whole history of the search. In standard TS terminology, residence frequencies count the times a variable "resides" in some value, in contrast to the transition frequencies which keep the number of changes in the values of variables. This long-term memory serves both the above strategic elements, primarily diversification and secondarily intensification, the latter through the use of "strong" (therefore influential) connections, in analogy with the notion of strong machines.

To this end, we keep a symmetric matrix $\mathbf{S}=(s_{ij})$ i,j=1,...,n initially with all elements zeros. At the end of each iteration, for every pair of machines (i,j) that belong to the same cell, the corresponding element in **S** is increased by one. We scale the values of this matrix to keep them comparable to those in the cost matrix **C**.

Therefore, at every stage of the search, values in S that are significantly smaller than their counterparts in **C** indicate connections which, being the more influential ones, were expected to appear more frequently than they actually did, that is these connections have not been explored much.

Therefore, we are interested in the values of the **C-S** matrix. If we select a move on the basis of this matrix instead of the normal cost matrix **C**, we give more weight to large cost - low frequency connections and less weight to small cost - high frequency ones. As a result, moves that lead to such connections are given priority/ are discouraged, respectively. This practice is in accordance with the use of long-term memory to provide incentives and penalties during the search and in fact matrix **-S** contains exactly the incentives/ penalties factors needed. Accordingly, this use of residence frequency-based long-term memory serves both the intensification and diversification targets, the first one through consideration of influential connections.

The second long-term memory is the one based on selected good solutions registered, the "elite" ones, and therefore completes the intensification target aspects.

To this end, we keep a symmetric matrix **B**, initially with all elements zeros. To take into account a solution for the update of matrix **B**, the following conditions must hold:

- the solution is a TS local optimum (not necessarily a strict local optimum, since the search is restricted by short-term memory), and
- the corresponding value does not exceed a predetermined upper bound set to: value of heuristic solution x 1.05.

In other words, only good TS local optima are recorded. After each iteration, if the current solution satisfies the above conditions, for every pair of machines (i,j) that belong to the same cell, the corresponding element in B is increased by one. Again this matrix is always scaled to have values comparable to those of the cost matrix **C**.

The heuristic used is the merge-and-swap one presented in [Spiliopoulos and Sofianopoulou, 1998]. Note that we allow the same solution to be recorded again if it re-appears. If the search insists on some areas, this must be taken into account.

If we compare moves on the basis of the **B** matrix (and not from the values of the cost matrix **C**), selection is based on attributes (connections) of high quality solutions already found. The introduction of such attributes helps towards focusing the search in good or high promising areas, therefore serving the intensification purpose.

An important detail is that not all values of the **B** matrix are used in this process. We are mainly interested in connections which systematically appear in many elite solutions, and impose this by filtering out values less than 0.8 of the maximum frequency observed so far, a threshold found to be adequate in practice. Smaller values would allow poor TS local optima that were visited only few times to contribute. On the other hand, a value of 1.0 would degenerate the process to depend on very few elite connections, especially during the later stages when the search has been adequately diversified.

4. THE SEARCH STRATEGY

4.1 Normal and Scheduled Iterations

The application of the long-term memory is based on a schedule. Certain conditions (explained later) trigger the following *scheduled* iterations sequence:

- At most 10 iterations mainly for diversification purposes, during which the move selection is based on the residence frequency-based long term memory, that is on the values of the **C-S** matrix. During these iterations all short-term memory restrictions (the tabu ones) are ignored. The iterations are stopped when the corresponding value can no more be improved, that is all candidate moves which involve high-cost connections have been visited enough and no further diversification can be achieved. This practice also prevents cycling in the absence of tabu states use.
- 10 iterations during which short-term recency memory is applicable, together with aspiration criteria.
- At most 10 iterations for intensification purposes, during which the move selection is based on the second long-term memory of elite solutions, that is on the values of the **B** matrix. Again during these iterations all short-term memory restrictions are ignored. In an analogous way with the diversification phase, these iterations are stopped when the corresponding value can no more be improved, that is candidate moves do not contain enough "good" connections, and intensification is no more served. This also prevents cycling in the absence of tabu states use.

The last intensification stage is not applied at all if the number of elite solutions has not reached a certain minimum, for reasons of reliability of the statistical information collected. This may happen in the earlier stages of the search for certain problems, depending on the quality of the heuristic solution and the search course. In such cases, the application of this stage is postponed and the threshold for a solution to enter the elite set is increased.

The logic for the above is the following: After a certain number of iterations has passed and if there is significant delay in global improvement, we first perturb the search to guide it towards unexplored areas, then impose a stabilization phase and finally introduce attributes of high quality solutions found, to correctly intensify the search in the new areas. This sequence is not repeated continuously of course. As it is explained in the next section, one condition for the start of scheduled iterations is delay in global improvement. Technically, the global improvement delay counter is also re-initialized at the end of each such sequence.

The iterations presented above are the *scheduled* ones. Any other, including the one triggering this sequence is called a *normal* iteration.

4.2 Outline of the Search Course

The algorithm starts with the most probable optimal shape, that is the one with the maximum number of cells of the maximum size M. The machines are randomly allocated to cells.

We then proceed with 500 iterations, starting with normal (i.e. not scheduled) ones. During the normal iterations, the cases are:

1. The best simple move leads to a solution equal to or better than the current one ⇒ this simple move is selected.
2. The best swap leads to a solution equal to or better than the current one and, either there are no candidate simple moves (all cells are saturated or all simple moves are irrevocably tabu), or the best simple move leads to a solution inferior to the current one ⇒ this swap is selected.
3. Both the best simple move and the best swap lead to solutions inferior to the current one ⇒ the best simple move becomes the candidate move but not yet carried out. We wait to check whether this iteration triggers the scheduled moves sequence.
4. The best swap leads to a solution inferior to the current one and there are no allowed simple moves ⇒ this swap becomes the candidate move but not yet carried out, as in case (3).

In (3), simple moves are preferred from swaps for reasons of diversification. In all cases, whenever there are no allowed swaps, all the non-zero elements of the tabu matrix **T** are reduced by one until at least one swap can be found.

The scheduled moves sequence is triggered in cases (3) and (4) if:

- the search has reached the 100th iteration, and
- at least 50 iterations have passed from the last global improvement or the last application of the scheduled moves sequence.

At the end of the scheduled iterations the search returns to normal ones. The search strategy is summarized in the following Figure 3 and also in the flowchart presented in Figures 4 and 5. For reasons of simplicity, the flowchart does not show the details in the application of short- and long-term memories described in sections 3.4 and 3.5 and the sampling sufficiency test for the intensification stage mentioned in section 4.1.

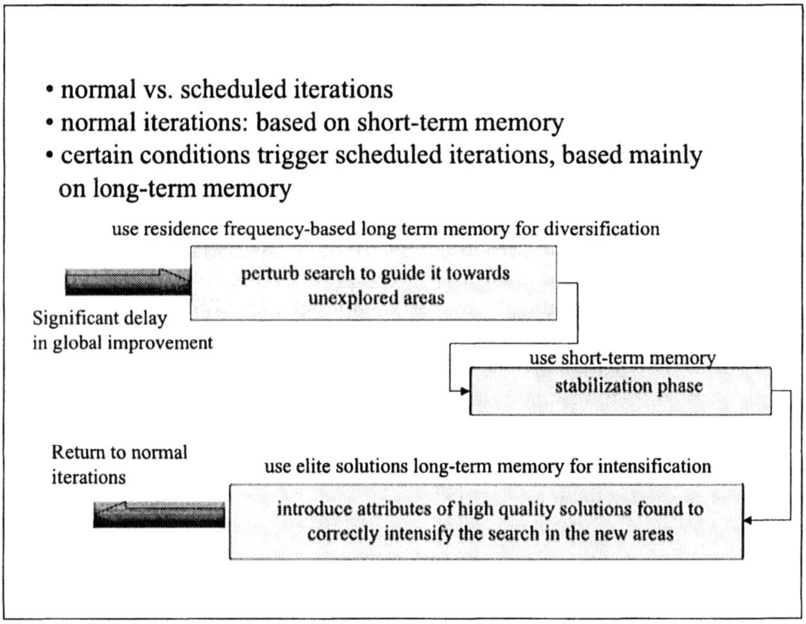

Figure 3. The core of the search strategy.

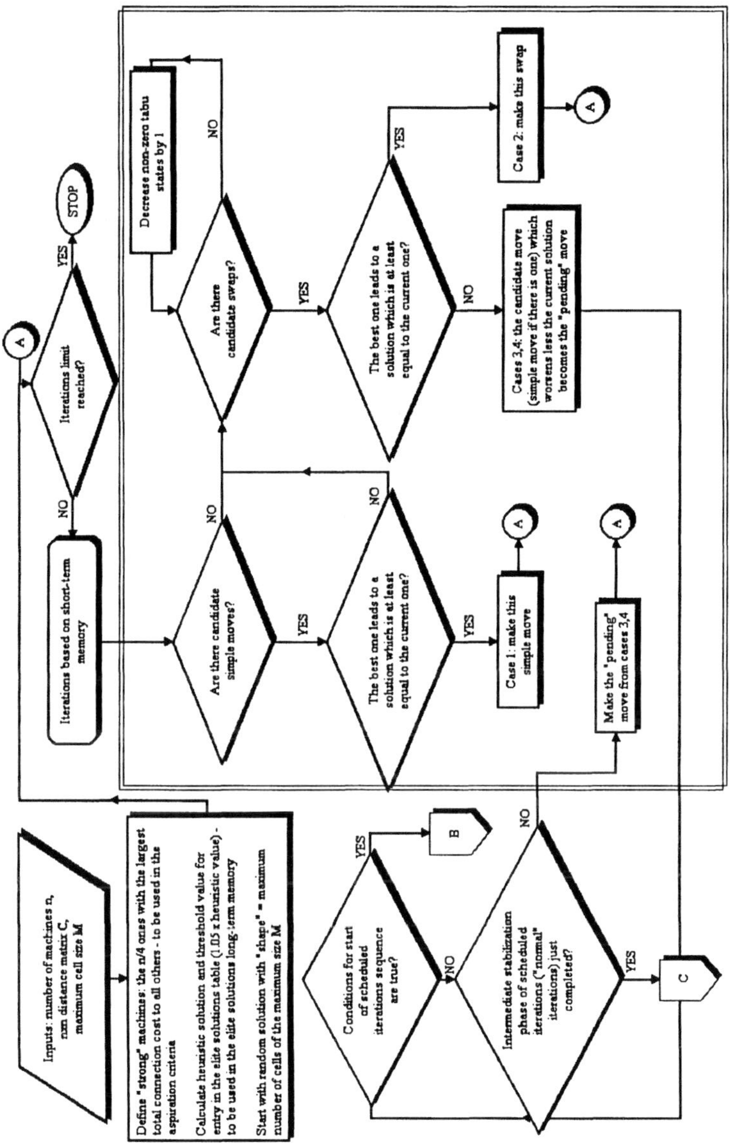

Figure 4. Outline of the Tabu Search algorithm.

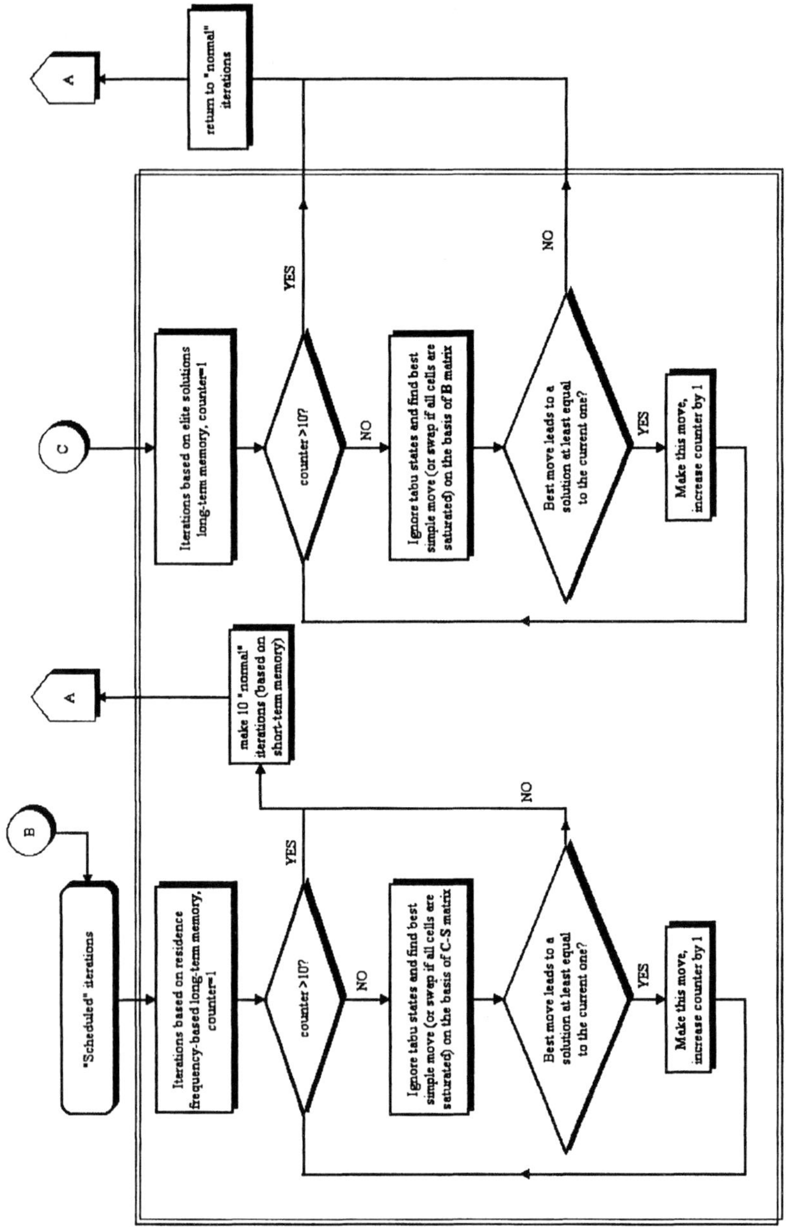

Figure 5. Outline of the Tabu Search algorithm (cont.)

5. COMPUTATIONAL RESULTS

We implemented a Fortran 77 code to test the algorithm on a series of problems with known optimal solutions. At the development stage, special care was taken to avoid occasional (non-cycling) re-visiting of past solutions, especially non "elite" ones. We monitored the occurrence of this effect with the use of a hash function to codify solutions and adjusted the parameters of the search by correlating their values with the degree of diversification. We also used the same technique to add or delete other search features (for example aspiration criteria).

The results were more than encouraging, in that the algorithm found the optimal solutions in all problems, in negligible times (on a Pentium 300 MHz 32MB PC). In fact, for most of these problems we did not even have to increase the maximum number of iterations which is quite low at the 500 level. In particular, we tested:

- The set of 80 problems published in [Sofianopoulou, 1997] and found the optimal solution in all. These are based on 10 cost matrices of dimension 16x16, with maximum cell size running over [5,12]. Total time was 140 seconds.
- The set of 50 problems published in [Spiliopoulos and Sofianopoulou, 1998] and found the optimal solution in all. These are based on 10 cost matrices of dimension 20x20, with maximum cell size running over [6,10]. Total time was 127 seconds.
- A set of 35 problems based on 7 cost matrices of dimension 25x25, with maximum cell size running over [8,12] and found the optimal solution in all. Total time was 148 seconds. The data for these problems is available by the authors and the optimal solutions were found by the exact algorithm presented in [Spiliopoulos and Sofianopoulou, 1998].
- A set of 45 problems based on 9 cost matrices of dimension 30x30, with maximum cell size running over [8,12] and missed only 3 optimal solutions (with 500 iterations), with mean deviation 0.05% and maximum 0.08% in these 3 instances. Total time was 292 seconds. For these problems we could no more use our exact algorithm, therefore we let the proposed algorithm itself run for 10,000 iterations, with zeroing of all tabu states information every 1000 iterations. These longer runs took a total time of 36 minutes on the same machine and produced conjectured optimal solutions which were very recently proved to be optimal by M. Sorensen (private communication) with the branch-and-cut exact algorithm presented in [Sorensen, 1995]. The 9 cost matrices for these problems were calculated from random parts routing data, combining 3 different probability distributions for the number of machines that each

part visits with 3 different numbers of parts. These are also available by the authors.

The only reason that did not enable us to test the proposed algorithm with problems of higher dimensionality was the unavailability of known optimal solutions.

6. CONCLUSIONS

A Tabu Search approach is presented for the solution of the manufacturing systems Cell Formation Problem. A simplified yet concise model is solved which takes into account the most important factors involved, namely the maximum size of the cells and the sequencing order of the parts in part routings. Other parameters can then be easily taken into account at a subsequent stage.

The Tabu Search approach consists of a proper framework for the short and long-term memory structures and the overall search strategy, so as to reach the benefits of intelligent search, namely adaptive memory and responsive exploration. To this end, we propose special long-term memory structures and use them in a scheduled way to achieve both the diversification and intensification Tabu Search targets. We give particular emphasis to influential moves by incorporating them also in the aspiration criteria. Oscillation between feasible and non-feasible areas is also used, to enhance diversification. As for the sensitivity of the results with regard to the various parameters, the fixed ones have been extensively tested and the non-fixed ones are sufficiently robust.

The algorithm produces most promising results for problems with up to 30 machines, this limit being imposed only by the absence of a benchmark with known optimal solutions for larger problems. We believe that the algorithm can give good quality solutions, without any changes, for problems up to dimension 50, but this remains to be tested. For higher dimensions, and in the case of deteriorating performance, there is much room for fine-tuning by making some of the static parameters dynamic. There is also the possibility of enhancements with probabilistic features (e.g. from Simulating Annealing) to introduce additional explorative capability, but this is not considered as the proper next step.

REFERENCES

Burbidge, J.L. (1993). Comments on clustering methods for finding GT groups and families. *Journal of Manufacturing Systems* 12(5), 428-429.

Cheng, C.H. (1992). Algorithms for grouping machine groups in group technology. *OMEGA International Journal of Management Science* 20(4), 493-501.

Cheng, C.H., Goh, C.-H., and Lee, A. (1996). Solving the generalized machine assignment problem in group technology. *Journal of the Operational Research Society* 47, 794-802.

Garey, M.R., and Johnson, D.S. (1979). *Computers and Intractability: A Guide to the Theory of NP-Completeness,* Freeman, New York.

Glover, F. (1990). Tabu Search: A tutorial. *Interfaces* 20(4), 74-94.

Glover, F., and Laguna, M. (1995). Tabu Search. In: Reeves, C.R. (ed) *Modern Heuristic Techniques for Combinatorial Problems*, McGraw-Hill, London.

Glover, F., and Laguna, M. (1997), *Tabu Search*, Kluwer, Boston.

Harhalakis, G., Nagi., R., and Proth, J.M. (1990). An efficient heuristic in manufacturing cell formation for group technology applications. *International Journal of Production Research* 28, 185-198.

Kandiller, L. (1994). A comparative study of cell formation in cellular manufacturing systems. *International Journal of Production Research* 32(10), 2395-2429.

Kern, G.M., and Wei, J.C. (1991). The cost of eliminating exceptional elements in group technology cell formation. *International Journal of Production Research* 29(8), 1535-1547.

Lozano, S., Guerrero, F., Eguia, I., and Onieva, L. (1999a). Cell design and loading in the presence of alternative routing. *International Journal of Production Research* 37(14), 3289-3304.

Lozano, S., Adenso-Diaz, B., Eguia, I., and Onieva, L. (1999b). A one-step tabu search algorithm for manufacturing cell design. *Journal of the Operational Research Society* 50, 509-516.

Seifoddini, H. (1989). Duplication process in machine cells formation in group technology. *IEE Transactions* 21(4), 382-388.

Skorin-Kapov, J., and Vakharia, A. (1993). Scheduling a flow-line manufacturing cell: A Tabu Search approach. *International Journal of Production Research* 31(7), 1721-1734.

Sofianopoulou, S. (1994). A mathematical programming approach to the manufacturing systems cell formation problem. Proceedings, *2nd Balkan Conference on OR*, October 1993, Salonica, Greece.

Sofianopoulou, S. (1996). Part and machine partitioning problem in cellular manufacturing: multiple machine environment. *Yugoslav Journal of Operations Research* 6, 183-191.

Sofianopoulou, S. (1997). Application of simulated annealing to a linear model for the formation of machine cells in group technology. *International Journal of Production Research* 35, 501-511.

Sofianopoulou, S. (1999). Manufacturing cells design with alternative process plans and/or replicate machines. *International Journal of Production Research* 37(3), 707-720.

Sorensen, M. (1995), "A polyhedral approach to graph partitioning ", Ph.D. Thesis, Dept. of Management Science and Logistics, The Aarhus School of Business, Aarhus, Denmark.

Spiliopoulos, K., and Sofianopoulou, S. (1998). An optimal tree search method for the manufacturing systems cell formation problem. *European Journal of Operational Research* 105, 537-551.

Wemmerlov, U., and Hyer, N.L. (1989). Cellular manufacturing in the U.S. industry: A survey of users. *International Journal of Production Research* 27(9), 1511-1530.

Chapter 15

State-Space Modeling and Analysis of Pull-Controlled Production Systems

Barış Tan
Graduate School of Business
Koç University

Key words: *Manufacturing design and control, State-space models, automated model generation, production line, Output processes, Production control policies.*

Abstract: A methodology that automatically generates the state space models of pull-controlled production systems is presented. The state space model generated by the algorithm is then used to evaluate the performance of these systems. In addition to long run measures such as the production rate, WIP levels, average backlog, fill rate, etc., the exact distribution of the number of parts produced in a short period of time, the distribution of the time to produce a given number of parts, and the distribution of the cycle time are also derived. This methodology can be applied to a wide range of discrete material flow production systems that can be modelled as Markov chains. The deterministic processing time model is considered to explain the methodology in detail. The state space models of unreliable automated production lines controlled by the kanban, constant WIP, Control Point Policy, Base Stock, and hybrids of these policies are generated and analysed.

1. INTRODUCTION

Developing analytical methods to design and control production systems has been the subject of numerous studies. Most of these studies have been focused on determining the long run average measures such as the production rate and expected WIP levels (Dallery and Gershwin, 1992).

Exact performance evaluation of manufacturing systems is limited to one- or two-station models. Therefore approximations that are usually based on decomposition are commonly used.

Furthermore, a wide range of production control policies can be analysed in the performance evaluation framework (Gershwin, 1997). In order to analyse the performance of production systems controlled by a given policy and, to determine the parameters of the control policy, efficient methods are needed. This study is motivated by this research direction.

In this study, a methodology that generates the state space models of pull controlled production systems automatically is presented. The state space model that is generated by the algorithm is then used in performance evaluation of these systems. In addition to long run measures such as the production rate, WIP levels, average backlog, fill rate, etc., the exact distribution of the number of parts produced in a short period of time, the distribution of the time to produce a given number of parts, and the distribution of the cycle time are also derived.

Exact analysis of production systems is useful to derive insights about the effects of control policies on the performance of a production system. Furthermore, the insights and observations obtained from the exact results can be used to develop efficient algorithms to determine the parameters of a control policy. These results can also be used to test the accuracy of the approximation methods.

State-space modelling can be used to evaluate the performance of "not-so-small" production systems exactly. It is known that the state-space based methods suffer from the rapid growth of states. This limits the applicability of these methods to a certain size. However, if the state-space modelling and analysis is done efficiently, the methodology can be useful to understand more complicated systems. The methodology presented here is not claimed to be an alternative to decomposition methods, rather it is presented as a useful tool that can be utilized to understand a system and to develop approximation methods.

In this study, a new numerical method that gives the exact distribution of the number of parts produced in a given time period and the time to produce a given number of parts is also presented. Analysing the variability of the output in a short period of time is crucial to achieve dependable deliveries.

There are two main contributions of this study. First, an algorithm to generate the state space models of production system controlled by various control policies automatically is developed. This algorithm is used to model production systems controlled with kanban, Control Point Policy (CPP), and CONWIP, and the hybrids of these policies. Second, a general state-space based methodology is given to determine the exact distribution of the number of parts produced in a given time period and the exact distribution of the time to produce a given number of parts.

The methodology can be applied to a wide range of discrete material flow production systems that can be modelled as Markov chains. The methodology is explained in detail by considering the deterministic processing time model described in (Gershwin, 1994). The state space models for unreliable automated production lines controlled by kanban, constant WIP (CONWIP), Control Point and hybrids of these three policies are generated and then analysed.

2. GENERAL STRUCTURE

The method presented in this study is consisted of two main parts: A *model builder* block and a *performance evaluation* block. The model builder block takes the *system structure, system parameters,* and *operational rules* as inputs. Given this information, the model builder generates the state space and the state-transition probability matrix automatically. Once the state space and the probability matrix are generated, the performance evaluation block determines the performance measures of interest by using the appropriate algorithms. Figure 1 below depicts the block diagram of the model.

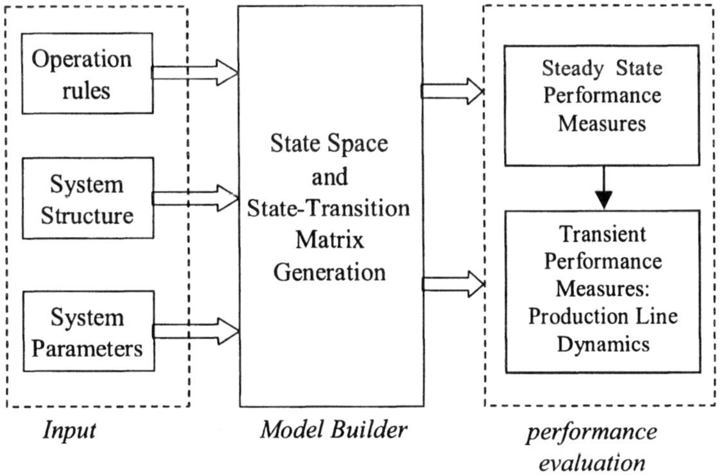

Figure 1. Block diagram of the methodology

The organization of the remaining part of this paper is as follows. The model builder block for a general Markov chain model is described in §3. In order to explain its operation in detail, a specific model is considered. The description and assumptions of this model are given in §4. The details of the algorithms, the system structure reduction, the state space generation, the state space reduction, and the transition matrix generation steps of the algorithm are presented in §5. Implementation of the kanban, constant WIP, Control Point and hybrid policies in the algorithm is explained in §6. Following the model builder block, the performance evaluation block is described in §7. Namely, the algorithms to determine the exact distributions of interest and the related performance measures are developed in this section. Some preliminary numerical results are given in §8. Finally, conclusions are given in §9.

3. MODEL BUILDER

The state-space based methods in the literature follow the approach of generating the probability matrix and then analysing the probability matrix for the performance measures of interest. For example, Miltenburg (1987) requires that the probability matrix should be given explicitly to determine the asymptotic variance rate of the output. However, it is rather cumbersome to generate the state-space and then to build the probability matrix manually. An early study by Schick and Gershwin (1978) discusses generation of the probability matrix of multistation production lines by using the nested tridiagonal structure of the probability matrix. Papadopoulos and O'Kelly (1989), Papadopoulos et. al. (1989) and Vidalis and Papadopoulos (1997) generate the transition matrix of a multistation production line recursively by exploiting the nested block tridiagonal structure of the underlying quasi birth and death process. Namely, the repeating structure of the submatrices is used and the probability matrix is formed by using the submatrices that are given explicitly.

In this study we follow an alternative route to generate the state space and the probability matrix automatically. More specifically, our objective is to develop a Model Builder Block that takes *system structure*, *system parameters*, and *operational rules* as inputs and then generates the state space, the probability matrix and the reward function. The system structure includes the information on how the stations are interconnected, where the buffers are located, etc. The system parameters contain buffer capacities, the number of stations, failure, repair probabilities, and control policy parameters. The *operational rules* include the control policy, blocking mechanism, failure mechanisms, and conventions used to update the system when a station fails, when the buffer level changes, etc. Given this information, the model builder generates the state-space and the state-transition probability matrix automatically.

A similar approach has been taken to develop the software tool SHARPE (Symbolic Hierarchical Automated Reliability and Performance Evaluator) that allows its users to construct and analyse performance, reliability, availability, and performability models, especially for computer systems (Sahner, Trivedi, and Puliafito, 1996). Stewart (1994) also presents a tool to generate and analyse Markov models of discrete-event systems. This study is a first step towards developing a similar environment for performance evaluation of production systems.

In the model builder block, state-space generation starts at a given state, say, at the state where the stations are up and the buffers are empty. Then for all possible changes in the machine states, the changes in the buffer levels are determined according to the operational rules. The next state is constructed by combining the new machine states and the buffer levels. Then the transition probability is calculated according to the operational rules. Finally, the index of the current state, the index of the new state, and the transition probability are stored. This procedure is continued with the next state in the state space until all the states in the state space are evaluated. As a result, the state space and the probability matrix are built. When the station of interest and the performance measure of interest are given, the reward matrix is formed from the state space. For example, if the production rate of the first station of a production line is of interest, then the states where the first station is up and not blocked are chosen. Denoting the chosen states with one and the others with zero yields the desired reward function.

4. MODEL

In order to describe the algorithm precisely, the deterministic processing time model described in (Gershwin, 1994) is used. This is the simplest model for studying failures, finite buffers, and various control policies. The assumptions and conventions of this model are summarized below. A more detailed description for production lines can be found in (Gershwin, 1994: pp. 71).

In this model, processing time of a part at each station is fixed and the same for all machines. This amount of time is taken as the time unit. All the stations start their operation at the same instant. A part starts its operation at machine M_1 and flows through buffer B_1 to downstream machines until it reaches the last machine M_k and after it finishes its last operation it flows out of the system. We assume that there is no scrapping of materials during operation. The transportation time from stations to buffers and from buffers to stations are assumed to be negligible compared to the service times. The first station is never starved and the last station is never blocked.

The repair state of machine M_i is $\alpha_i \in \{0,1\}$ where $\alpha_i=1$ if the machine is operational and $\alpha_i=0$ if the machine is under repair. Therefore the repair state of k machines is a k-tuple $(\alpha_1, \ldots, \alpha_k) \in A = \{0,1\}^k$. Repairs and failures take place at the beginnings of time units. The amount of material in Buffer i at time t is $n_i(t)$, $0 \leq n_i(t) \leq N_i$. The buffer state of a production system with k-1 buffers is $(n_1, \ldots, n_{k-1}) \in B=\{0, 1,\ldots,N_1\} \times \{0, 1,\ldots,N_2\} \times \ldots \times \{0, 1,\ldots,N_{k-1}\}$. Then the state of the system at time t, $X(t)$ is described as a tuple of size $2k$-1,

$$X(t) = (n_1, \ldots, n_{k-1}, \alpha_1, \ldots, \alpha_k), \quad X(t) \in S' = A \times B \tag{1}$$

The state space S' has

$$\mathrm{n}(S') = 2^k \prod_{i=1}^{k-1} (N_i + 1) \tag{2}$$

states. In the Markov chain model, there is a set of transient states. Furthermore, due to the control policies and operational rules, some of the states are never visited. Therefore the state space S' can be reduced in size by eliminating the states that are never visited. The state space after reduction is referred as S. Obviously, the number of states in the reduced state space n(S) is less than n(S').

The states in S are partitioned into two disjoint sets: U and $\overline{\mathrm{U}}$. The set U is the set of *productive* states in which a part is produced at the end of that cycle. For example, for a two-station production line with a finite buffer, all the states where the second station is operational and the buffer is not empty, and the first and the second stations are operational and the buffer is empty constitute U. The set $\overline{\mathrm{U}}$ is the complement set of U, i.e., $\overline{\mathrm{U}} = S \setminus \mathrm{U}$.

Let Φ_i be an indicator variable that describes the effect of a *control policy* and *operational rules* on machine M_i. If the control policy does not restrict the flow out of Machine M_i, Φ_i is 1, otherwise it is 0. For example, for production lines

$$\Phi_1(t) = \begin{cases} 1 & \text{if } n_1(t) < N_1 \\ 0 & \text{otherwise} \end{cases} \tag{3}$$

$$\Phi_i(t) = \begin{cases} 1 & \text{if } n_{i-1}(t) > 0, n_i(t) < N_i \\ 0 & \text{otherwise} \end{cases}, \quad i = 2,\ldots,k \tag{4}$$

$$\Phi_k = \begin{cases} 1 & \text{if } n_{k-1} > 0 \\ 0 & \text{otherwise} \end{cases} \tag{5}$$

For other control policies, Φ_i is set accordingly. In Section 5, Φ_i i=1, 2,…,k are given explicitly for the kanban, constant WIP, Control Point , and hybrid policies.

Buffer i gains or loses one unit depending on the repair state of its upstream and downstream stations and control policies. According to the convention, buffer levels are updated at the end of the time units. Thus, if blocking and starvation do not affect buffer i,

$$n_i(t+1) = n_i(t) + I_i(t+1) - I_{i+1}(t+1) \tag{6}$$

where $I_i(t)$ indicates whether a part arrives at buffer i from upstream

$$I_i(t+1) = \begin{cases} 1 & \text{if } \alpha_i(t+1) = 1 \text{ and } \Phi_i(t) = 1 \\ 0 & \text{otherwise} \end{cases} \tag{7}$$

It is assumed that a station that is not processing a part cannot fail, i.e., operational failures are assumed. The probability that a working station fails during a time unit is p_i.

The probability that a failed station is repaired during a time unit is r_i.. When a machine fails, the part it was processing is taken to the upstream buffer to wait for until the machine is repaired.

The state transition probability matrix, or the probability matrix, $Q = \{q_{i,j}\}$ $i,j \in S$ completely describes the transition probabilities between states, i.e.,

$$q_{i,j} = P[X(t+1)=j \mid X(t) = i] \ \forall \ t, i, j \tag{8}$$

If at state s, machine i is operational and the control policy allows flow into downstream buffer B_i, the reward of this machine at this state $w_{s,i}$ is 1, otherwise it is 0. That is

$$I_i(t+1) = \begin{cases} 1 & \text{if } \alpha_i(t+1) = 1 \text{ and } \Phi_i(t) = 1 \\ 0 & \text{otherwise} \end{cases} \tag{9}$$

The rewards are stored in n(S) × k matrix $W=\{ w_{s,i}\}$ $s\in S$, i=1,2,..,k. Note that $s \in U$ if $w_{s,k}$=1.

5. AN ALGORITHM TO GENERATE THE STATE SPACE MODELS

The state transition matrix Q and the reward function W defined in the previous section are generated automatically when the system parameters, the system structure, and the operational rules, more specifically Φ_i $i=1, 2,\ldots,k$, are given. The methodology that generates Q and W consists of four main parts: *system structure reduction*, *state space generation*, *state space reduction*, and *transition matrix and reward function generation*. In this section, these steps are described for a production line controlled by general pull policies. In the next section, the implementation issues for the kanban, CONWIP, Control Point Policy and hybrid policies are discussed in detail.

5.1 System structure reduction

When a production line with a control policy is given, a number of different state space models that describe the same system can be developed. Since the size of the state space determines the computational requirements, the objective of this step is to reduce the system structure to an equivalent representation that conveys the same information with a fewer number of states.

The equivalence theorem for assembly/disassembly networks (Gershwin, 1994), loop invariants in loop networks, and special properties of synchronization stations are utilized in order to reduce the system structure.

For example, Figure 2 depicts a production line with a loop. This system can be used to evaluate the performance of a production line with the CONWIP policy.

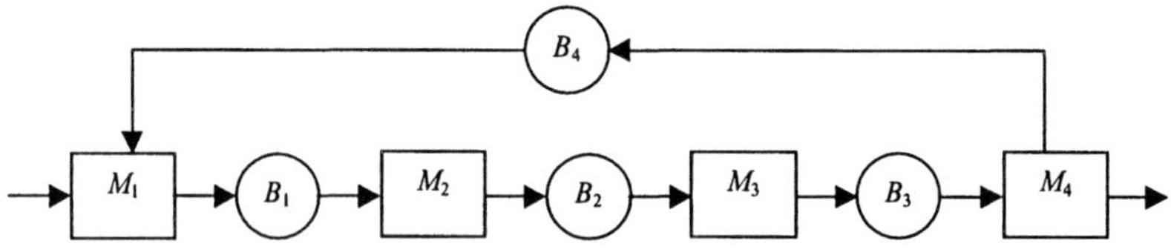

Figure 2. A production line with a loop

The number of parts in the loop network is constant because there is no way for parts to arrive or depart. That is, for the system shown in Fig 2

$$n_1 + n_2 + n_3 + n_4 = L \tag{10}$$

where L is the population size, i.e., the maximum number of parts allowed in the system. Equation (10) yields the number of parts in B_4 when L, n_1, n_2, and n_3 are given. Therefore it is not necessary to include n_4 in the system state. This loop invariant reduces the number of states in the state space by a factor of (N_4+1).

Now consider a station with backlogged demand shown in Figure 3. Station H is an infinitely fast and perfectly reliable synchronization station and station D is a station that represents the demand. Buffer B_3 holds the parts processed at M_3 and buffer B_4 holds backlogged demand.

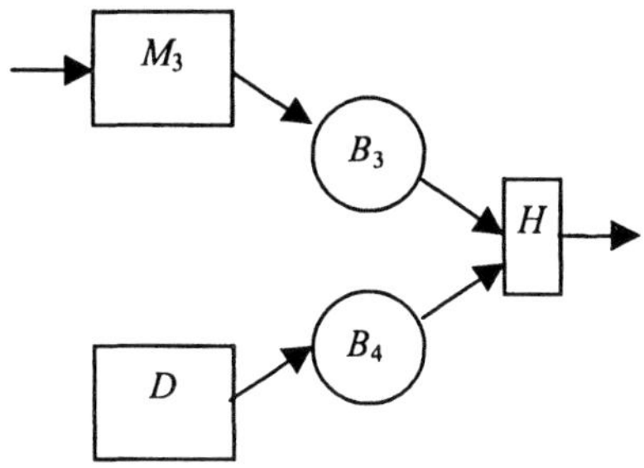

Figure 3. A station with backlogged demand and equivalent representation

Since the synchronization station is infinitely fast, at any given time either B_3 or B_4 is empty. Then the same dynamics can be captured by an equivalent two-station line where buffer B'_3 has a capacity of $N_3 + N_4$ and $n'_3 = N_4 + n_3 - n_4$. If the equivalent system is used, n_3 and n_4 can be determined from n'_3 as $n_3 = \max\{ n'_3 - N_4, 0\}$ and $n_4 = \max\{ N_4 - n'_3 , 0\}$. Bonvik (1996) uses a similar approach to analyse a continuous material flow system with backlogged demand. This transformation reduces the number of states by a factor of $(N_3+1)(N_4+1)/(N_3+N_4+1)$.

Furthermore, the reductions yield a series arrangement of stations. Accordingly, a basic production line model can be used to analyse production lines controlled by kanban, CONWIP, Control Point, and hybrid policies. The state space after reduction is denoted by S'.

5.2 State Space Generation

The reduced state space of a series arrangement of workstations includes all possible combinations of repair states of machines and buffer states of interstation buffers as shown in equation (1).

Let all the states in S' be stored an $n(S') \times (2k-1)$ matrix Γ. Furthermore let Γ be partitioned as $[\Gamma_M : \Gamma_B]$ where matrix Γ_M is an $n(S') \times k$ matrix that includes the machine states and matrix Γ_B is an $n(S') \times (k-1)$ matrix and it includes the buffer states. These matrices are generated automatically by using two functions `repmat` and `reshape`.

The function G = `repmat` (F, i, j) replicates and tiles the matrix F to produce the $i \times j$ block matrix G. The function G =`reshape`(F, i, j) returns the $i \times j$ matrix whose elements are taken columnwise from F that has $i \times j$ elements. These functions are available in MATLAB 5.1 as built-in matrix functions.

For example, let F=[1 2 3 4],

$$\texttt{repmat}\ (F, 2, 1) = \begin{bmatrix} 1 & 2 & 3 & 4 \\ 1 & 2 & 3 & 4 \end{bmatrix} \text{ and } \texttt{reshape}(F, 2, 2) = \begin{bmatrix} 1 & 3 \\ 2 & 4 \end{bmatrix}.$$

Matrix Γ_M is generated as

$$\Gamma_M = \texttt{repmat}\,(A, \mathrm{n}(S')/2^k, 1) \tag{11}$$

where n(S') is given in equation (2) and the ith column of A, $A^{<i>}$, i=1, 2,..k-1, is given as

$$A^{<i>} = \texttt{repmat}\,(\texttt{reshape}\,(\texttt{repmat}(\,[0, 1]\,, 2^{k-i}, 1), 2^{k-i+1}, 1), 2^{i-1}, 1) \tag{12}$$

Similarly the ith column of matrix Γ_B, $\Gamma_B{}^{<i>}$, i=1,2,..k-1 is generated as

$$\Gamma_B{}^{<i>} = \mathrm{repmat}(\mathrm{reshape}(\texttt{repmat}\,(\,[0,\ldots,N_i]\,, 2^k \prod_{j=i+1}^{k-1}(N_j+1), 1), 2^k(N_i+1)\prod_{j=i+1}^{k-1}(N_j+1), 1\,), C(i), 1) \tag{13}$$

where

$$C(i) = \begin{cases} 1 & i = 1 \\ \prod_{j=i-1}^{k-1}(N_j+1) & i = 2,\ldots k-1 \end{cases} \tag{14}$$

In order to refer to a state easily, a unique index number for each state is required. This index number can be determined by following a base conversion method that yields the equivalent number in base 10 when the tuple $(n_1, \ldots, n_{k-1}, \alpha_1, \ldots, \alpha_k)$ is treated as a 2k-1 digit number in a specific number base. The function $\varphi(s)$ given as

$$\varphi(s) = 2^k\left[n_{k-1} + \sum_{j=1}^{k-2} n_j \prod_{i=j+1}^{k-1}(N_i+1)\right] + \sum_{j=1}^{k}\alpha_j 2^{k-j} + 1 \tag{15}$$

yields a unique index for state $s = (n_1, ..., n_{k-1}, \alpha_1, ..., \alpha_k)$. Note that, the latter part of the right hand side of the above equation is simply the base 10 equivalent of the binary number $(\alpha_1, ..., \alpha_k)$. The first part is derived in a similar way. Note also that,

$$\varphi(0, \ldots 0, 0, \ldots, 0) = 0 \quad \text{and} \quad \varphi(N_1, \ldots N_k, 1, \ldots, 1) = 2^k \prod_{i=1}^{k-1} (N_i + 1).$$

5.3 System State Reduction

As mentioned before, some of the states in S' will never be visited and some of them will be transient. Since the transient analysis may also be of interest to study the production line dynamics, in this step we focus on reducing the state space by eliminating the states that will never be visited.

When the control policy and the operational rules are given, they describe a certain kind of relationship between the number of parts in the buffers. If this relationship is not observed in a specific state of S', it cannot be a member of S. Therefore, each state is checked to see whether it satisfies the requirements of the control policy. If it satisfies, it is included in S.

For example, the total number of parts in a production line controlled by CONWIP policy cannot exceed the population size L. That is,

$$(n_1, \ldots, n_{k-1}, \alpha_1, \ldots, \alpha_k) \in S \text{ if } (n_1, \ldots, n_{k-1}) \in \left\{ (n_1 \quad \cdots \quad n_{k-1}) \mid \sum_{i=1}^{k-1} n_i \le L \right\} \tag{16}$$

This condition can easily be checked to reduce S'. When Φ_i i=1, 2,...,k are given for a control policy, they also indicate how the state space can be reduced. We present state space reduction conditions for each control policy in Section 6.

5.4 Transition Matrix and Reward Function Generation

Once the state space is finalized, the transition matrix and the reward function are generated by calculating the transition probability from a given state to all the other possible states accessible from this state.

Let the current state of the system be $s' = (n_1, ..., n_{k-1}, \beta_1, ..., \beta_k)$ and the next state be $s' = (n'_1, ..., n'_{k-1}, \alpha_1, ..., \alpha_k)$. Note that, when $(n_1, ..., n_{k-1}, \beta_1, ..., \beta_k)$ and $(\alpha_1, ..., \alpha_k)$ are given, $(n'_1, ..., n'_{k-1})$ is completely determined by the operational rules and conventions used in the model. For example, for the deterministic processing time model equations (6) and (7) yield $(n'_1, ..., n'_{k-1})$ when $(n_1, ..., n_{k-1}, \beta_1, ..., \beta_k)$ and $(\alpha_1, ..., \alpha_k)$ are given. Then by generating possible machine states $(\alpha_1, \ldots, \alpha_k) \in \{0, 1\}^k$ and calculating $(n'_1, \ldots, n'_{k-1})$ give 2^k possible transitions from state s to s'.

For the model described in Section 4, the transition probabilities $Q=\{q_{s',s}\}$ $s,s' \in S$ can be calculated as

$$q_{s',s} = \prod_{i=1}^{k} \left(f_i^{1-\alpha_i}(1-f_i)^{\alpha_i} \right)^{\beta_i} \left((1-r_i)^{1-\alpha_i} r_i^{\alpha_i} \right)^{1-\beta_i} \quad s' \neq s \tag{17}$$

$$q_{s',s'} = 1 - \sum_{s \in S, s \neq s'} q_{s,s'} \tag{18}$$

where $f_i = p_i \Phi_i$ reflects the operational failures assumption. Equation (17) is repeated 2^k times for all possible repair states, $(\alpha_1, \ldots, \alpha_k)$ to obtain all the transitions from a given state.

In order to store Q in a compact form, only the non-zero elements are stored. Note that if Φ_i=0, the transition probability calculated by equation (17) is zero indicating that this transition is not possible. The index of the current state, the index of the state that will be visited next, and the transition probability is stored only for non-zero transition probabilities. Equation (15) gives the index of a given state. Due to the system state reduction and state space reduction steps, the list of indices obtained from equation (15) may include gaps between indices. In order to make it sequential, the list is renumbered after these steps. Finally, these steps are repeated for all the states in the state space to form Q. Since Q and W are stored as sparse matrices, the algorithms used in the performance evaluation block are written in a way to handle sparse matrices efficiently. For more information about sparse matrix representations and other computational issues, the reader is referred to (Stewart, 1994).

Table 1 shows the number of states in the state space before and after reduction and the number of non-zero elements for a specific four station line controlled by the kanban, CONWIP, Base Stock, and hybrid policies. Table 1 indicates that depending on the control policy parameters, the system structure reduction and state space reduction steps decrease the number of non-zero elements in the final transition matrix substantially.

Figure 4 also gives information about the effectiveness of the system state, state space reduction, and transition matrix generation steps for a three-station production line controlled with a CONWIP policy. In Figure 4, the number of elements in the initial state space and the number of elements in the reduced state space are plotted for different values of L in the above part. The number of elements in the state transition matrix if the initial, and the reduced state spaces were used, and the number of non-zero elements generated and used in the performance evaluation are shown in the below figure. Since the computational complexity depends on the number of non-zero elements, these reduction steps improve the computational requirements of the algorithm. The implementation issues related to kanban, CONWIP, CPP, Base Stock, and hybrid policies are discussed in the next section.

Table 1. Performance of the Reduction Step

Control Policy	Control Policy Parameters	n(S')	Reduction n(S)	No. of non-zero elements in Q
CONWIP	$L = 10$	21296	4576	57360
Base stock	$V = [10, 8, 7, 6]$	12672	4032	50720
Kanban	$N = [5, 5, 5]$	3456	3456	43120
CONWIP/Kanban	$L = 10$ $N = [5, 5, 5, 10]$	3456	2896	36000
Base stock/Kanban	$V = [10, 8, 7, 6]$, $N = [5, 5, 5]$	3456	2816	35072

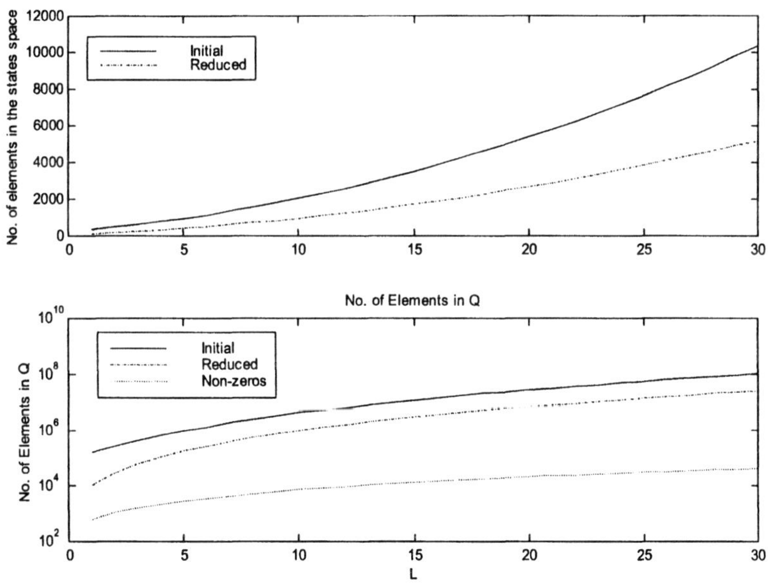

Figure 4. Number of elements in the state space and the number of elements in the state transition matrix of a three-station production line controlled with a CONWIP policy

6. AUTOMATED TRANSFER LINES CONTROLLED BY PULL POLICIES

Kanban based production control methods are commonly used in lean manufacturing. However, it has been recently shown that some new control policies make it possible to reach high service levels with significantly lower inventories than kanban (Bonvik, 1997; Frein et. al., 1995). These policies are hybrids of existing policies that limit the local inventories (e.g. kanban) and policies that provide a global flow of information (e.g. CPP, Base Stock, CONWIP). In this section, the algorithm explained in the preceding section is implemented to model production lines controlled by pull policies.

6.1 Transfer Lines Controlled with Kanban-type Policies

The operation of a finite buffer production line is similar to a production line controlled by a kanban-type policy. Figure 5 depicts a production line with finite buffers. The last station in this line represents the demand. When there is a demand, a part is pulled from B_3. If there are no parts in B_3, the demand is lost.

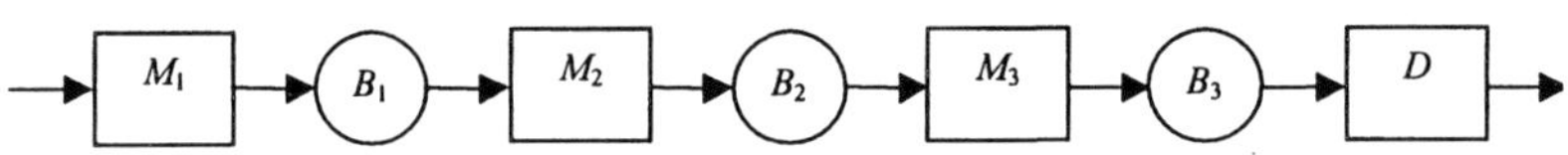

Figure 5. A production line controlled with a kanban-type policy: the lost sales case

When there is no space for a part that has just finished processing on a machine, that machine is blocked. This blocking controls the WIP in the system. Similarly a machine is starved when there are no parts in the upstream buffer. By selecting the interstation buffer capacities appropriately, a production system can be controlled by a kanban-type policy. Note that there are many different kanban system implementations (Liberopoulos and Dallery, 1998). This is a simple single parameter per stage policy.

The system structure is a series arrangement of workstations. Therefore, the structure cannot be reduced further. The state space can be reduced by eliminating the transient states. However if the transient analysis is also of interest, $\{0,1\}^k \times\{0, 1,\ldots,N_1\}\times\{0, 1,\ldots,N_2\}\times\ldots\times\{0, 1,\ldots,N_{k-1}\}$ is the state space used. Then once the system parameters and operational rules are given, the algorithm described in the previous section yields the transition matrix and the reward function. The system parameters are simply the failure and repair probabilities and the buffer capacities, i.e., p_i, r_i

i=1,2,..,k N_i, i=1,2,..,k-1 Operational rules are summarized in Φ_i i=1, 2,...,k which a re given in equations (3), (4), and (5).

In order to model a production line controlled by a kanban policy with backlogged demand, the last station that represents the demand is replaced with the one depicted in Figure 3 and the system structure is reduced to a series arrangement as described in Section 5.1. The remaining implementation follows directly the case with lost sales. When considering backlog demand case, a limit is imposed to the number of backlogged demand stored in B_4. If n_4=N_4, station D is blocked to produce additional demand. In other words, if the backlog is large, customers choose not to wait and the order is lost. Although N_4 can be set to a large number to approximate infinite backlog case, this will increase the computational complexity. Furthermore, if the production system is designed properly, backlog should not reach such high levels. If it reaches such high levels, it is plausible to assume that new customers choose not to place orders.

6.2 Production Lines Controlled with CONWIP policies

Limiting the WIP inventory and maintaining a constant WIP to control a production system received some attention in recent years (Duenyas et.al., 1993) . This is, may be, the simplest control policy, since it requires only one parameter that is the population size L.

A production line with constant WIP (CONWIP) control policy can be represented as a loop network as depicted in Figure 2. In this case the last buffer that connects the last station to the first one can represent the information about whether the population size is reached. For example, if a card is attached to each part, whenever a part leaves the production line, it is placed in this buffer. Then implementing this policy only requires starting production at M_1 only if there is a card in buffer B_k. Then the population size can be set by having the same number of cards circulating in the system. Similarly, pallets can be used for this purpose.

In Section 5.1, it is explained that a production line with a loop can be reduced to a series arrangement of stations. Furthermore equation (16) shows how the state space is reduced by imposing constant WIP restriction.

Note that, in CONWIP, there is no restriction for the interstation buffer capacities. However, since only L parts are allowed in the system, each buffer can hold at most L parts. In other words, we set the interstation buffer capacities to L initially, i.e., $N_i = L$ i=1, 2, ...,k-1. The only required system parameters are failure and repair probabilities, p_i, r_i i=1,2,..,k, and the population size L. Operational rules for the CONWIP control policy are summarized in Φ_i i=1, 2,...,k given below

$$\Phi_1 = \begin{cases} 1 & \text{if } \sum_{j=1}^{k-1} n_j < L \\ 0 & \text{otherwise} \end{cases} \tag{19}$$

$$\Phi_i = \begin{cases} 1 & \text{if } n_{i-1} > 0 \\ 0 & \text{otherwise} \end{cases} \quad i = 2,\ldots,k \tag{20}$$

It is very easy to implement a CONWIP / kanban hybrid policy by setting interstation buffer capacities externally. In other words, the required system parameters are L, p_i, r_i i=1,2,..,k, and N_i, i=1,2,..,k-1. The system structure and state space reduction steps are identical to CONWIP case. Operational rules Φ_i i=1, 2,…,k combines equations (3-5) with equations (19-20). The operational rules for the CONWIP/kanban control policy are given below

$$\Phi_1 = \begin{cases} 1 & \text{if } \sum_{j=1}^{k-1} n_j < L\,,\ n_1 < N_1 \\ 0 & \text{otherwise} \end{cases} \tag{21}$$

$$\Phi_i = \begin{cases} 1 & \text{if } n_{i-1} > 0\,,\ n_i < N_i \\ 0 & \text{otherwise} \end{cases} \quad i = 2,\ldots,k-1 \tag{22}$$

$$\Phi_k = \begin{cases} 1 & \text{if } n_{k-1} > 0\,,\ L - \sum_{i=1}^{k-1} n_i < N_k \\ 0 & \text{otherwise} \end{cases} \tag{23}$$

Note that equation (23) indicates that the last station is blocked if B_k is full. This case can be avoided by setting N_k equal to L. However, if this cannot be done due to space limitations or availability of pallets, the last station will be blocked occasionally.

Figure 6 shows the effect of L on the system performance, namely production rate, expected inventory, and service level for a three-station production line controlled by a CONWIP policy (p_i = [0.01,0.01,0.007]; r_i = [0.09,0.09,0.05], i=1,2,3). Figure 7 depicts the same system for the CONWIP/Kanban case with N=[10, 10, 10]. In the first case B_3 is large enough to hold L parts, therefore the last station is never blocked. In this case, the production rate, the total WIP, and the service levels are increasing function of L. When $N_3<L$ as in Figure 7, the system is blocked frequently when L is close to $N_1+N_2+N_3$. As a property of closed queueing networks, the production rate of this system is maximized when L is close to the half of total buffer capacity which is 15 in Figure 7.

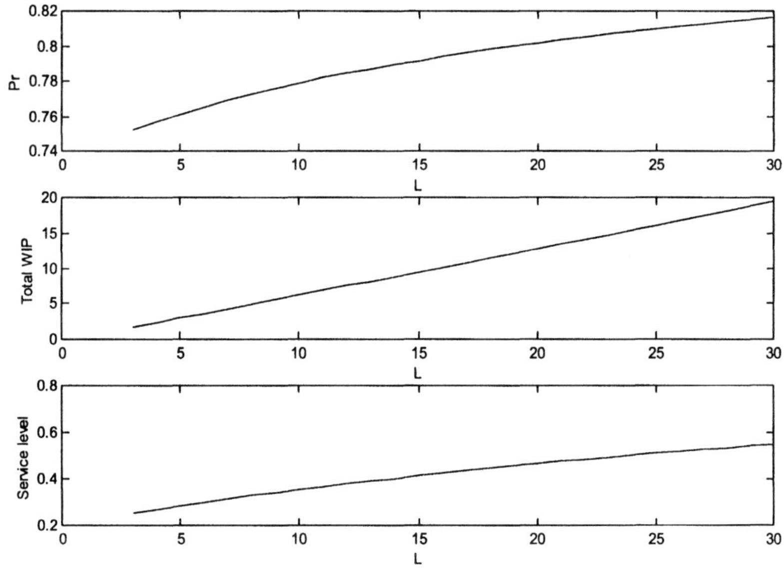

Figure 6. Production rate, Total WIP, and Service level in a three-station production line controlled with CONWIP policy with parameter L.

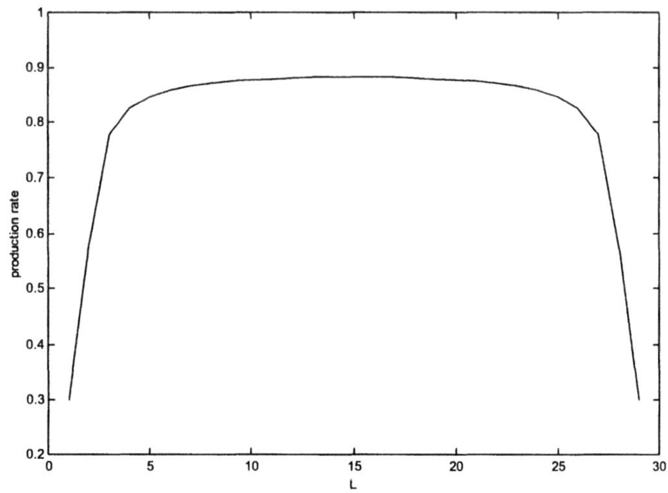

Figure 7. The Production rate in a three-station production line controlled with CONWIP/kanban hybrid policy with parameter L.

6.3 Transfer Lines Controlled with Control Point Policies

Gershwin (1997, 2000) introduced the Control Point Policy (CPP) as an effective class of policies that optimises the tradeoff between inventory build up and backlogging by communicating demand information directly to all stations when a new demand arrives. Bonvik (1996) shows that it performs well compared to other policies.

Figure 8 depicts a three-station production line controlled with a Control Point Policy. Demand is represented by station D. Buffers G_i i=1, 2, 3 hold backlogged demand information. Buffers B_1 and B_2 hold information about the number of parts produced at M_1 and M_2. The finished goods inventory is B_3. Station H represents infinitely fast and perfectly reliable synchronization stations.

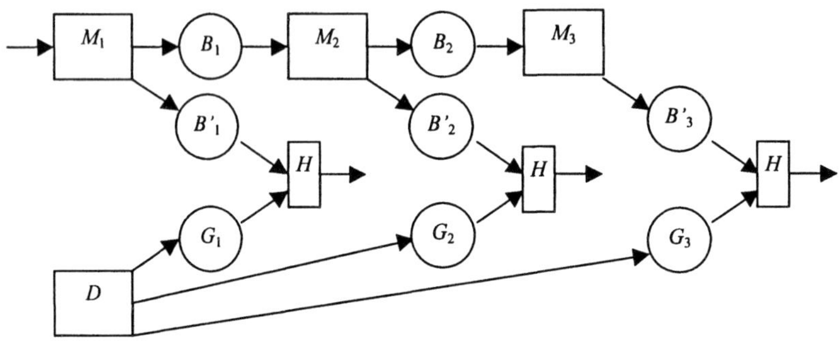

Figure 8. A three-station production line controlled by a Control Point Policy

The first step in system structure reduction is replacing synchronization stations and its upstream buffers with a single buffer and rearranging the flow of materials. Applying the system structure reduction techniques explained in Section 5.1 yields the production system shown in Figure 9 where the capacities of buffers B^*_1, B^*_2, and B^*_3 are adjusted accordingly.

The production system depicted in Figure 9 includes loops. Utilizing loop invariants in loop networks reduces the system structure further. In this system, loop 1 shown in Figure 9 yields

$$(n_1 + n_1 + n_3^*) - n_1^* = \text{constant} \tag{24}$$

Furthermore loop 2 yields

$$(n_2 + n_3^*) - n_2^* = \text{constant} \tag{25}$$

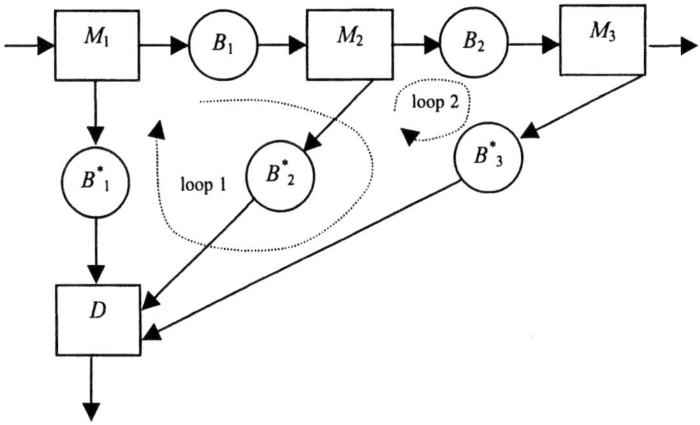

Figure 9. An equivalent representation of a production line controlled by a Control Point Policy

Equations (24) and (25) indicate that, at any given state, the number of parts in buffers B^*_1 and B^*_2 can be calculated if the number of parts in buffers B_1, B_2, and B^*_3 are given. This observation suggests that a production line controlled by a Control Point Policy can be reduced to a series arrangement of stations that includes stations M_1, M_2, M_3, D and buffers B_1, B_2, and B^*_3.

The system parameters for the Control Point Policy are the interstation buffer capacities N_i, hedging points Z_i, p_i, and r_i i=1,2,..,k. Hedging point Z_i is the size of buffer B'_i in Figure 8. The state space is reduced according to the following condition:

$$(n_1,\ldots, n_{k-1}, \alpha_1,\ldots, \alpha_k) \in S \text{ if } (n_1,\ldots, n_{k-1})$$
$$\in \{(n_1 \quad \cdots \quad n_{k-1}) \mid n_i \leq \min\{N_i, Z_i\},\ i = 1,\ldots,k-2;\ n_{k-1} \leq Z_{k-1}\} \quad (26)$$

Finally, the operational rules for the Control Point Policy are given as

$$\Phi_1 = \begin{cases} 1 & \text{if } n_1 < Z_1 \text{ , } n_1 < N_1 \\ 0 & \text{otherwise} \end{cases} \quad (27)$$

$$\Phi_i = \begin{cases} 1 & \text{if } n_{i-1} > 0 \text{ , } n_i < Z_i \text{ , } n_i < N_i \\ 0 & \text{otherwise} \end{cases} \qquad i = 2,\ldots,k-1 \quad (28)$$

$$\Phi_k = \begin{cases} 1 & \text{if } n_{k-1} > 0,\ n_k < Z_k \\ 0 & \text{otherwise} \end{cases} \quad (29)$$

6.3.1 Base Stock Policy

Liberopoulos and Dallery (1998) prove that the operation of the Control Point Policy is identical to the base stock policy. In the Base Stock Policy, a base stock level is given to each station. Then the system parameters are base stock level for station i V_i, p_i, and r_i i=1,2,..,k. Base Stock level V_i is the maximum number of parts allowed in all the downstream buffers of M_i. If no interstation buffer capacity is set for the buffers, we set N_i=V_i, i=1,2,..,k-1 initially and then reduce the state space according to the following restriction

$$(n_1,\ldots,n_{k-1},\alpha_1,\ldots,\alpha_k)\in S \text{ if } (n_1,\ldots,n_{k-1})$$

$$\in\left\{(n_1 \quad \cdots \quad n_{k-1}) \mid \sum_{i=1}^{k-1} n_i \le V_1, \sum_{i=2}^{k-1} n_i \le V_2, \cdots, n_{k-1} \le V_{k-1}\right\} \tag{30}$$

The operational rules for the base stock policy, are given below:

$$\Phi_1 = \begin{cases} 1 & \text{if } \sum_{j=1}^{k-1} n_j < Z_1 \\ 0 & \text{otherwise} \end{cases} \tag{31}$$

$$\Phi_i = \begin{cases} 1 & \text{if } n_{i-1} > 0\,,\ \sum_{j=i}^{k-1} n_j < Z_i \\ 0 & \text{otherwise} \end{cases} \qquad i = 2,\ldots,k-1 \tag{32}$$

$$\Phi_k = \begin{cases} 1 & \text{if } n_{k-1} > 0 \\ 0 & \text{otherwise} \end{cases} \tag{33}$$

Once these steps are completed, it is straightforward to impose limits on interstation buffer capacities by combining the above conditions with the ones given in equations (3-5).

In this section, only specific policies are considered. However, once a control policy is given, the operational rules can be written in terms of the system state. Then these rules can be used to reduce the state space and then to generate the transition matrix and the reward function. The performance evaluation block that used the transition matrix and the reward function as inputs is discussed next.

7. PERFORMANCE EVALUATION BLOCK

The structure of the performance evaluation block is depicted in Figure 10. The performance evaluation is classified according to the time period of interest: the steady state evaluation for the performance in the long run and the transient analysis for the performance of the system in a given time interval.

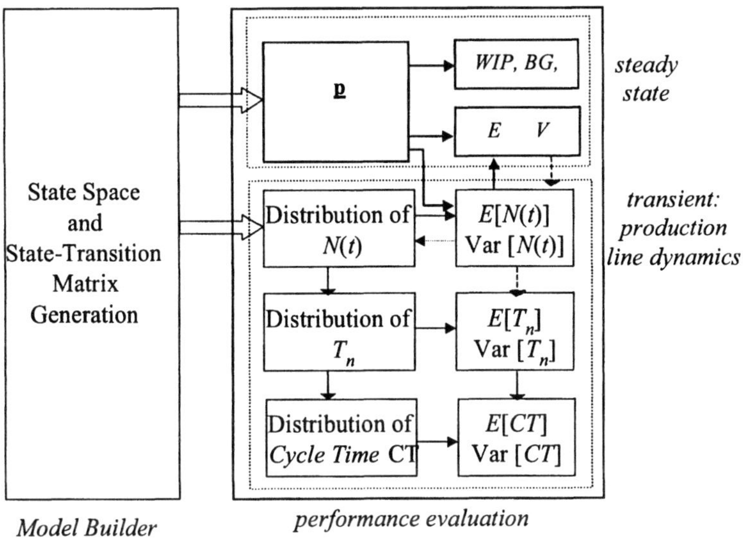

Figure 10. Structure of the performance evaluation block

7.1 Steady State Performance Measures

The steady state results have been used in performance evaluation extensively. The most commonly used performance measures are the production rate, average inventory and backlog levels and the service level. All the steady-state performance measures are derived from the steady state probabilities.

The steady state mean, i.e., the production rate, and the steady state variance rate of the number of parts produced during $[0, t)$, $N(t)$ are defined as,

$$E = \lim_{t \to \infty} \frac{E[N(t)]}{t} \qquad V = \lim_{t \to \infty} \frac{Var[N(t)]}{t} \tag{34}$$

respectively. Tan (1999) presents an efficient algorithm to determine E and V directly from the state space model. Furthermore, it is shown that the distribution of the $N(t)$ is asymptotically normal.

There are various efficient methods to determine the steady state probabilities when a state transition matrix is given. For the pertinent literature and the methods, the reader is referred to (Stewart, 1994). Therefore, we now focus on the performance of the system in a given time interval.

7.2 Transient Performance Measures: Production Line Dynamics

The steady-state analysis is more of a static nature while the transient analysis captures the dynamics of the behaviour. Therefore, the study of transient behaviour of the output variability in production systems is referred as *production system dynamics* (Tan, 1999). The study of production system dynamics supports operational decision making in a shorter planning horizon and supplements the steady-state analysis used in the design phase. Therefore combining both the transient and steady state analysis provides a more powerful decision support tool.

In the transient analysis part of the performance evaluation block, the exact distribution of $N(t)$, the exact distribution of the time to produce n parts, T_n, and the distribution of cycle time, CT, are determined. After determining these distributions, a number of performance measures such as the mean and the variance of these random variables, e.g., the mean and variance of $N(t)$, $E[N(t)]$ and $Var[N(t)]$, and the probabilities of certain events such as the probability of meeting the due date, etc. can be derived exactly from these distributions.

By using the asymptotic normality of $N(t)$, one can use E and V to approximate both $E[N(t)]$ and $Var[N(t)]$ as $E \cdot t$ and $V \cdot t$ and also to approximate the distribution with a normal distribution with mean $E \cdot t$ and variance $V \cdot t$. Note that this approximation will be accurate as t approaches infinity.

In a recent study, Tan (2000) presents a state-space based algorithm to determine $E[N(t)]$ and $Var[N(t)]$ directly without determining the exact distribution of $N(t)$. By using the asymptotic normality of $N(t)$, the distribution of $N(t)$ can be approximated with a normal distribution with mean $E[N(t)]$ and variance $Var[N(t)]$. Note that both of these methods yield approximate distributions for $N(t)$ and T_n.

The relationship between direct calculation and then obtaining approximations is depicted in Figure 10 by showing the exact links with a solid line and approximations with a broken line. In this study, we derive the exact distributions for $N(t)$ and T_n which are not reported before.

There is a tradeoff between the accuracy of the results and the computational complexity. If the performance of the system in a short time period is desired, then the exact distributions should be obtained. If the performance of the system in the long run is of interest then, E and V should be obtained first. Similarly, if the performance of the systems in a medium range is to be analysed, then mean and variance at time t should be obtained first. Then the other performance measures of interest can be obtained from these results.

7.2.1 Distribution of the Number of Parts Produced in [0, *T*)

The approach we use to determine the distribution of the number of parts produced during a fixed time interval is similar to the one given in (Gershwin, 1993). Gershwin (1993) derives and solves a set of difference equations analytically for a model of an unreliable machine. We extend this approach to a general model of a production system and solve the resulting set of difference equations iteratively.

Let us define $\pi(n, i, t)$ as the probability that the state of the system is i at time t and n parts have been produced during $[0, t)$:

$$\pi(n, i, t) = P[\, X(t) = i, N(t) = n] \quad i \in S, \quad n \in \{0, 1, 2, \ldots, t\} \tag{35}$$

Note that, since at most one part can be produced in one time unit, $N(t) \in \{0, 1, 2, \ldots., t\}$.

If the system is in a productive state at time t and n parts have been produced during $[0,t)$, i.e., $X(t) = i$ and $i \in \mathrm{U}$ and $N(t)=n$, then a part is produced at the end of that cycle and $N(t+1) = n+1$. That is, for $n \in \{0, 1, 2, \ldots, t\}$ and $j \in S$:

$$P[X(t+1) = j, N(t+1) = n+1 \mid X(t) = i, N(t) = n] = q_{i,j} \quad i \in U \tag{36}$$

$$P[X(t+1) = j, N(t+1) = n \mid X(t) = i, N(t) = n] = 0 \quad i \in U \tag{37}$$

Otherwise, if $X(t) = i$ and $i \in \overline{\mathrm{U}}$, then

$$P[X(t+1) = j, N(t+1) = n+1 \mid X(t) = i, N(t) = n] = 0 \quad i \notin U \tag{38}$$

$$P[X(t+1) = j, N(t+1) = n \mid X(t) = i, N(t) = n] = q_{i,j} \quad i \notin U \tag{39}$$

The above equations yield the following difference equations

$$\pi(0,j,t+1)=\sum_{i\notin U}\pi(0,i,t)q_{i,j} \quad j\in S \tag{40}$$

$$\pi(n,j,t+1)=\sum_{i\notin U}\pi(n,i,t)q_{i,j}+\sum_{k\in U}\pi(n-1,k,t)q_{k,j} \quad n=1,2,\ldots,t+1,\ j\in S \tag{41}$$

In order to rewrite the above equations in matrix notation let us define $\pi^n(t)$ which is a $1\times n(S)$ row vector whose ith element is $\pi(n, i, t)$:

$$\pi^n(t)=\{\pi(n,i,t)\} \quad i\in S \tag{42}$$

and the diagonal matrix $G=\{g_{i,j}\}$ where

$$g_{i,j}=\begin{cases}1 & \text{if } w_{i,k}=1 \text{ and } i=j\\ 0 & \text{otherwise}\end{cases} \tag{43}$$

Then equations (40) and (41) can be written as

$$\pi^0(t+1)=\pi^0(t)(I-G)Q \tag{44}$$

$$\pi^n(t+1)=\pi^n(t)(I-G)Q+\pi^{n-1}(t)GQ \quad n=1,2,\ldots t+1 \tag{45}$$

where I is an $n(S)\times n(S)$ identity matrix.

Let the initial state of the production system at t=0 be given as x_0, i.e., $X(0)=x_0$. Then

$$\pi(n,i,0)=\begin{cases}1 & \text{if } n=0 \text{ and } i=x_0\\ 0 & \text{otherwise}\end{cases} \tag{46}$$

Alternatively, the production system can be assumed to be in steady state at t=0. In this case, the steady-state probabilities must be determined initially.

Now, if the distribution of $N(\tau)$ is of interest, the difference equations (44) and (45) can be solved iteratively starting at the initial condition $\pi^0(0)$ until $\pi^n(\tau)$ is obtained. Finally,

$$P[N(\tau)=n]=\sum_{i\in S}\pi(n,i,\tau)=\pi^n(\tau)\underline{u} \tag{47}$$

where $\underline{u}$ is a column vector of ones with an appropriate dimension.

From the distribution function of $N(\tau)$, all the necessary performance measures can be determined. The most important performance measures are the mean and variance of the number of parts produced in $[0,\tau)$, $E[N(\tau)]$ and $Var[N(\tau)]$, are

$$E[N(\tau)] = \sum_{n=0}^{\tau-1} nP[N(\tau) = n] \tag{48}$$

$$Var[N(\tau)] = \sum_{n=0}^{\tau-1} n^2 P[N(\tau) = n] - E^2[N(\tau)] \tag{49}$$

Similarly, higher moments of $N(\tau)$ can also be determined if they are of interest. In addition to the moments of $N(t)$, covariance and correlation coefficient of $N(t)$ and $N(t\text{-}1)$ can also be obtained from $\pi(n,i,t)$.

Figure 11 shows the exact distribution of $N(T)$ for a four station production line controlled by a kanban-type policy for T=10, 50, 75, and 100. In the system, initially all the stations are up and there is one part in each buffer and p_i=0.1, r_i=0.9, N_i=15, i=1, 2, 3, 4. In this specific model, there are 65536 states in the state space and the number of non-zero elements in the state-transition matrix is 954800. The effect of interstation buffer capacity on variability of $N(t)$ for a three station production line with p_i = 0.1, r_i=0.5, N_i=N, i=1,2,3 is shown in Figure 12. In the example, initially all the stations are up and there is one part in each buffer.

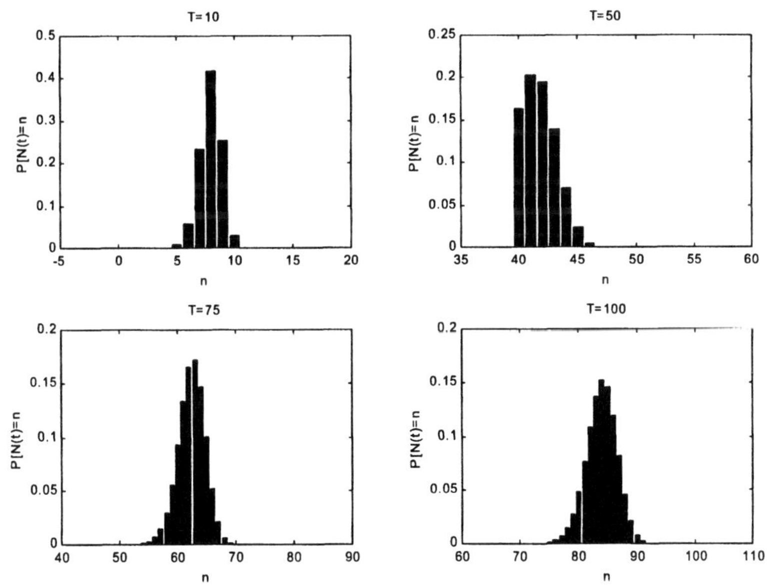

Figure 11. Distribution of $N(t)$ for a four-station production line

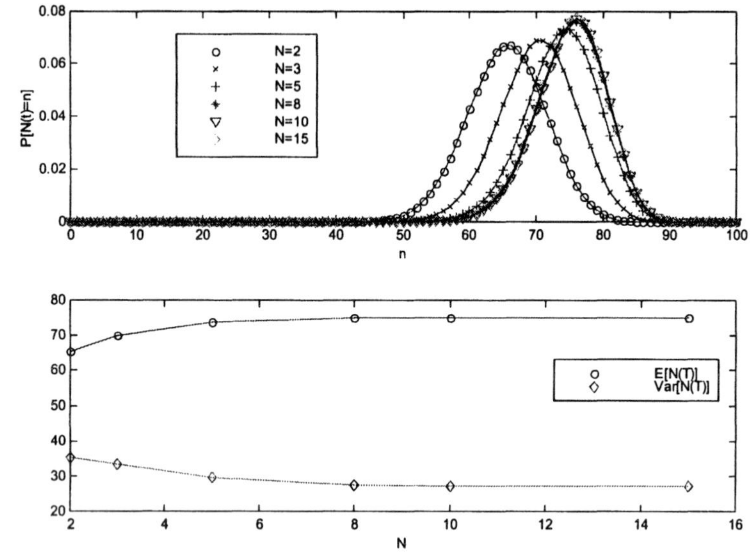

Figure 12. Effect of interstation buffer capacity on the variability of $N(T)$, T=100. Three station production line with kanban control.

7.2.2 Time to Produce a Given Order

In addition to the number of parts produced in a given time interval, the time to produce a given number of parts is also of interest. For example, determining the variability of the time to produce a number of parts allows one to set due-dates according to a desired level of service. It is possible to determine the due-time performance from the distribution of $N(t)$ directly by determining the probability of completing an order before its due-date. However, determining the distribution and variance of the time to produce n parts is not straightforward. In this section, we present a method that can be used to determine the distribution, mean, and variance of the time to produce a fixed lot.

The relationship between the processes $\{N(t), n=0, 1, 2,\ldots\}$ and $\{T_n, t\geq 0\}$ yield

$$P[N(t)\leq n-1]=P[T_n > t] \tag{50}$$

Furthermore $T_n = t$ if and only if $N(t) = n$ and $N(t\text{-}1) = n\text{-}1$. That is if the time to produce, say, 5 parts is 10 time units, then it means that 4 parts have been produced in [0,9) and the 10th part has been produced during [9, 10). Then

$$P[T_n = t] = P[N(t) = n \text{ and } N(t-1) = n-1] \tag{51}$$

Now, the above equation yields

$$P[T_n = t] = \begin{cases} 0 & \text{if } t < n \\ \pi^{n-1}(t-1)GQ\underline{u} & \text{if } t \geq n \end{cases} \tag{52}$$

In equation (52), since t takes infinite number of different values, the calculations can be terminated at $t=\tau$ where $P[T_n=\tau]<\varepsilon< P[T_n=\tau-1]$ where ε is the tolerance level. In other words, the upper tail of the distribution of T_n is truncated at $t=\tau$ and then the resulting distribution is normalized.

If only the mean and variance of T_n is to be determined, then we can directly work with equation (50). Equation (50) can be rewritten as

$$P[T_n \leq t] = \begin{cases} 0 & \text{if } t < n \\ \sum_{m=n}^{t} P[N(t) = m] & \text{if } t \geq n \end{cases} \tag{53}$$

Then the expectation and variance of the time to produce n parts, $E[T_n]$ and $Var[T_n]$ can be evaluated from the distribution of $N(t)$ as

$$E[T_n] = \sum_{t=0}^{\infty} P[T_n > t] = \sum_{t=0}^{\infty} P[N(t) \leq n-1] = \sum_{t=0}^{\infty} \sum_{m=0}^{n-1} P[N(t) = m] \tag{54}$$

Similarly,

$$Var[T_n] = E[T_n^2] - (E[T_n])^2 \tag{55}$$

where

$$E[T_n^2] = \sum_{t=0}^{\infty} 2tP[T_n > t] = \sum_{t=0}^{\infty} 2tP[N(t) \leq n-1] = \sum_{t=0}^{\infty} 2t \sum_{m=0}^{n-1} P[N(t) = m] \tag{56}$$

The above equations can be evaluated numerically after further simplifying them by noting that $P[N(t)=m] = 0$ if $t<m$, i.e., it is not possible to produce more than t parts in $[0,t)$, and terminating the summation at $t=\tau$ where $P[N(\tau)]<\varepsilon<P[N(\tau-1)]$.

Figure 13 gives the distribution of T_n for a three-station line with $p_i=0.1$, $r_i=0.9$, $N_i=5$, $i=1, 2, 3$. Initially all the stations are up and there is one part in each buffer. In this example, there are 288 states in the state space and the number of non-zero elements in the state-transition matrix is 1948. By using the distribution of T_n, the cycle time distribution conditioned on the initial state of the production system can also be determined easily.

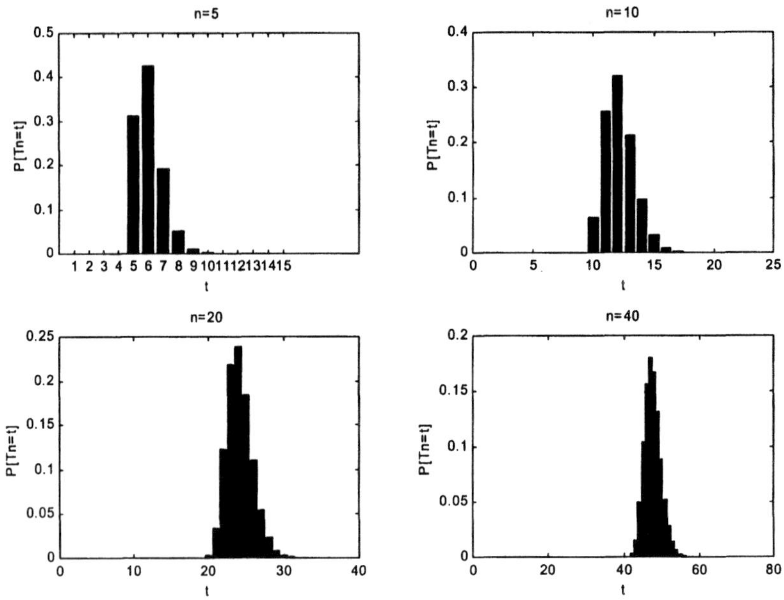

Figure 13. Distribution of the time to produce n parts, T_n, for n=5, 10, 20, 40 for a three station production line

Note that, if the WIP inventory is m at a given time, the time a new part will spend in the system, i.e., the cycle time, is equal to the time to produce m+1 parts T_{m+1}. Thus the distribution of the cycle time can be obtained from the distribution of the time to produce a given number of parts. Figure 14 below depicts the conditional cycle time distribution for a specific four-station production line with p_i=0.1, r_i=0.9, N_i=5, i=1, 2, 3, 4.

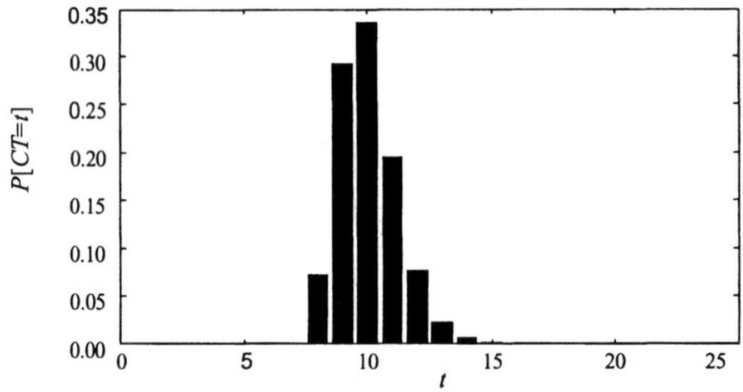

Figure 14. Distribution of the conditional cycle time for a four-station production line

8. NUMERICAL RESULTS AND EXTENSIONS

The objective of this section is to introduce the capabilities of the method through a number of representative cases. A thorough numerical study that focuses on performance evaluation of pull-controlled production systems will be presented in a further study. First, we discuss how this method can be used to develop efficient algorithms to determine the parameters of a given control policy. Then, a number of performance evaluation cases are given. Finally, we focus on buffer allocation and its effects on due-date performance of a production system.

8.1 Determination of the Control Policy Parameters

When design and performance evaluation are treated in the same framework, parameter determination problem shows some similarities to the buffer allocation problem in finite buffer production lines. Note that determining the parameters of a production line controlled by a kanban-type policy described in this study is similar to this problem.

Figure 15 shows iso-production rate lines, i.e. the lines that yield the same production rate, for a three-station production line controlled by a base stock policy with parameters V_1 and V_2.

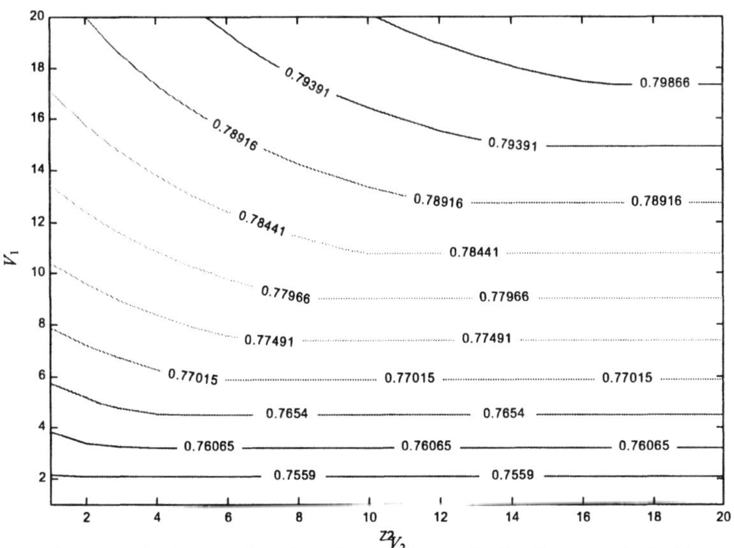

Figure 15. Iso-production rate lines for a three-station production line controlled with a Base Stock Policy.

A similar iso-production rate figure is observed in the buffer allocation problem in production lines (Schor,1995). This figure suggests that efficient algorithms based on gradient search and one-dimensional search similar to the one given in (Schor, 1995) can be used to determine the parameters of the Base Stock and CPP. Figure 16 depicts iso-service level lines for the same system. Therefore, determining the control policy parameters that maximize the service level can be done by using a similar algorithm.

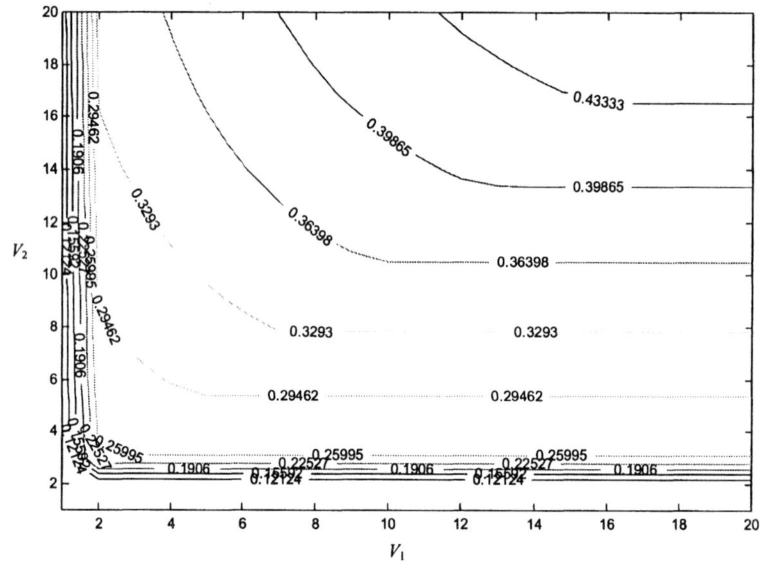

Figure 16. Iso-service level lines for a three-station production line controlled with a Base Stock Policy.

8.2 Performance Evaluation and Comparison of Control Policies

The tool described in this study can be used to determine the performance measures of interest for a production system controlled by a given policy. Once this can be done efficiently, the performance measures can be compared. A thorough numerical experiment is required to claim some general remarks about the control policies. Here, we simply present some representative cases.

The performance of kanban, CONWIP, and Base Stock policies are compared in Figures 17, 18, and 19 by considering a three-station line with backlogged demand where the last station represents the demand and p_i =

[0.01,0.01,0.007], r_i = [0.09,0.09,0.05], N_{BG}=5, i=1,2,3. For all three policies, the optimal parameters of the control policy that minimize the total WIP inventory subject to a desired service level are determined by total enumeration.

Figure 17 shows the results for the total WIP required to achieve a desired service level. The figure suggests that the kanban control minimizes the WIP required to achieve a desired service level. However, Figure 18 depicts the production rate obtained to achieve a desired service level while minimizing the total WIP. While kanban policy minimizes the total WIP, it also yields a low production rate. Finally Figure 19 shows the average backlog for the same case. These figures indicate that a thorough comparison of control policies require a careful look at the objective function used. Furthermore, a given policy may yield good results for a wide range of system parameters and various objective functions, while another policy is better than this one at only a specific set of parameters. In this case, the *robustness* of a control policy is a relevant property that needs to be investigated.

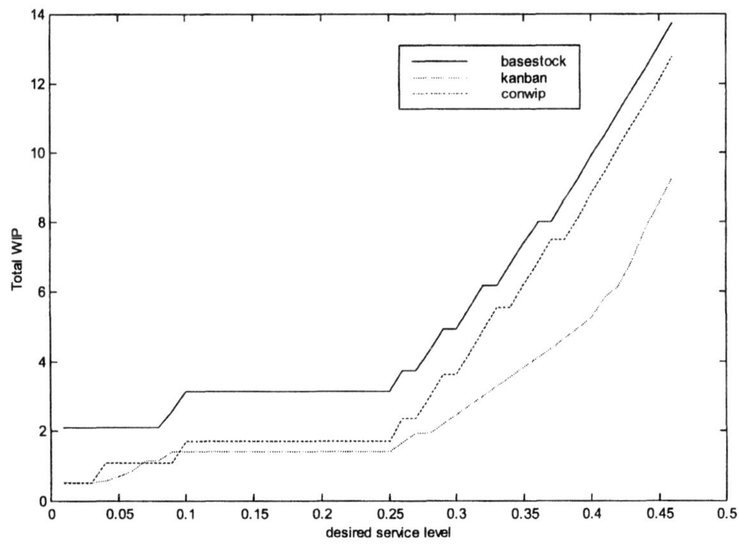

Figure 17. The minimum total WIP levels required to achieve a desired service level in a three station production line controlled with kanban, base stock, and CONWIP policies

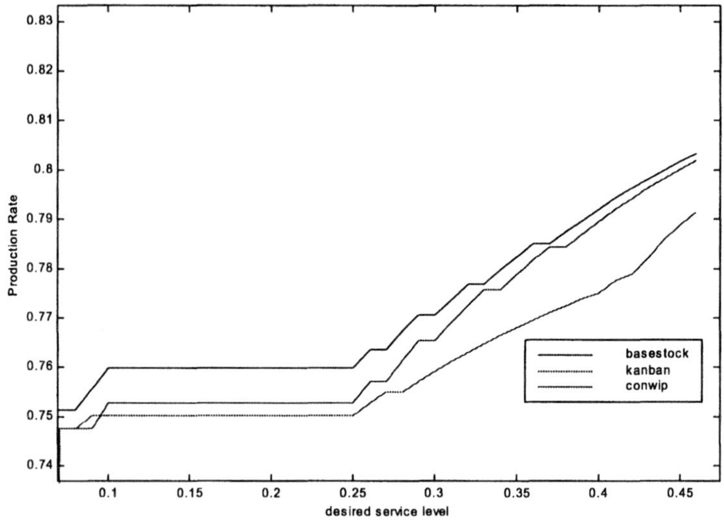

Figure 18. The production rates for the minimum total WIP levels required to achieve a desired service level in a three-station production line

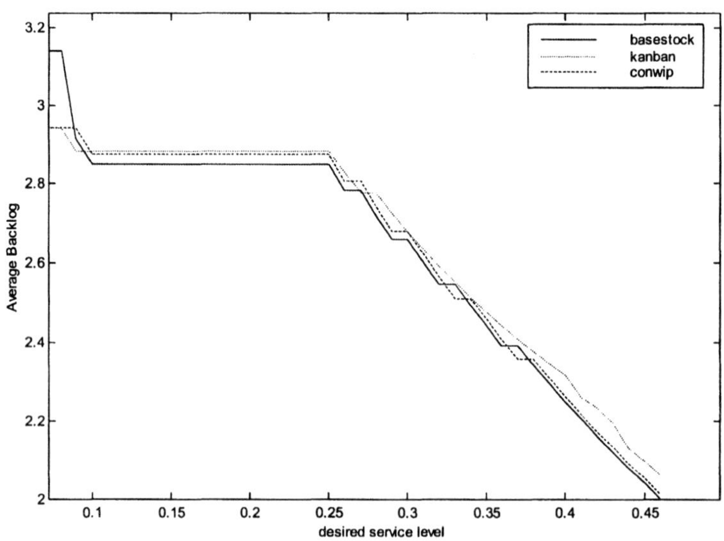

Figure 19. The average backlog for the minimum total WIP levels required to achieve a desired service level in a three-station production.

8.3 Buffer Allocation and Variability

We also investigate the effect of buffer allocation in a production line controlled by a kanban-type policy on the variability of $N(t)$ and the due-date performance. The probability of meeting a customer order on time is taken as a performance measure for the due-date performance. If the number of parts to be produced by the promised due-date is greater than or equal to the number of parts ordered, then the order is met on time. Therefore the probability of meeting a customer order on time, or the ratio of orders met on time, is $P[N(T)\geq Q]$ where T is the due date and Q is the number of parts ordered. Figure 20 shows that buffer allocation affects both the mean and also the variance of $N(T)$. Therefore, the best buffer allocation that maximizes the due-date performance is of interest.

Although, there are many studies on buffer allocation, these studies aim at either maximizing the production rate for a given buffer capacity or minimizing the total buffer capacity for a desired production rate. Figure 21 depicts the maximum $E[N(T)]$ and the best due-date performance that can be attained for a given total buffer capacity. The best buffer allocation for this case is found by total enumeration.

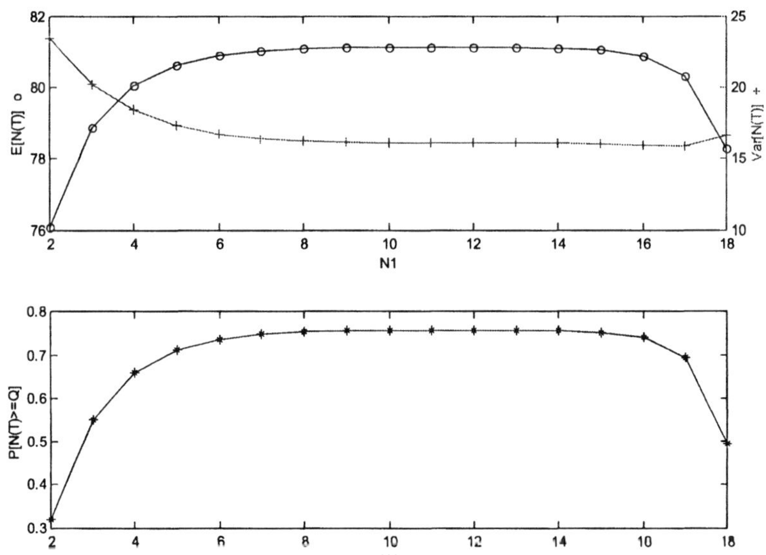

Figure 20. Effect of buffer allocation on the due-date performance of the line and variability of N(T), T=100. Three station production line:

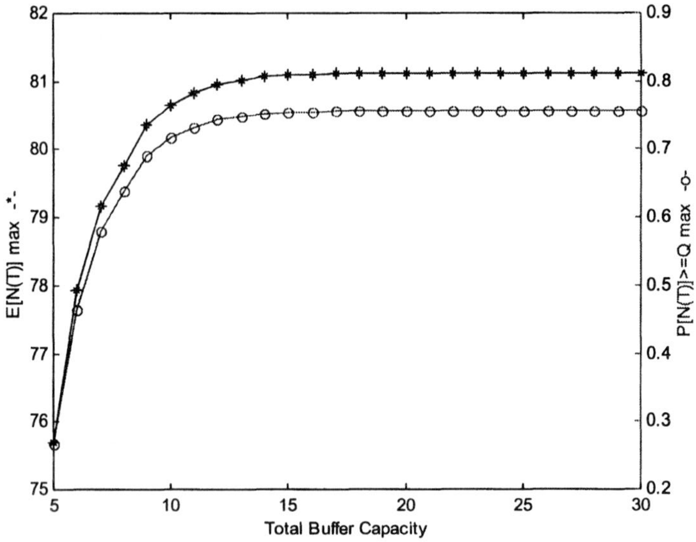

Figure 21. Effect of the total buffer allocation on $E[N(T)]$ and due-date performance. The best allocation that maximizes $E[N(T)]$ and $P[N(T) \geq Q]$ are reported. T=100, Q=80

9. CONCLUSIONS

In this study, a tool that is used to generate the state-space models of production systems controlled by a given policy is presented. The performance evaluation algorithms that work with the transition matrix and the reward function can be used to analyse a given system in the long run and also in a short period of time. The MATLAB codes to generate the state space models of pull-controlled production lines are available at http://home.ku.edu.tr/~btan/statespace.html.

A new result that yields the exact distribution of the number of parts produced during a given time period and also the distribution of the time to produce a given number of parts is also presented. These distributions can be used to answer questions such as *"How dependable our deliveries are,"* *"How variable our output is,"* *"How variable our cycle time is,"* *"What is the probability that the current orders will be shipped by their due-dates"*, *"What is the expected time for a new order to be completed?,"* etc. depending on the current status of the production system.

Similar to all state-space methods, computational requirements suffer from rapid increase of the states depending on the number of stations and buffer capacities. Therefore, the proposed method cannot be used effectively to evaluate the performance of, say, a 50-station production line. However, note that single buffer production lines have been analysed thoroughly to understand the operation of the production system and also to use this easy-to-solve subsystem to decompose more complex production systems. The method presented in this study extends the same approach to more generalized systems controlled by kanban, CONWIP, CPP, and hybrid policies. We can gain insights from the operation of more complicated production systems. Moreover, approximation methods that evaluate the performance of a manufacturing system by decomposing the system into tractable subsystems can be developed.

Furthermore, if a short time period is of interest, then only the part of a production system that currently contains the number of parts that can be produced during this time can be analysed in isolation. For example, consider a 50-station production line. If the performance of this 50-station production line during $[0, T)$, T=30 is of interest then the maximum number of parts that can be produced in this time period is 30. Let us assume that stations 47, 48, 49, 50 are up and Buffers 47, 48, and 49 currently contain 12, 6, and 14 parts respectively. Then analysing a subsystem of a four-station line which is consisted of Station 47, 48, 49, and 50 during [0, 30) yields the desired results for the whole system.

ACNOWLEDGEMENT

The author acknowledges the support of the TUBITAK-NATO Science Fellowship program. The author also thanks two anonymous referees for their comments and suggestions .

REFERENCES

Bonvik, A.M. (1996), Performance Analysis of Manufacturing Systems Under Hybrid Control Policies, Ph.D. Thesis, MIT.

Dallery, Y., and Gershwin, S.B. (1992), "Manufacturing Flow Line Systems: A Review of Models and Analytical Results," *Queueing Systems Theory and Applications*, Special Issue on Queueing Models of Manufacturing Systems, Vol. 12, No. 1-2, December, 1992, pp. 3-94.

Duenyas, I., Hopp, W.J., and Spearman, M.L.(1993), "Characterizing the Output Process of a CONWIP Line with Deterministic Processing and Random Outages," *Management Science*, Vol. 39, No. 8, pp. 975-988.

Frein, Y., Di Mascola, M., Dallery, Y. (1995), "On the Design of Generalized Kanban Control Systems," *International Journal of Operations&Production Management,* Vol. 15, No. 9, pp.158-184

Gershwin, S. (1993), "Variance of the Output of a Tandem Production System," *Queueing Networks with Finite Capacity.* Proceedings of the Second International Conference on Queueing Networks with Finite Capacity, Edited by R. Onvural and I. Akyildiz, Elsevier, Amsterdam.

Gershwin, S.B. (1994), Manufacturing Systems Engineering, Prentice Hall, Englewood Cliffs, NJ 07632.

Gershwin, S.B. (1997), "Design and Operation of Manufacturing Systems: A Control-Based Synthesis," *European Journal of Control*, Vol. 3, pp. 166-181

Gershwin, S.B. (2000), "Design and Operation of Manufacturing Systems: The Control Point Policy," *IIE transactions on Design and Manufacturing, forthcoming.*

Liberopoulos, G. and Dallery, Y. (1998), "A Unified Framework for Pull Control Mechanisms in Multi-Stage Manufacturing systems," unpublished manuscript

Miltenburg, G.J. (1987), "Variance of the Number of Units Produced on a Transfer Line with Buffer Inventories During a Period of Length T" *Naval Research Logistics*, Vol. 34, pp. 811-822.

Papadopoulos, H.T., and M.E.J. O'Kelly (1989), "A Recursive Algorithm for Generating the Transition Matrix of Multistation Series Production Lines," *Computers in Industry*, Vol. 12, pp.227-240

Papadopoulos, H.T., C. Heavey, and M.E.J. O'Kelly (1989), "Throughput Rate of Multistation Reliable Production Lines with Interstation Buffers (I) Exponential Case" *Computers in Industry*, Vol. 13, pp.229-244

Sahner, R, Trivedi, K.S., and Puliafito, A. (1996), Performance and Reliability Analysis of Computer systems: an Example-based Approach Using the SHARPE Software Package, Kluwer Academic Publishers, Massachusetts 02061.

Schick, I. C. and Gershwin, S. (1978), Modelling and Analysis of Unreliable Transfer Lines with Finite Interstage Buffers, Electronics Systems Laboratory Report, ESL-FR-834-6, Massachusetts Institute of Technology, Cambridge, Massachusetts 01239.

Schor, J.E. (1995), "Efficient Algorithms for Buffer Allocation," M.S. thesis, MIT, Cambridge MA.

Stewart, W.J. (1994), Introduction to the Numerical Solution of Markov Chains, Princeton University Press, New Jersey.

Tan, B. (1999), "Variance of the Output as a Function of Time: Production Line Dynamics," *European Journal of Operational Research*, Vol. 117, No.3, pp. 470-484.

Tan, B. (2000), "Asymptotic Variance Rate of the Output in Production Lines with Limited Buffers," *Annals of Operations Research*, Special Issue on Performance Evaluation and Optimization of Production Lines, Vol. 93, pp. 385-403.

Chapter 16

USING FLUID SOLUTIONS IN DYNAMIC SCHEDULING

Michael H. Veatch
Department of Mathematics
Gordon College
Wenham, MA 01984
veatch@gordon.edu

Abstract We review some recent work on fluid approximation of queueing network control problems and present new results. The relationship between the optimal control of a queueing network and the optimal control of the associated fluid model is investigated. A general theory, a number of examples, and a few numerical results are presented. The fluid model replaces stochastic counting processes by their mean rates. Surprisingly, although the fluid model is deterministic and transient, it often contains valuable information about the original problem. We show that, under some technical assumptions, the asymptotic slope (normal vector in higher dimensions) of the switching surfaces for large queue lengths can be found from the fluid model. The implications of the fluid policy for designing manufacturing scheduling and control policies are discussed. Unlike the queueing network control problem, which is usually an intractable dynamic program, very efficient algorithms exist to solve the fluid problem from a given initial buffer state. Fluid solutions are presented for a number of examples, some of them new.

Keywords: fluid models, dynamic scheduling, switching point controls, fluid limits

1. Introduction

Considerable attention has been paid recently to fluid models of stochastic network control problems. In the fluid model, stochastic counting processes are replaced by their mean rates. The fluid model provides a highly simplified framework for analysis. It has deterministic, continuous trajectories. For the models we consider–mostly open multiclass queueing networks with sequencing, arrival routing, and input control–the original problem has an average cost criterion but the fluid model is transient, minimizing the "cost to drain." In light of these differences, fluid models have often been viewed as being at a much higher level in the hierarchy of models. In this view, the fluid model is useful when a highly aggregated model is needed. If large time scales and queue lengths are of interest then fluid models are a natural choice because these are precisely the quantities that are scaled in fluid limit theorems. [Che94] proposes that these fluid models are suitable for long-range strategic planning, while queueing networks are appropriate for short-range, e.g., weekly, control of production operations.

This paper seeks to demonstrate another role for fluid models. Surprisingly, their solution often contains valuable information about the original problem, almost regardless of the "level of aggregation" at which decisions are to be made. Call the optimal policy for the original network the *discrete policy* and the optimal policy for the fluid model the *fluid policy.* We present an introduction to minimum cost draining of fluid models, review the existing theory relating the discrete and fluid policies, build on this theory to establish characteristics of the fluid policy which are shared by the discrete policy, and provide several examples to illustrate these connections. Two of the examples compare the fluid policy with the policy obtained from a diffusion approximation. Most of the fluid solutions are taken from earlier papers. In presenting the case for using fluid models, much of the work on fluid optimality is reviewed; however, no attempt is made to systematically survey the field and many papers are not mentioned. For a survey of fluid models and limit theorems, mostly without optimization, see [Che94]. A compilation of diffusion solutions is given by [Kel93]; [Wil96] gives a broader review of the Brownian control literature, emphasizing limit theorems.

Fluid models under a given policy are, under very general conditions, the weak limit of a sequence of systems where time and queue length are scaled by the same large factor; see [Dai95]. Thus, queue length vectors in the fluid model correspond to large vectors with the same proportions in the discrete system. Intuitively, when queue lengths are large, due to an unusual event such as a burst of arrivals, the optimal policies should be similar because of this scaling argument. We give some general results on the similarity between the discrete and fluid policy and explain the implications for designing manufacturing scheduling and control policies through a number of examples. For an important class of problems, the fluid model is (i) tractable (when the discrete system is not), (ii) insightful, in that the fluid policy provides understanding of the discrete policy, and (iii) useful, in that the fluid policy–possibly after translation or correction back to the discrete system–is stable, near optimal, and relatively simple to implement.

First, an efficient algorithm exists for a very general class of fluid models with linear costs and dynamics that are linear in the control. The algorithm of [Luo98] uses a separated continuous linear programming formulation, time discretization, and the fact that the problem admits piece-wise constant optimal controls to find them exactly. There are no restrictions on the network topology–reentrant lines and multiclass sequencing problems are included. In fact, the algorithm applies to all our examples. This algorithm solves for the (deterministic) trajectory from a given initial state. For smaller or specially structured problems, the control *policy* can be found; more examples are given in [ABR95], [Weis95], and [Set99]. Both of these methods are used in this paper. It should be noted that the fluid model is more tractable than the Brownian diffusion model, where solutions can usually be found only if a state space collapse condition is met or the problem admits pathwise optimal solutions. Flow control models, which have fluid dynamics in time intervals punctuated by Markovian changes in the dynamics (e.g., machine failures), are also not as tractable as fluid models.

The insights drawn from fluid models begin by dividing them into two types. Greedy solutions are discussed in Section 3. If costs are linear, these solutions use static priorities when all buffers are nonempty. In this case the discrete and fluid policy use the same priorities, differing only on the boundary of the state space, if at all. A single station with feedback is used to illustrate these connections. In this case the fluid model retains most or all the information needed to solve the discrete problem. Make-to-stock queues are also considered. They demonstrate a general limitation of fluid policies: They do not use safety stock.

Solutions containing switching curves are discussed in Section 4. For systems with linear costs, the fluid policy has linear switching curves (surfaces if more than two dimensions) through the origin. Meyn [Mey97], [Mey98], [Mey01a] has established that optimality of the discrete policy is preserved in the fluid limit. We use his result to show that, under some technical assumptions, the discrete switching curves (surfaces) have the same asymptotic slope (normal vector) as in the fluid. Thus, when the fluid policy switches in the interior it captures the main feature of the discrete policy. However, if the fluid policy switches on the boundary of the state space (e.g., when a buffer empties), the discrete policy will generally switch in the interior and the fluid fails to capture the essential characteristic of the discrete policy. An arrival routing example is used to numerically verify this relationship and demonstrate that the fluid policy can be near-optimal for the discrete system. Fluid solutions are also given for a series queue [ABR95] and a series make-to-stock system. Finally, fluid trajectories are generated for the input control problem of [Wein90] and compared with his diffusion solution.

For a fluid policy to be useful, it must be translated into a discrete policy that is stable, near optimal, and reasonably easy to implement. Usually, adjustments must be made at the boundary of the state space, where the fluid policy moves material through empty buffers. Unfortunately, naive translations, such as serving the next lower priority class when a buffer is empty, might not even be stable; an example is given in Section 2. [Mag00] constructs discrete policies by tracking the fluid policy but enforcing safety stocks. These policies are stable and asymptotically optimal under fluid scaling. However, this translation introduces some artificial policy features, such as discrete review, that may not be desirable in the actual system. The nonidling property is also violated. [Bau00] shows that implementing the fluid policy in an open-loop fashion, based only on the initial state, is asymptotically optimal with discounting. This result illustrates that asymptotic optimality is a very mild criterion. It has the same fundamental limitations as the fluid model. It appears that translating a policy in a near-optimal fashion is largely an open question. See [Eng96] for work on one example and [Mey01a] for some general proposals. However, some of our examples avoid this difficulty. For the arrival routing example of Section 4.2, the fluid policy can be directly applied to the discrete system. For the examples in Section 3, a naive translation is either optimal or has little impact.

The next section introduces the discrete stochastic network control problem and its fluid model. After discussing the greedy and switching curve policies, conclusions are given in Section 5.

2. The Discrete and Fluid Models

We present the discrete stochastic network in the Markov decision process context. However, more general processes also have fluid limits; see [Dai95]. Consider a network containing m single-server stations indexed by $j = 1, \ldots, m$ and n classes of jobs indexed by $i = 1, \ldots, n$. Although the underlying model is continuous time, we will consider the uniformized discrete-time process. The state is $\Phi_k \in Z_+^n$, where $[\Phi_k]_i$ denotes the number of class i jobs in the system after k transitions. All transitions–arrivals and service completions–are potentially controlled. Let $\mathcal{P}$ be the routing matrix, so that after service completion at class i, a job is routed to class j with probability p_{ij}. The transition probabilities are

$$P(\Phi_{k+1} = x + e_i \mid \Phi_k = x) = \lambda_i \widetilde{v}_i(k)/\Lambda \tag{1}$$

$$P(\Phi_{k+1} = x + e_j - e_i \mid \Phi_k = x) = \mu_i p_{ij} \widetilde{u}_i(k)/\Lambda, \tag{2}$$

where e_i is the ith basis vector and we use class 0 and the convention $e_0 = 0$ to represent departures. The exogenous arrival rate of class i customers is λ_i with control $\widetilde{v}_i(k)$. The service rate for class i customers is μ_i with control $\widetilde{u}_i(k)$. The tilda denotes the discrete model. Let $\widetilde{u}(k)$ and $\widetilde{v}(k)$ be vectors defined in the natural way. The appropriate policy will be a stationary, state-feedback control, so that Φ is a Markov chain. Let $\mathcal{P}$ be its transition probability matrix, defined by (1), (2), and $\mathcal{P}_{xx} = 1 - \sum_{y \neq x} \mathcal{P}_{xy}$. The uniformization constant

$\Lambda = \sum_{i=1}^{n}(\lambda_i + \mu_i)$ makes $\mathcal{P}_{xx} \geq 0$. Class i is served by station $s(i)$. The standard constraints on the controls in state $\Phi_k = x$ are

$$\sum_{i:s(i)=j} \widetilde{u}_i(k) \leq 1 \quad \text{for } j = 1, \ldots, m \tag{3}$$

$$\widetilde{u}_i(k) = 0 \text{ if } x_i = 0, \quad \widetilde{u}_i(k) \geq 0, \quad 0 \leq \widetilde{v}_i(k) \leq 1.$$

Here $\widetilde{u}_i(k)$ is the fraction of time server $s(i)$ allocates to class i. A server's allocations cannot exceed one. Other linear constraints can be added to reflect uncontrolled arrivals, arrival routing, and uncontrolled servers. (Without additional constraints, the optimal policy will have no arrivals.) The general form of constraints we will consider is

$$\begin{aligned} D\widetilde{u}(k) &\leq (\geq)\ e \\ H\widetilde{v}(k) &\leq (\geq)\ e \\ \widetilde{u}_i(k) &= 0 \text{ if } x_i = 0 \\ \widetilde{u}(k) &\geq 0, \quad \widetilde{v}(k) \geq 0. \end{aligned} \tag{4}$$

In a slight abuse of notation, let u denote the state-feedback control $(\widetilde{u}(x), \widetilde{v}(x))$. There is a linear holding cost $c'x$ in state x. The objective is to minimize average cost

$$J(x, u) = \limsup_{T\to\infty} \frac{1}{T} E_x \sum_{k=0}^{T-1} c'\Phi_k, \tag{5}$$

where E_x denotes expectation given $\Phi_0 = x$. Assuming a stable policy exists, there will typically be a solution to the average cost optimality equation

$$\eta_*/\Lambda + h(x) = \min_u c'x/\Lambda + \mathcal{P}_u h(x), \tag{6}$$

where the minimum is taken over state-feedback policies satisfying (4). In (6) the dependence of $\mathcal{P}$ on the policy has been made explicit and $\mathcal{P}h$ is function composition, i.e., $\mathcal{P}h(x) = \sum_{y\in Z_+^n} \mathcal{P}_{xy}h(y)$. The solution gives the optimal average cost, $\eta_* = \min_u J(x, u)$ for all x, and the relative value function $h(x)$. An extreme point control will achieve the minimum in (6), so the action space can be considered finite.

A fluid model is obtained when all transitions are replaced by their mean rates and a continuous state is used; $x_i(t)$ is the length of the class i queue at time t, with $x(t) \in R_+^n$. The relevant stability criterion for most fluid models is that they drain from any initial state; i.e., for each initial state we can choose a time horizon T such that $x(t) = 0$ for all $t \geq T$ for some policy. Stable models that do not drain still move to a "best state" in finite time. The fluid control

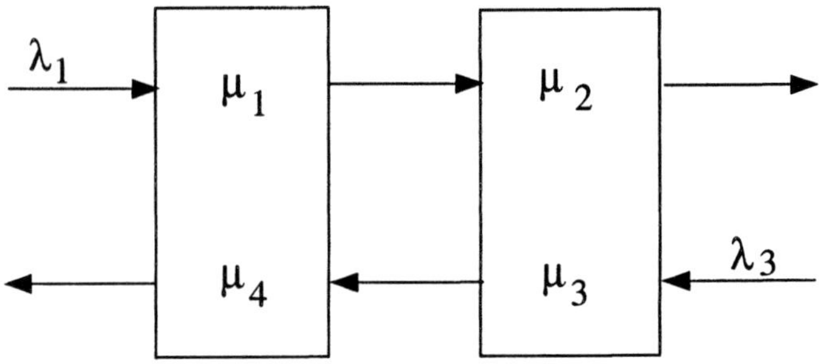

Figure 1. The Rybko-Stolyar network.

problem corresponding to (4) and (6), but without arrival control, is

$$\text{(FNET)} \quad \min \int_0^T c' x(t) dt$$
$$\dot{x}(t) = Bu(t) + b$$
$$Du(t) \le (\ge)\ e$$
$$x(0) = x$$
$$x(t) \ge 0, \quad u(t) \ge 0,$$

where $b = (\lambda_1, \ldots, \lambda_n)^T$, $B = (P^T - I)M$, and $M = \text{diag}(\mu_1, \ldots, \mu_n)$. The function $x(t)$ is continuous. The optimal $u(t)$ is piece-wise constant, making $x(t)$ piece-wise linear. Let $J^*(x)$ denote the optimal fluid cost from initial state x.

The feasible u in state $x > 0$ in Z_+^n are the same as the feasible $\widetilde{u}$. If $x_i(t) = 0$, however, (FNET) requires

$$(Bu(t) + b)_i \ge 0. \tag{7}$$

If (7) does not imply $u_i(t) = 0$, then the fluid constraints are weaker than the discrete constraints (4) and we say there is a *translation problem.* We will use only simple translations that enforce $u_i(t) = 0$. However, there are many examples of non-acyclic networks where these translations are unstable. The network of [Ryb92] shown in Figure 1 is a simple example.

Consider the data $\lambda_1 = \lambda_3 = 1, \mu_1 = \mu_3 = 6,\ \mu_2 = \mu_4 = 1.5$, and $c = 1$. The fluid policy is last-buffer-first-served (LBFS), i.e., classes 2 and 4 have priority, except that the server is split to avoid starvation whenever the exiting class at the other server is empty. Any translation that only makes adjustments on the boundary is inadequate for the discrete system; see [Mag00]. The difficulty is that the servers wait too long to switch from the exiting to the entering class, causing undesirable idleness. The translation that waits the longest to switch, pure LBFS, is actually unstable.

Two variations of (FNET) will occur in the examples. First, arrival or input control can be added by replacing b with $bV(t)$, where $V(t) = \text{diag}(v_1(t), \ldots, v_n(t))$ and placing appropriate constraints on $V(t)$. However, an equivalent formulation is to simply redefine b and D. Second, make-to-stock systems are modeled by removing the nonnegativity constraints on certain classes. Other extensions, such as controlled routing, can also be accommodated.

3. Greedy Solutions

For some well-known examples the greedy policy is discrete optimal and for others it provides an attractive approximate policy. After some general observations about the simplicity of greedy policies, we give an example in Section 3.1 where it is discrete optimal and in 3.2 where it is only fluid optimal. Section 4.1 gives an example where the greedy policy is fluid optimal in sufficiently heavy traffic. Interestingly, there seems to be a general connection between heavy traffic and effectiveness of the greedy policy; it is explored in [Mey01b].

Similar greedy solutions can be found using either the fluid or the discrete system. For the fluid, the greedy control in state x minimizes the time derivative of the cost rate. It is a solution to the LP

$$\begin{aligned} \text{(LPNET)} \quad & \min c'Bu \\ & \dot{x} = Bu + b \\ & Du \le e \\ & x_i \dot{x}_i \ge 0, \quad i = 1, \ldots, n \\ & u \ge 0. \end{aligned}$$

Chen and Yao (1993) give a procedure to generate a trajectory $x(t)$ for the greedy policy: Solve (LPNET) with $x = x(0)$. Apply this u until some $x_i(t)$ hits 0, then resolve. If there is no translation problem, greedy policies for the fluid problem translate directly into greedy policies for the discrete problem:

$$\widetilde{u}(x) = u(x) \text{ for all } x \in Z_+^n. \tag{8}$$

To see this, replace the relative cost function in (6) with the cost rate $c'x$. Setting $v_i = 1$ to eliminate arrival control, the function to be minimized is

$$\begin{aligned} \mathcal{P}_u c' x &= c' x + \sum_{i=1}^{n} c_i \lambda_i / \Lambda + \sum_{i=1}^{n} \sum_{j=1}^{n} (c_j - c_i) \mu_i p_{ij} u_i / \Lambda \\ &= c'(x + \dot{x}/\Lambda). \end{aligned} \tag{9}$$

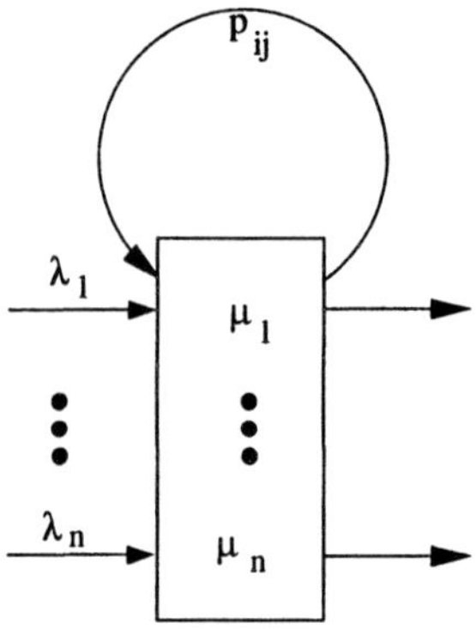

Figure 2. Klimov's problem.

Minimizing (9) subject to (4) is equivalent to (LPNET). If there is a translation problem, then the greedy control for the discrete problem will generally differ on the boundary.

Because of the linear objective function, the fluid policy is greedy in the interior of the state space ($x > 0$) exactly when it is static priority. Clearly, a greedy policy can always be chosen to be static priority because the constraints in (LPNET) depend only on which $x_i = 0$. For $x > 0$, minimizing (9) is accomplished by serving at each station the class with minimal index

$$f_i = \sum_{j=1}^{n} (c_j - c_i)\mu_i p_{ij}, \tag{10}$$

where we use class 0 and the convention $c_0 = 0$ to represent departures. If all indices are positive, a station idles. The indices (10) are static and we say the greedy policy has static priorities on the interior of the state space. [ABR95] shows that the fluid policy in general has dynamic indices. For these indices to be static, it can be shown that they must agree with (10). (Here we use the fact that the state space is unbounded.) Thus, if the fluid policy is static priority on $x > 0$, then it is also greedy on $x > 0$.

3.1 Single Station with Feedback

When there is only one station, the discrete control problem is tractable. Still, it is interesting to note that the fluid and discrete policies are essentially identical. Although there is a translation problem, i.e., they differ on the boundary of the state space, the naive translation gives the correct discrete policy. For a single station without feedback, the familiar $c\mu$ rule is optimal for both. A single station with feedback is known as Klimov's problem (Figure 2).

A static priority, greedy policy is optimal for both. [Che93] gives a simple algorithm for computing the priorities based on (LPNET) and uses duality to show that this greedy solution is also optimal for the fluid model. [Luo95] gives a dual algorithm to solve (FNET) directly, obtaining a static priority policy. He proves that the fluid priorities are consistent with the discrete priorities by showing that his dual algorithm also solves the discrete problem. Here we illustrate the simplicity of finding the priorities.

For $x > 0$, (LPNET) can be written

$$\begin{aligned} &\min\ c'Bu \\ &\sum_{i=1}^{n} u_i \le 1 \\ &u \ge 0, \end{aligned}$$

the solution of which is to serve the class with smallest

$$[c'B]_i = \left(\sum_{j=1}^{n} p_{ij}c_j - c_i\right)\mu_i.$$

If $x_i = 0$ for $i \in R$ and $x_i > 0$ for $i \in N$, and all classes in R have higher priorities than all classes in N, then it can be shown that the solution to (LPNET) has $\dot{x}_i = 0$ for $i \in R$, i.e., buffers do not refill once they are emptied. Thus, (LPNET) only needs to be solved for R containing the k highest priority classes, $k = 0, \ldots, n-1$. Using this fact, (LPNET) can be written

$$\begin{aligned} &\min\ c'Bu \\ &[Bu + b]_R = 0 \\ &\sum_{i=1}^{n} u_i \le 1 \\ &u \ge 0. \end{aligned} \tag{11}$$

For $i \in N$, let $\overline{u}_i$ be the maximum feasible value of u_i (the rest of the server's time is allocated to R to keep $x_R = 0$). [Luo95] proves that $\overline{u}_i$ is well-defined. Then the solution to (11) allocates $\overline{u}_i$ to class $i \in N$ with minimal

$$\left(\sum_{j=1}^{n} p_{ij}c_j - c_i\right)\mu_i\overline{u}_i. \tag{12}$$

The priorities can be found by repeatedly applying (12), adding the next priority class to R each time. The references above give explicit formulas for (12).

3.2 Make-to-Stock Queues: Fluid and Diffusion Models

Now consider a single-station production system that operates in make-to-stock fashion. Class i items are produced from an unlimited supply of raw material and placed in a finished goods inventory, x_i. This inventory is depleted by demand at rate λ_i. Unmet demand is backordered and recorded as negative inventory. The fluid model is

$$\begin{aligned} \min \int_0^T \sum_{i=1}^n \left(c_i^+ x_i^+(t) + c_i^- x_i^-(t)\right) dt \\ \dot{x}_i(t) = \mu_i u_i(t) - \lambda_i \\ \sum_{i=1}^n u_i(t) \le 1 \\ x_i(t) = x_i^+(t) - x_i^-(t) \\ x(0) = x \\ x^+(t) \ge 0, \quad x^-(t) \ge 0, \quad u(t) \ge 0. \end{aligned}$$

Decomposing x into x^+ (inventory) and x^- (backorders) allows holding and backorder cost to be written as a linear function of the state variables.

This example illustrates a fundamental limitation of fluid models: They do not use safety stock. The discrete policy, described in [deV00], consists of switching surfaces and a hedging point at which the server idles. When $x(t) \le 0$ a $c\mu$ rule is used. In contrast, the fluid policy from initial $x \le 0$ is the same as for the single-station queue, namely, the $c\mu$ rule. It operates the system without safety stock, ignoring the make-to-stock capability and the data c^+. Generally, not holding finished goods inventory in virtual buffers is suboptimal. In contrast, the fluid policy's lack of safety stock in real buffers can be unstable, as discussed at the beginning of Section 3.

Diffusion models, with their recognition of randomness, do considerably better than fluid models for this problem. They set reasonable hedging points and recover the $c\mu$ rule; see [Wein92] and [Vea96]. In the single-class case, the diffusion model gives a hedging point (and, hence, a policy) which is asymptotically optimal as $\rho \to 1$. The fluid policy is only optimal when ρ is sufficiently small and holding costs are large. Similar conditions for optimality of the fluid policy hold for two classes; see [Vea01b].

4. Fluid Limits and Asymptotic Switching Curves

The solution to many stochastic network control problems has a threshold form–below some switching surface (switching curve if $n = 2$), one extreme point is used and above it another is used. Based on scaling arguments and numerical results, it has been conjectured that the fluid policy gives the correct asymptotic orientation of these surfaces. In other words, the fluid tells us what to do when queues are large. This section presents a general theory, culminating in a corollary that establishes this connection between the fluid and discrete switching surfaces. Several examples are given in the following subsections. To simplify the exposition, consider the sequencing problem (4) without arrival control and when the extreme points have $\widetilde{u}_i = 0$ or 1. In this section we refer to $\Phi(t)$ and $\widetilde{u}(t)$ in continuous time. They can be constructed from the discrete-time versions as piece-wise constant, right-continuous left-limit functions or can be thought of as the continuous-time MDP before uniformization.

The most common justification for using fluid models is the existence of fluid limits when time and queue length are scaled by the same factor. The following result, called a functional strong law of large numbers, is proven in more general form in [Dai95]. Consider a sequence of discrete systems indexed by N such that $\Phi^N(0) = \lfloor Nx(0) \rfloor$, $x(0) \in R^n_+$. All use the same policy. Introduce

$$\widetilde{T}_i(t) = \int_0^t \widetilde{u}_i(s)ds$$

to represent the cumulative time that server i has devoted to class i customers by time t.

Theorem 1 (Dai) *For almost all sample paths there is a subsequence $\{N_r\}$ such that, as $N_r \to \infty$,*

$$\begin{aligned} &\frac{1}{N_r}(\Phi^{N_r}(N_r t), \widetilde{T}^{N_r}(N_r t)) \to (x(t), T(t)), \\ \textit{where} \quad & x(t) = x(0) + BT(t) + bt \\ & T(t) = \int_0^t u(s)ds \qquad (13) \\ & Du(t) \le e \\ & x(t) \ge 0, \quad u(t) \ge 0. \end{aligned}$$

The limit exists uniformly on compact sets. The processes $(x(t), T(t))$ are called the *fluid limit.* Note that each initial state and policy has at least one fluid limit and that they satisfy the same constraints as the fluid model. Thus, the fluid model includes all fluid limits and possibly more trajectories.

Theorem 1 does not guarantee a *unique* fluid limit for a given initial state and policy. Different sample paths might have different fluid limits; even for one sample path, different subsequences might have different fluid limits. Because the theory of unique fluid limits is incomplete, we will simply *assume* that the policies we consider have unique fluid limits from all initial states, except possibly a set of lower dimension. We conjecture that this is not a restrictive assumption. The fact that these fluid limits satisfy the fluid model constraints (13) and that they are all generated by the same discrete policy severely limits how nonuniqueness can arise. A trivial nonuniqueness occurs if the discrete policy is not unique and ties are not broken in a consistent fashion. This difficulty can be avoided by appropriate selection from among the optimal policies. The only examples of reasonable state-feedback policies giving non-unique fluid limits that we are aware of are policies such as "shortest queue first" that make the fluid trajectory map discontinuous on switching surfaces. We also note that [Mag00] makes a similar assumption, namely, that the trajectory map is continuous. However, some skepticism about our assumption may be warranted in light of the nonunique fluid limits and other unexpected behavior encountered in unstable systems.

Theorem 1 also does not give a method for computing fluid limits. Again the theory is incomplete–in fact, constructing the fluid limit is known to be complicated in some cases–but we will proceed in a pragmatic manner. At all differentiable points of $T(\cdot)$, (13) defines a fluid control $u(t)$ in state $x(t)$ and $\dot{x}(t)$ is the velocity in that state. Under mild conditions (see the semigroup property of [Mag00]), the collection of fluid limits $x(\cdot)$ from all initial states, called the trajectory map, defines a fluid policy $u(x)$ (except possibly on a set of lower dimension). Call this policy the *fluid limit policy* of $\tilde{u}$.

In order to establish a connection between the discrete and fluid policies, we make the following additional assumption about the discrete policy, which we call a *scalable policy*. Define the *scaled policy* associated with $\tilde{u}$ as $\overline{u}(x) = \lim_{N\to\infty} \tilde{u}(\lfloor Nx_1 \rfloor, \ldots, \lfloor Nx_n \rfloor)$ when the limit exists. Assume that $\overline{u}(x)$ exists except on a set of lower dimension and consists only of extreme points of (4). Let $\overline{u}^1, \ldots, \overline{u}^K$ be the values taken on by $\overline{u}(\cdot)$ and $S_k = \{x : \overline{u}(x) = \overline{u}^k\}$, $k = 1, \ldots, K$. Call the S_k control switching sets (CSSs). By construction, the CSSs are scale-invariant: If $x \in S_k$ then $\alpha x \in S_k$ for $\alpha > 0$.

To illustrate the general result to follow, consider a small example. The scaled policy associated with the following scalable policy for two queues in series (Section 4.2) with $\mu_1 < \mu_2$ is shown in Figure 3.

Always serve at the downstream queue: $\tilde{u}_2(x) = 1$ if $x_2 > 0$ and 0 otherwise. Use a switching curve at the upstream queue: $\tilde{u}_1(x) = 1$ if $x_2 < s(x_1)$ and 0 otherwise. Let $\gamma = \lim_{x\to\infty} s(x)/x$. We will show that the fluid limit policy has the switching curve $x_2 = \gamma x_1$ and that if $\tilde{u}$ is discrete optimal then its fluid limit policy is fluid optimal. If $\gamma > 0$, the CSSs are $S_0 = \{0\}$, $S_1 = \{x :$

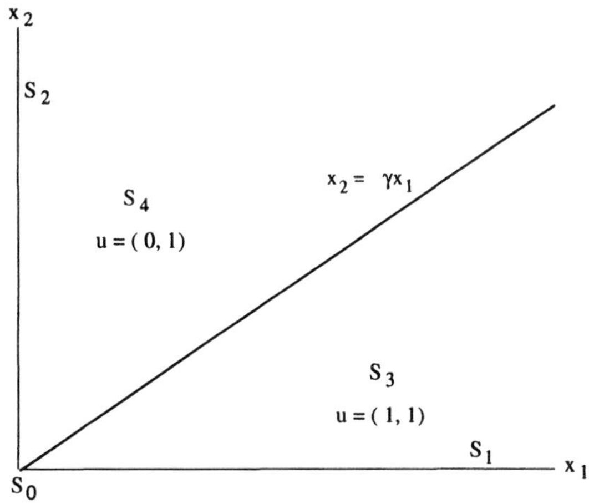

Figure 3. Control switching sets for two queues in series.

$x_1 > 0, x_2 = 0\}$, $S_2 = \{x : x_1 = 0, x_2 > 0\}$, $S_3 = \{x : 0 < x_2 < \gamma x_1\}$ and $S_4 = \{x : x_2 > \gamma x_1\}$. Two adjustments to $\overline{u}$ are needed to obtain the fluid limit policy. The limit $\overline{u}(1,\gamma)$ may not exist. However, the switching curve is *deflective*, i.e., trajectories do not remain on the curve, and we can arbitrarily set $u(1,\gamma) = \overline{u}^3 = (1,1)$. In contrast, S_1 is *attractive* (trajectories remain in S_1) and the scaled control $\overline{u}^1 = (1,0)$ needs to be adjusted by "time averaging" over the time the discrete system spends in S_1 and S_3 when $x_1 \gg 0$. On S_1, the fluid limit control is the convex combination of $\overline{u}^1$ and $\overline{u}^3$ that remains in S_1: $u(1,0) = \alpha\overline{u}^1 + (1-\alpha)\overline{u}^3$ and $\dot{x}_2(t) = 0$, leading to $u(1,0) = (1, \mu_1/\mu_2)$. For $x \in S_k$, $k \neq 1$, set $u(x) = \overline{u}^k$. Theorem 2 below shows that u is the fluid limit policy associated with $\overline{u}$ in S_3 and S_4. The fluid limit is not defined by (13) on the transient S_2 and the switching curve (nondifferentiable points), and we have defined u on S_1 in the only way to construct trajectories that are smooth there and are consistent with the nonidling nature of $\overline{u}$. As u is the only fluid policy that satisfies (13) and is consistent with $\overline{u}$, it must contain the (unique) fluid limit policy of $\overline{u}$.

The following theorem establishes that all switching surfaces in the interior of the state space are the same for the fluid limit policy and the scaled policy.

Theorem 2 *At z in the interior of a CSS S_k of full dimension, the fluid limit policy matches the scaled policy,* $u(z) = \overline{u}^k$.

Proof. We proceed as in, e.g., [Dai95, proof of Theorem 7.1]. Let $B_i = \{\alpha x \in R_+^n : 0 \le \alpha \le 1 \text{ and } \widetilde{u}_i(x) > 0\}$ and define

$$f_i(x) = \min\{\min_{y \in B_i} |x - y|, 1\},$$

where $|x| \equiv \sum_i |x_i|$. For all x with $\widetilde{u}_i(x) > 0$, $f_i(x) = f_i(x/N) = 0$ so that

$$\int_0^t f_i(\frac{\Phi^N(s)}{N}) d\widetilde{T}_i^N(s) = 0. \tag{14}$$

Because f_i is bounded and continuous, Lemma 4.4 of [Dai95] applies and (14) passes to the fluid limit:

$$\int_0^t f_i(\frac{\Phi^N(s)}{N}) d\widetilde{T}_i^N(s) \to \int_0^t f_i(x(s)) dT_i(s). \tag{15}$$

Here $x(\cdot)$ is the fluid limit and the convergence is uniformly on compact sets. Now, (14) and (15) imply that $f_i(x(t)) = 0$ or $u_i(t) = 0$ for almost every t along a trajectory. We can select a fluid limit policy for which this is true at all t, since doing so will not change $T(\cdot)$. Let z be in the interior of a CSS S_k of full dimension with $\overline{u}_i^k = 0$ so that $f_i(x) > 0$. Consider the fluid trajectory with $x(0) = z$. It remains in S_k for a nonzero time; hence, $u_i(z) = 0 = \overline{u}_i^k$.

A similar argument applies to CSSs where $\overline{u}_i^k = 1$ By assumption, $\overline{u}_i^k = 0$ or 1, so these are the only two cases. □

The key theoretical result is that optimality is preserved in the fluid limit. It is proven in [Mey01a, Theorem 6] using the "fluid scale asymptotic optimality" result of [Mey97, Theorem 7.2]. Here is the part of his theorem that we will use.

Theorem 3 (Meyn) *If a stable policy exists for the discrete system, then there exists a discrete optimal policy whose fluid limit is optimal for the fluid model, in the sense that, for any x, almost every sample path from x has a fluid limit that achieves the optimal fluid cost $J^*(x)$.*

The following corollary establishes a relationship between the discrete policy (through the scaled policy $\overline{u}$) and the fluid policy.

Corollary 1 *If a stable policy exists for the discrete system, then there exists a discrete optimal policy $\widetilde{u}$ whose scaled policy $\overline{u}$ matches some fluid optimal policy in the interior of CSSs of full dimension.*

Proof. Consider a discrete optimal policy that satisfies Theorem 3. The assumption that the fluid limit is unique implies that its fluid limit policy u

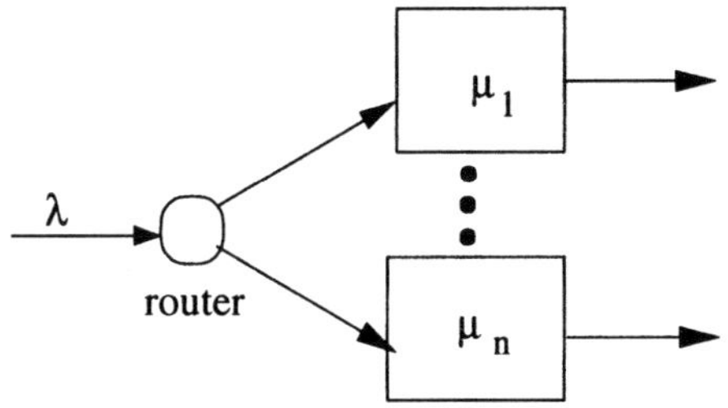

Figure 4. Arrival routing.

matches some fluid optimal policy (except on a set of lower dimension). In particular, $u(x)$ is a fluid optimal control for x in the interior of CSSs of full dimension because, for any fluid limit with $x(t) = x$, $T(\cdot)$ is differentiable at t. Theorem 2 guarantees that $\overline{u}$ matches u in these interiors. □

The fluid policy is known to consist of CSSs bounded by linear switching surfaces through the origin. By the corollary, some discrete policy, when scaled, has the same switching surfaces as some fluid policy. If we avoid the particular parameter values that lead to ties, these switching surfaces will be unique. Also, it appears that Meyn's proof applies to *any* discrete policy. Thus, we can say that any (scaled) discrete and fluid policy have the same switching surfaces. More precisely, the asymptotic slopes (unit normal vectors for $n > 2$ dimensions) of the discrete switching surfaces as $|x| \to \infty$ are the same as the slopes of the corresponding fluid switching surfaces.

4.1 Arrival Routing

The arrival routing problem analyzed in [Haj84] (Figure 4) provides a good illustration of how switching curves are related in the discrete and fluid models.

The fluid solution is presented in more detail in [Vea01a]. Customers arrive at rate λ and are routed to one of n servers to minimize holding cost. Server i has service rate μ_i and holding cost rate $c_i > 0$. Its fluid model can be written in the form of (FNET):

$$
\begin{aligned}
J(x) = \min \int_0^T & c' x(t) dt \\
\dot{x}_i(t) &= \lambda v_i(t) - \mu_i u_i(t) \\
\sum_{i=1}^{n} v_i(t) &= 1 \qquad\qquad (16) \\
x(0) &= x \\
x(t) \geq 0, \quad v(t) &\geq 0, \quad 0 \leq u(t) \leq 1.
\end{aligned}
$$

If $v_i(t) = 1$, arrivals are routed to class i at t. The control u_i has been added to enforce idling at empty queues. It is optimal to set $u_i(t) = 1$ if $x_i(t) > 0$ and $u_i(t) = \lambda v_i(t)/\mu_i$ otherwise. The greedy policy will avoid starving servers whenever possible and then route to the server with smallest holding cost. In light of this, we order the classes $c_1 \leq \ldots \leq c_n$.

For $n = 2$ servers, we consider two cases. If $\lambda \geq \mu_2$, the fluid policy is greedy. It routes to 1 whenever $x_2 > 0$. When $x_2 = 0$, it avoids starving server 2 by setting

$$v_1 = 1 - \mu_2/\lambda \text{ and } v_2 = \mu_2/\lambda. \tag{17}$$

The second case, $\lambda < \mu_2$, has a switching curve: Route to server 2 if

$$x_2 \leq \gamma x_1 = \frac{c_1}{c_2}\left(\frac{\mu_2 - \lambda}{\mu_1}\right)x_1, \tag{18}$$

otherwise route to server 1. Note that this policy makes a tradeoff between the higher short term cost of routing to server 2 and postponing the starvation of server 2. The optimality of (17) and (18) is shown in [Vea01a].

The fluid policy for n servers makes similar tradeoffs. Let M be the integer satisfying

$$\sum_{i=M}^{n} \mu_i > \lambda \geq \sum_{i=M+1}^{n} \mu_i.$$

The fluid policy avoids starving whenever possible and can be chosen so that servers $M + 1, \ldots, n$ never starve. Let $E(x) = \{i : x_i = 0\}$. Avoiding starving requires routing $\lambda_0(t) = \sum_{i \in E(x(t))} \mu_i$ to empty buffers. The remaining decision is how to route $\lambda_1(t) = \lambda - \lambda_0(t)$. We will describe policies by how they route $\lambda_1(t)$; essentially, this is the routing to nonempty buffers. We consider only policies that do not split $\lambda_1(t)$ between servers, since splitting is not needed to achieve optimality. This part of the problem ends at the starvation time $\tau_S = \min\{t : \lambda_1(t) < 0\}$. No servers are starved before τ_S. Let $x(t)$ be an optimal trajectory and $t_i = \liminf \{t > 0 : x_i(t) = 0\}$. Note that if $x_i(0) = 0$ but the initial routing makes $\dot{x}_i(0) > 0$, then t_i is the *next* time buffer i empties.

The fluid policy has the following properties, proven in [Vea01a].

(a) Never route to servers $M + 1, \ldots, n$.

(b) Never switch to an empty buffer at $t > 0$.

(c) Only switch to higher-cost servers (from i to j, $j > i$).

(d) If $t_i < \tau_S$, then never route to servers i or higher.

It is optimal to route to the server with the minimal index $k_i(x) = D_{e_i}J(x)$. This directional derivative will exist for all x and can be thought of as the incremental cost per unit of fluid initially in buffer i. Furthermore, [ABR95] shows that $k_i(x)$ is continuous along a trajectory. Thus, if the optimal routing switches from server i to server j at x, then

$$k_i(x) = k_j(x). \tag{19}$$

Consider a small amount of additional fluid initially in buffer i. Relative to $x(t)$, the additional fluid stays in buffer i until t_i. If $t_i \geq \tau_S$ then the additional fluid leaves the system at t_i. If $t_i < \tau_S$ and routing is to l at t_i, then the additional fluid is in buffer l from t_i until t_l. This shifting continues until τ_S, giving a sequence of cost terms

$$k_i(x) = c_i t_i + c_l(t_l - t_i) + \dots . \tag{20}$$

A consequence of the properties above is that trajectories are acyclic, in the sense that once a buffer empties it remains empty. If a trajectory switches from server i to j at x, then, by Property 4, $t_i \geq \tau_S$. Hence, (20) contains only the first term and (19) simplifies to

$$c_i t_i = c_j t_j. \tag{21}$$

The optimal policy never routes to i again, so $t_i = x_i/\mu_i$. Because t_j depends on the policy, we do not have a simple algorithm for computing an optimal trajectory. Let τ be the time of the next switch after switching to j at time 0. If there are no more switches, set $\tau = t_j$. Again using Property 3,

$$t_j = \frac{x_j + \int_0^\tau \lambda_1(s)ds}{\mu_j}. \tag{22}$$

Note that τ and $\lambda_1(\cdot)$ depend on x. These conditions are consistent with those for two servers, as can be seen by setting $i = 1$, $j = 2$, $\tau = t_2$ and $\lambda_1(s) = \lambda$. For $i < j$, (21) has a positive solution if and only if $j \leq M$. The parameter M specifies which servers have switching surfaces in the interior of the state space.

In principle, the following approach can be used to compute the switching surfaces. However, the number of cases to be computed is exponential in n. First, find the i, M switching surfaces. Setting $j = M$, there are no more switches, so $\tau = t_j$. Knowing the future control, find $\lambda_1(s)$ in terms of x and solve (22) and (21). Next, find the $i, M-1$ switching surfaces (set $j = M-1$). To find τ, consider the trajectory that continues routing to $M-1$. If it intersects the $M-1, M$ switching surface, the intersection is at τ. If not, set $\tau = t_j$. Again, the future routing is known and we can find $\lambda_1(s)$. Continue in this fashion, decreasing j to find all of the switching surfaces.

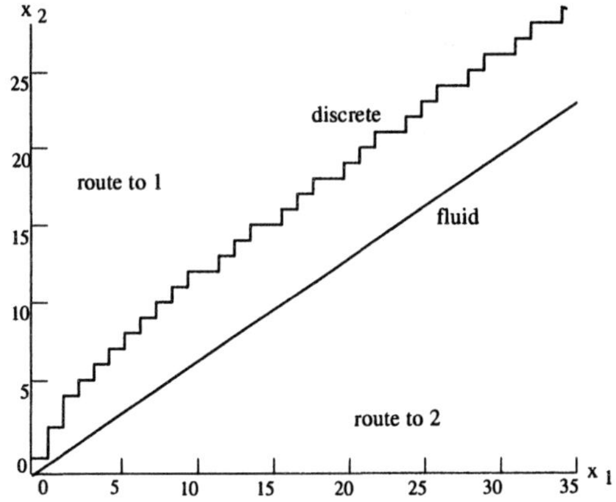

Figure 5. Fluid and discrete switching curves for arrival routing, $\lambda = 1$, $\mu = (0.5, 1.5)$, $c = (1, 1.5)$.

Figure 5 compares the fluid and discrete policies for an example with two servers and a switching curve.

The discrete policy was found using dynamic programming value iteration on a truncated state space. The policies were compared by computing their average cost in the discrete model. The fluid policy was translated to the discrete model by applying (18) directly except at the origin, where routing is to the server with minimum index c_i/μ_i. This index gives the optimal control in light traffic and matches the discrete policy. Routing is to server 1 above the curve and server 2 below it. Average cost is 1.18 for the discrete policy and 1.36 for the fluid policy (15% suboptimal). This and other examples suggest that the fluid policy is similar to the discrete policy when there are switching surfaces on the interior of the state space, but far from optimal when it switches on the boundary.

4.2 Series Systems

Consider a system of n stations in series, each serving a single class, with arrivals at rate λ to the first class. The fluid model is

$$\begin{aligned}
&\min \int_0^T c' x(t) dt \\
&\dot{x}_1(t) = \lambda - \mu_1 u_1(t) \\
&\dot{x}_i(t) = \mu_{i-1} u_{i-1}(t) - \mu_i u_i(t), \quad i = 2, \ldots, n \\
&x(0) = x \\
&x(t) \geq 0, \quad 0 \leq u(t) \leq 1.
\end{aligned}$$

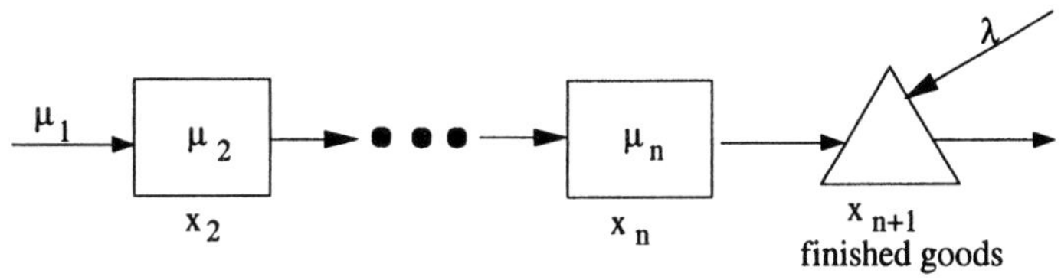

Figure 6. Series make-to-stock system.

If $c_i \geq c_{i+1}$ (or $i = n$), it is optimal to serve at the maximum possible rate at station i. We restrict our attention to the more interesting case where value is added, $c_1 \leq \ldots \leq c_n$. If $\mu_1 \geq \ldots \geq \mu_n$, starvation can be avoided until $x(t) = 0$ and the fluid policy simply avoids starving station n. More specifically, $u_i(t) = \mu_n/\mu_i$ if $x_j(t) = 0$, $j = i+1, \ldots, n$ and zero otherwise.

For the two-station system, [ABR95] show that the remaining case, $c_1 < c_2$ and $\mu_1 < \mu_2$, has a switching curve: idle station 1 if

$$x_2 \geq \frac{c_1}{c_2}\left(\frac{\mu_2 - \mu_1}{\mu_1 - \lambda}\right)x_1, \tag{23}$$

otherwise $u_1 = 1$. For n stations, [Luo95] establishes that there exist optimal fluid solutions with the following properties.

(a) Once a station starts working, it works at the maximum rate such that all empty queues will remain empty.

(b) $u(t)$ is piece-wise constant with at most $2n + 2$ pieces. It changes only when a station starts working or a queue empties.

The small number of break points suggests that there is an efficient algorithm to find the fluid policy for this model. See [Set99] for progress on this problem.

The discrete problem has only been solved using iterative dynamic programming methods on small systems. Most of the work on this problem assumes a finite buffer control, such as a kanban mechanism, then evaluates performance and optimizes within this class of controls. Several aspects of the fluid solution are reasonable for the discrete problem. First, the discrete policy is nonidling exactly when the fluid policy is, namely, when $c_1 \geq \ldots \geq c_n$. Also, replacing "empty" in property 1) with "a small buffer" we obtain, roughly, a kanban blocking mechanism upstream of a bottleneck.

As in Section 3.2, the series system can be extended to a make-to-stock setting with backordering as shown in Figure 6.

Backorders, with cost c_{n+1}^{-}, are only allowed in class $n + 1$, representing finished goods. The state x_1 can be eliminated, since μ_1 functions as arrival control. The fluid model dynamics are unchanged for $i = 2, \ldots, n$ and

$$\dot{x}_{n+1}(t) = \mu_n u_n(t) - \lambda.$$

The fluid policy again neglects the finished goods hedging point. However, it does produce ahead of demand at the upstream stations while there are backorders. For two stations, station 2 works if and only if $x_3 \le 0$. Station 1 has a switching curve in the case $\mu_1 < \mu_2$: Idle station 1 if

$$x_3 \ge -\left[\frac{(c_2/c_3^{-})(\mu_1 - \lambda) + \mu_2 - \lambda}{\mu_2 - \mu_1}\right] x_2, \tag{24}$$

otherwise $u_1 = 1$. Note that to have $u_1 = 1$ there must be backorders ($x_3 < 0$). It is interesting to compare the fluid policy with base stock and kanban policies. Stability requires the slope in (24) to be strictly less than -1. Base stock policies have a slope of -1; hence, they cannot be optimal or even asymptotically optimal under fluid scaling. Kanban, or any finite buffer policies, have an asymptotic slope of $-\infty$, which can only be optimal when $\mu_1 \ge \mu_2$. Similarly, (23) shows that kanban type policies can only be optimal for a series queue when $\mu_1 \ge \mu_2$. The discrete policy is computed numerically in [Vea94]. For the data $\lambda = 1$, $\mu_1 = 1.2$, $\mu_2 = 2$, $c_2 = 1$, and $c_3^{-} = 4$, (24) gives a slope (dx_3/dx_2) of -1.31 for the station 1 switching curve, while their policy (case 3) has a slope of about -1.05. The discrepancy is small enough that it could be due to state space truncation. The zero slope of the station 2 switching curve matches the computed discrete policy even more closely.

Returning to make-to-stock systems with n stations, [Per94] and [Luo95] give the following properties of fluid solutions.

(a) Once a station starts working, it works at the maximum rate such that all empty queues will remain empty until backorders are removed.

(b) Once backorders are removed, there will be no future backorders.

(c) Once backorders are removed and a station starts working, it works at the maximum rate such that all empty queues will remain empty.

(d) $u(t)$ is piece-wise constant with at most $4n + 4$ pieces. It changes only when a station starts working, a queue empties, or backorders are removed.

4.3 Input Control

Next we consider a two-station, six-class network with controllable inputs studied in [Wein90]. We are particularly interested in comparing his diffu-

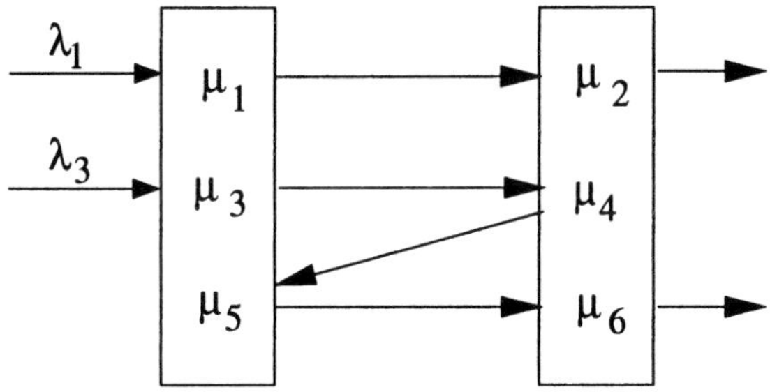

Figure 7. Wein's input control example.

sion solution with the fluid solution. This example illustrates that the fluid model can give useful information that is in some ways better than the diffusion model. The timing of arrivals is controlled but not the mix. In Wein's formulation, the time of each arrival is specified by the control. An approximate MDP formulation is to use a scalar arrival control $0 \le \tilde{v}(k) \le 1$ and make λ_i large in (1). Average cost is minimized subject to the average throughput constraint

$$\limsup_{T\to\infty} \frac{1}{T} E_x \sum_{k=0}^{T-1} \sum_{i=1}^{n} \lambda_i \tilde{v}(k) \ge \bar{\lambda}. \tag{25}$$

The dynamic programming formulation (6) does not apply to this problem because of (25).

It would not be appropriate to use a time-averaged throughput constraint in a fluid model, since the fluid model is transient. Instead, we apply the more stringent constraint that the output rate must be at least $\bar{\lambda}$ at every t. Then the fluid input control problem corresponding to (FNET) is

$$\begin{aligned} \text{(FIC)} \quad \min \int_0^T & c' x(t)dt \\ \dot{x}(t) &= Bu(t) + bv(t) \\ Du(t) &\le e \\ a'u(t) &\ge \bar{\lambda} \\ x(0) &= x \\ x(t) \ge 0, \ u(t) &\ge 0, \ 0 \le v(t) \le 1, \end{aligned}$$

where $a_i = \mu_i p_{i0}$ so that $a'u(t)$ is the flow out of the network. [Ric95] analyzes (FIC) and proves that, given the optimal sequencing control, the optimal input control is the minimal $v(t)$ that maintains $x(t^+) \ge 0$.

The network is shown in Figure 7.

The data are $\mu = (1/4, 1, 1/8, 1/6, 1/2, 1/7)$, $c = (1, 1, 1, 1, 1, 1)$, $\bar{\lambda} = 0.127$, and $\lambda_1 = \lambda_3$ (input is equally divided between classes 1 and 3). The utilization required to meet (25) is $\rho_1 = \rho_2 = 0.9$. To describe the input control based on the diffusion model, let M_{ji} equal the expected time that station j must devote to a class i customer before it exits the network. Define the two-dimensional workload process $w(t) = Mx(t)$ and interpret $w_j(t)$ as the expected amount of work for station j represented by customers in the network at time t. The diffusion control only accepts input if

$$w_1 < 19 \text{ and } w_2 - \frac{1}{4}w_1 \leq 0 \text{ or} \tag{26}$$

$$w_2 < 62 \text{ and } w_1 - \frac{2}{13}w_2 \leq 0, \tag{27}$$

where

$$w_1 = 4x_1 + 10x_3 + 2x_4 + 2x_5$$
$$w_2 = x_1 + x_2 + 13x_3 + 13x_4 + 7x_5 + 7x_6.$$

We have not found the complete fluid policy. However, fluid trajectories can be used to compare the fluid and diffusion input controls in the workload space (Figure 8).

The constraints $w_1 < 19$ and $w_2 < 62$ are not shown. They are based on variance and cannot be captured by the fluid model. When only $x_i > 0$, the workload lies on a line through the origin with slope M_{2i}/M_{1i}. The diffusion control (26) and (27) accepts input when on or above the line for class 4, or when on the line for class 1. In contrast, the computed trajectories and analysis suggest that the fluid control accepts input if and only if $x_1 = x_2 = x_5 = x_6 = 0$. Thus, input *may* occur between the lines for class 3 and 4. For example, the optimal fluid trajectory from $x = (0, 0, 1, 1, 0, 0)$ accepts input when $x_3 > 0$, disagreeing with the diffusion control, while the trajectory from $x = (1, 1, 1, 1, 1, 1)$ does not.

It is peculiar that the diffusion control accepts input *above* the line for class 4, since there is no need to accept input when x_2 or x_6 is large. The difficulty arises from translating the diffusion policy to the discrete model. Under heavy traffic scaling, the workload can be instantaneously shifted between classes. Let $\widehat{w}$ denote the diffusion scaled version of w. Input control allows $\widehat{w}_1$ and $\widehat{w}_2$ to be instantaneously *simultaneously* reduced, but not rebalanced. The *workload imbalance,* $\widehat{w}_1(t) - \widehat{w}_2(t)$, is the only independent quantity that cannot be immediately eliminated. All workload is held in class 1 (by giving it the lowest priority) when station 1 is behind and in class 4 when station 2 is behind. Hence, $x_2 = x_6 = 0$ is maintained and input is accepted *on* the line for class 4.

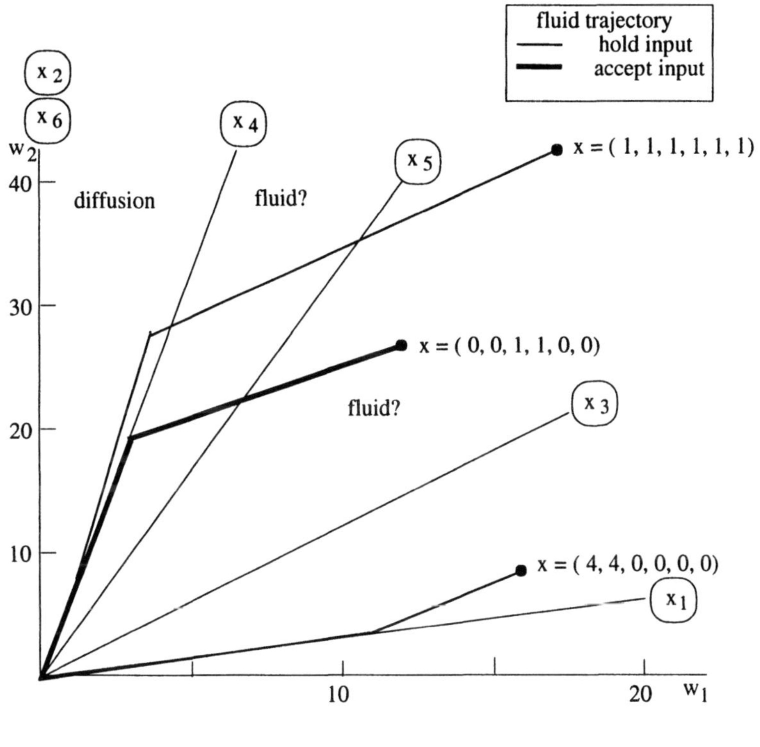

Figure 8. Regions where input is accepted for Wein's example.

Wein interprets this as *on or above* the line for the discrete model. The example suggests that it should be interpreted as *on* the line, with equality replacing less than or equal to in (27) (or within some ϵ of the line to allow safety stock).

Sequencing priorities are compared in Table 1. The diffusion priorities are one of many that are consistent with the diffusion solution; they reverse when station 1 has more work in the network. The fluid sequencing at station 2 begins serving each class in the same order as the diffusion, but splits the server between classes. In the second time interval, classes 6 and 2 are served. In the last time interval, class 4 is the only non-zero class.

Table 1. Priorities (highest to lowest) for Wein's input control example

	Diffusion(DRC)*	Greedy	Fluid from $x = (1, 1, 1, 1, 1, 1)$
Station 1	$1, 3, 5$	$1, 5, 3$	$1, 5, 3$
Station 2	$2, 6, 4$	$2, 6, 4$	2, 6/2, 6/4/2, 4

* When there is more work in the network for station 2

To summarize, the policies differ significantly in their input control and sequencing. The diffusion accepts input when the ratio w_1/w_2 of workloads for the two stations is far from one. The fluid accepts input only when needed for feasibility; roughly, when classes that have small M_{ji} (the "downstream" queues) are empty. We find the fluid input control, with its recognition of the importance of small or empty queues, to contain more potentially useful information. Much of the difference between the sequencing policies appears to be due to the different throughput constraint in the fluid model. The constraint *at each time* in (FIC) causes the output to be spread over time so that input can be avoided for as long as possible. The server splitting shown in Table 1 accomplishes this. In light of the formulation difference, it would be even more surprising if, as suggested by our qualitative analysis, the fluid policy performs better than the diffusion in the discrete model.

5. Conclusion

Numerous examples have demonstrated that the fluid model can contain valuable information about the optimal control of a stochastic network, including sequencing, scheduling, arrival routing, and work release decisions. The first connection between the fluid policy and the discrete policy (the optimal policy for the stochastic network) is their *priorities*. For problems that admit greedy solutions, the discrete and fluid policy use the same priorities when all buffers are nonempty. This correspondence suggests that the fluid priorities may be useful for other, more difficult problems.

When policies exhibit switching surfaces, a second, more powerful connection holds. The asymptotic slopes (normal vectors in higher dimensions) of the switching surfaces can be obtained from the fluid policy. We have provided a general theory for this equivalence, albeit with some difficult to check assumptions. Our numerical examples, combined with other numerical studies, suggest that the equivalence is robust.

We have also demonstrated that the fluid policy can often be found analytically; if not, the optimal trajectory from an initial state can be efficiently computed using the algorithm of [Luo98]. In fact, fluid solutions can be obtained in many more cases than diffusion solutions. Establishing a connection between the fluid and discrete models through limit theorems is also more straightforward than establishing a heavy traffic limit theorem. In light of this simplicity, it is appealing to use the fluid approximation whenever it gives comparable or better information than the diffusion approximation. The make-to-stock example illustrates that for systems where safety stock is important, the diffusion policy is better–though it may be possible to synthesize a good policy from the fluid policy and some other stochastic method. The input control example suggests that the fluid policy can be better than the diffusion, particularly when there are difficulties translating the diffusion policy back to the discrete model.

When considering the relationship between the discrete and fluid policy, it is important to distinguish between three cases.

(a) *Scalable interior switching.* If the fluid policy switches in the interior of the state space, then the discrete policy does also with the same asymptotic slopes.

(b) *Nonscalable interior switching.* If the discrete policy switches in the interior of the state space, but scales to a boundary switching policy (because the asymptotic slopes are zero), then the fluid policy switches on the boundary.

(c) *No interior switching.* If the discrete policy does not switch in the interior of the state space (it either switches on the boundary or has static priorities), then neither does the fluid policy.

In cases 1) and 3), the fluid policy is in some sense close to the discrete policy, while in case 2) the fluid policy fails to capture a primary feature of the discrete policy. It would be of interest to classify problems as belonging to 2) or 3), since solving the fluid model does not distinguish between these cases. It would also be of interest to know, as has been conjectured, whether whenever the fluid policy is static priority the discrete policy is also static priority.

Finally, fluid policies need to be adjusted near the boundaries, or *translated* back to the original problem. A major question for future research is whether general methods can be developed to adjust the fluid policy to give a good policy for the original problem (stable for all problems and robustly near optimal). However, our numerical tests on the arrival routing problem demonstrate that, even without any adjustments, the fluid switching curve policy can perform well. For some problems, then, the fluid switching surfaces are useful without any adjustments.

6. Acknowledgments

I would like to thank Dimitris Bertsimas and Costis Maglaras for their many helpful suggestions and Jay Sethuraman and Michael Yee for assisting me in performing the numerical work. I would also like to recognize the work of Michael Ricard and Andy Luo in their theses, which I have drawn on heavily.

References

[ABR95] Avram F., Bertsimas D., Ricard, M. "Fluid models of sequencing problems in open queueing networks: An optimal control approach." In *Stochastic Networks,* F. Kelly and R. Williams, eds., vol. 71 of the Proceedings of the IMA, pp. 199-234. New York: Springer-Verlag, 1995.

[Bau00] Bäuerle, N. Asymptotic optimality of tracking policies in stochastic networks. *Ann. Appl. Prob.* 2000; 10:1065-1083.

[Che93] Chen H. and Yao D.D. Dynamic scheduling of multiclass fluid networks. *Oper. Res.* 1993; 41:1104-1115.

[Che94] Chen H., Mandelbaum A. "Hierarchical modeling of stochastic networks; Part I: Fluid models." In *Stochastic modeling and analysis of manufacturing systems,* D. D. Yao, ed. New York: Springer-Verlag, pp. 47-105, 1994.

[Dai95] Dai, J.G. On the positive Harris recurrence for multiclass queueing networks: A unified approach via fluid limit models. *Ann. Appl. Prob.* 1995; 5:49-77.

[deV00] de Véricourt F., Karaesman F., Dallery Y. Dynamic scheduling in a make-to-stock system: A partial characterization of optimal policies *Oper. Res.* 2000; 48:811-819.

[Eng96] Eng D., Humphrey J., Meyn S.P. Fluid network models: Linear programs for control and performance bounds. In *Proceedings of the 13th IFAC World Congress.*, J. Cruz, J. Gertler and M. Peshkin, eds., vol. B, pp. 19-24, San Francisco, CA, 1996.

[Haj84] Hajek B. Optimal control of two interacting service stations. *IEEE Trans. Automat. Control* 1984; AC-29:491-499.

[Kel93] Kelly F.P., Laws, C.N. Dynamic routing in open queueing models: Brownian models, cut constraints and resource pooling. *Queueing Systems* 1993; 13:47-86.

[Luo95] Luo X.D. Continuous linear programming: Theory, algorithms, and applications. Ph.D. thesis, MIT Operations Research Center, Cambridge MA, 1995.

[Luo98] Luo X.D., Bertsimas D. A new algorithm for state-constrained separated continuous linear programming. *SIAM J. Control and Optim.* 1998; 37:177-210.

[Mag00] Maglaras C. Discrete-review policies for scheduling stochastic networks: Fluid-scale asymptotic optimality. *Adv. Appl. Prob.* 2000; vol. 10, no. 3.

[Mey97] Meyn S.P. The policy improvement algorithm for Markov decision processes with general state space. *IEEE Trans. Automat. Control* 1997; AC-42:191-197.

[Mey98] ______ Stability and optimization of multiclass queueing networks and their fluid models. Vol. 33 of *Lectures in Applied Mathematics*, pp. 175-17; American Mathematical Society, 1998.

[Mey01a] ______ Sequencing and routing in multiclass queueing networks. Part I: Feedback regulation. 2000 IEEE International Symposium on Information Theory, Sorrento, Italy, June 25 - June 30, 2000, and *SIAM J. Control and Optimization*, 2001; 40:741-776.

[Mey01b] ______ Sequencing and routing in multiclass queueing networks. Part II: Workload relaxations. To appear, *SIAM J. Control and Optimization*, 2001.

[Per94] Perkins J.R., Kumar P.R. Optimal control of pull manufacturing systems. Technical report University of Illinois, Urbana, IL, 1994.

[Ric95] Ricard M.J. Optimization of queueing networks: An optimal control approach. Ph.D. thesis, MIT Operations Research Center, Cambridge MA, 1995.

[Ryb92] Rybko A.N., Stolyar A.N. Ergodicity of stochastic processes describing the operations of open queueing networks. *Problems of Information Transmission,* 1992; 28:199-220.

[Set99] Sethuraman J. Scheduling job shops and multiclass queueing networks using fluid and semidefinite relaxations. Ph.D. thesis, MIT Operations Research Center, Cambridge MA, 1999.

[Vea01a] Veatch M.H. Fluid analysis of arrival routing. *IEEE Trans. Automat. Contr.* 2001; 46:1254-1257.

[Vea01b] Veatch M.H., de Véricourt F. Zero-inventory conditions for a two part-type make-to-stock production system. Submitted for publication, 2001.

[Vea94] Veatch M.H., Wein L.M. Optimal control of a two-station tandem production/inventory system. *Oper. Res.* 1994; 42:337-350.

[Vea96] Veatch M.H., Wein L.M. Scheduling a make-to-stock queue: Index policies and hedging points. *Oper. Res.* 1996; 44:634-647.

[Wein90] Wein L.M. Scheduling networks of queues: Heavy traffic analysis of a two-station network with controllable input. *Oper. Res.* 1990; 38:1065-1078.

[Wein92] ______ Dynamic scheduling of a multiclass make-to-stock queue. *Oper. Res.* 1992; 40:724-735.

[Weis95] Weiss G. On optimal draining of re-entrant fluid lines. In *Stochastic Networks,* F. Kelly and R. Williams, eds., vol. 71 of the Proceedings of the IMA, pp. 91-103. New York: Springer-Verlag, 1995.

[Wil96] Williams R.J. On the approximation of queueing networks in heavy traffic. In *Stochastic Networks: Theory and Applications,* F. Kelly, S. Zachary, and I. Ziedeins, eds., Oxford University Press, pp. 35-56, 1996.

Index